Clinical Xenotransplantation

David K. C. Cooper • Guerard Byrne
Editors

Clinical Xenotransplantation

Pathways and Progress in the Transplantation of Organs and Tissues Between Species

Springer

Editors
David K. C. Cooper
Xenotransplantation Program
University of Alabama at Birmingham
Birmingham, AL
USA

Guerard Byrne
Department of Surgery
University of Alabama at Birmingham
Birmingham, AL
USA

ISBN 978-3-030-49126-0 ISBN 978-3-030-49127-7 (eBook)
https://doi.org/10.1007/978-3-030-49127-7

This Springer imprint is published by the registered company Springer Nature Switzerland AG
The registered company address is: Gewerbestrasse 11, 6330 Cham, Switzerland

Foreword

Organ transplantation was one of the great surgical advances of the second half of the twentieth century and has become one of the greatest success stories in medical history. The University of Alabama at Birmingham was early into this field when Dr. Arnold Diethelm initiated our kidney transplant program in 1968. Subsequent programs of heart, liver, pancreas, and lung transplantation followed. We have now carried out almost 15,000 organ transplants at UAB.

Advances in immunosuppressive therapy have contributed to continuing improvement in the results of organ transplantation, but the field remains limited by the inadequate number of organs from deceased donors that become available each year. Xenotransplantation, using organs from genetically engineered pigs, offers a solution to the problem of donor organ availability. In collaboration with our colleagues at United Therapeutics and its subsidiary, Revivicor, UAB is playing an active role in moving the field forward. In addition to the experimental work continuing in our laboratories, we have established a "clean" pig facility that we anticipate meets US Food and Drug Administration guidelines, which will enable us to initiate a clinical trial of pig kidney transplantation in the near future.

To review progress in xenotransplantation, and gain insight into current advances in clinical allotransplantation that may impact xenotransplantation, Herbert Chen, the chairman of the department of surgery at UAB, proposed a one-day conference that was held at UAB in March 2019. This brought together experts in most aspects of xenotransplantation research and leaders in the field of clinical allotransplantation. Topics discussed ranged from progress in experimental models, preparations to enable a clinical trial to proceed, and selection of patients for the first clinical trials to the impact of xenotransplantation on health-care economics. The conference attracted participates from around the globe and proved very successful in defining some of the parameters for a clinical trial.

Members of the UAB department of surgery have compiled edited versions of the presentations made at the conference together with other contributions relating to xenotransplantation, and these are published in this book as a source of information for those who are interested in the topic but were unable to attend the conference.

My colleagues and I at UAB believe that xenotransplantation will have a major impact on medicine, and we are pleased to be playing a part in its development.

University of Alabama at Birmingham, School of Medicine Selwyn M. Vickers
Birmingham, AL, USA

Welcome from the University of Alabama at Birmingham (UAB)

It is my privilege to welcome you to the University of Alabama at Birmingham (UAB), and to what we believe is the world's first conference on advancing to clinical trials of pig organ xenotransplantation.

I particularly wish to welcome the members of the US Food and Drug Administration (FDA) who are participating in this conference, and also those from the National Institute of Allergy and Infectious Diseases (NIAID) of the US NIH, who have joined us today. The NIAID has strongly supported the field of xenotransplantation research for many years, and I know that the scientists in this field remain most indebted to the institute and its staff. The input of all of these scientific representatives from the FDA and NIH will be most welcome. I also particularly wish to welcome our invited speakers, both those from UAB and those from other distinguished universities, who are all experts in fields that are important in one way or another to our goal of moving towards clinical xenotransplantation.

UAB has one of the busiest organ allotransplantation programs in the country, if not in the world. We also have one of the most active xenotransplantation research programs in the world. The grants that we have been awarded for xenotransplantation research have contributed to UAB's rise in the ranking of US academic institutions that receive federal funding. We are fortunate to have as our research partners the scientists of Revivicor and its parent company, United Therapeutics, and we welcome them here today. Without the advances that they and others have pioneered in the genetic engineering of pigs, we would not have advanced towards the clinic so quickly.

The potential clinical impact of xenotransplantation is immense, encompassing as it does not only organ transplantation, but tissue and cell transplantation. Although today's conference concentrates attention largely on the transplantation of pig kidneys and hearts, clinical xenotransplantation may ultimately play a role in the treatment of conditions as diverse as (i) diabetes, where pig pancreatic islet transplantation may be life-saving, (ii) Parkinson's disease, where the transplantation of pig neuronal cells may be largely curative, (iii) corneal blindness, where pig corneas will resolve the shortage of deceased human corneas for transplantation, and (iv) life-threatening trauma, where pig red blood cell transfusion is likely to become important.

The topics we will discuss in the next two days will be of immense importance to the field of organ transplantation, and will throw light on how we and other groups should proceed in the future. We hope you will find the conference of great interest, and we encourage you to participate fully in the discussion sessions. All of the presentations and discussions will be recorded, and the proceedings of the conference will be published in book form, and so your questions and comments are important.

Finally, it is our honor to have with us today Professor Leo Buhler from the University of Geneva, who is not only the current president of the International Xenotransplantation Association (IXA), but also the Editor-in-Chief of its official journal, '*Xenotransplantation*'. We greatly appreciate his personal support for this conference. I invite him to welcome you on behalf of the IXA.

Department of Surgery, University of Alabama at Birmingham Herbert Chen
Birmingham, AL, USA

Welcome from the President of the International Xenotransplantation Association

First, I would like to thank Dr. Chen and his colleagues at the University of Alabama at Birmingham (UAB) for organizing this conference, which is highly relevant and timely for our field. Indeed, we are on a pathway that will advance xenotransplantation to clinical reality, and this path will certainly bring new questions and problems that we have not yet faced. The UAB has become a key spot on the map of xenotransplantation and has concentrated a high density of specialists, both researchers and clinicians.

Recently, I discussed intensively with my son, Benjamin, who is 11 years-old, the progress that has been made in space exploration. Benjamin taught me a great deal about that topic. I learned that space exploration began just after the Second World War and that it took approximately 20 years to send humans to the moon. Benjamin also told me that the distance between our planet and the moon is exactly 384,400 km, or 238.855 miles, which seems a long way.

So how come that, after almost 40 years of effort, xenotransplantation has not yet progressed into the clinic? Here, I think we should remember a point made by the late Claus Hammer, a pioneer in our field, who defined our research as a battle against evolution. We must fight millions of years of divergent evolution to overcome all of the biological differences that have developed between pigs and humans during this period of time. So perhaps the passage of time is more difficult to overcome than the distance of space.

Our field has benefited from more than four decades of research, involving many attempts and many failures, as well as from the collaboration of many scientists and clinicians of all continents and numerous specialties. In recent years, we have seen an acceleration of progress, mainly thanks to the development of new tools for genetic-engineering of organ-source animals and the introduction of novel forms of immunosuppressive therapy. The survival of porcine organs and tissues in nonhuman primates has been prolonged from days to months, or even years in some cases.

The initiation of clinical trials is around the corner, and we must anticipate problems that may arise, such as early acute humoral rejection and infectious complications. For example, what tests will be indicated if a pig heart recipient develops a fever on post-operative day 30? Today''s conference is key to our preparation to initiate these new trials, and the uncertainties they will bring.

I once again thank Dr. Chen and the UAB team for hosting this conference, and for inviting many of the top specialists in the field to participate. I am sure the conference will be of interest to us all.

University of Geneva
Geneva, Switzerland

Leo Buhler

Acknowledgment

The editors are most grateful to Pat Kaley of Patricia A. Kaley Secretarial Services, Bridgeville, PA, for her assistance in the preparation of this book.

Introduction

The first series of organ (kidney) transplants was carried out by Yu Yu Voronoy (Fig. 1) in the Ukraine in the 1930s and 1940s, without success. Further series were carried out in Paris (by the groups of René Küss (Fig. 2) and Jean Hamburger (Fig. 3)) and Boston (by the groups of David Hume (Fig. 4) and Joseph Murray (Fig. 5)) in the 1950s, with occasional success. The conditions under which these pioneering transplants were undertaken were primitive in the extreme. René Küss

Fig. 1 Yu Yu Voronoy (1895–1961)

Fig. 2 René Küss (1913–2000)

Fig. 3 Jean Hamburger (1909–1992)

Fig. 4 David Hume (1917–1972)

Fig. 5 Joseph Murray (1919–2012) Nobel Prize 1990

reported waiting in a Paris prison for a prisoner to be guillotined, and then removing the kidneys from the corpse on the prison floor. A few years later, Roy Calne in the UK remembers not being allowed to transfer a potential donor to the operating room, but having to remove the kidneys on the donor's hospital bed, shielded only by curtains from the view of the other patients in the ward.

Progress began to be made with the introduction of pharmacologic immunosuppressive therapy in the form of azathioprine and corticosteroids, largely through the work of Roy Calne (Fig. 6) and Tom Starzl (Fig. 7) in the early 1960s. The subsequent introduction of cyclosporine (by Calne in the late 1970s) and tacrolimus (by Starzl in the 1990s) were major advances allowing 1-year kidney graft survival to increase from approximately 50% to >80%. Immunosuppressive regimens based on cyclosporine or tacrolimus also allowed more successful transplantation of livers (by Starzl and Calne), hearts (by Shumway (Fig. 8) and Barnard (Fig. 9)), and other organs. Further novel agents have improved graft survival even more.

The one major problem that remains is an inadequate supply of organs from deceased human donors, which severely limits the number of organ transplants that can be performed each year. In the USA alone, approximately 120,000 patients

Fig. 6 Roy Calne (1930–) Lasker Prize 2012

Fig. 7 Thomas Starzl (1926–2017) Lasker Prize 2012

Fig. 8 Norman Shumway (1923–2006)

Fig. 9 Christiaan Barnard (1922–2001)

await an organ transplant, yet only approximately 40,000 transplants are carried out each year (with organs from approximately 20,000 deceased donors). This reflects the situation worldwide. Immense efforts have been made over the past 70 years to increase organ donation, with only partial success. The increase in the number of organs that become available each year has to some extent been related to a decision to accept organs of less-than-perfect "quality" (e.g., extended criteria donors, donation after cardiac death).

Xenotransplantation, using pigs as sources of organs, if successful, would resolve this continuing problem. However, the primate immune response to a pig organ proved to be rapid and severe, and overcoming this barrier has taken the combined efforts of many groups worldwide over approximately 35 years. Nevertheless, immense progress has been made, largely through the availability of increasingly sophisticated genetically engineered pigs and the introduction of novel immunosuppressive agents. The results in life-supporting pig-to-nonhuman primate models have steadily improved, and now pig kidney and heart graft survival is measured in many months rather than minutes or days.

This has encouraged preparations to be made to initiate clinical trials of organ xenotransplantation. This has necessitated attention to be directed to such widely differing topics as (i) the building of facilities that will house pigs under biosecure conditions to minimize the risk of infections in the pig that could be transferred to the human recipient, (ii) selection of potential patients for the first clinical trials, and (iii) the regulatory, ethical, legal, and financial aspects of xenotransplantation.

What is clearly obvious is that, when pig organ xenotransplantation is introduced into the clinic, the state of the science will be far in advance of that when organ allotransplantation was first attempted in the 1950s and 1960s. In those days, there was very limited experience in dogs before clinical attempts were made, and hardly any effective method of immune suppression. In contrast, today (i) we benefit from 70 years of clinical experience with allotransplantation, (ii) we have more than three decades of experience of organ xenotransplantation in pig-to-nonhuman primate models, and (iii) we have a wide range of genetically engineered pigs and novel immunosuppressive agents available to us.

To consider many of these topics, the department of surgery at UAB organized a conference in March 2019, held in Birmingham, at which many experts were invited to present their data and opinions on how xenotransplantation can move forward into the clinic. Attention was concentrated on pig kidney and heart transplantation as it is in regard to these organs that most progress has been made. Pig liver and lung transplantation, and cell transplantation, e.g., islets, neuronal cells, were not discussed.

Towards the end of the conference, the audience of approximately a hundred people was asked whether patients in need of a kidney or those in need of a heart should be the first to be included in a clinical trial. A point of considerable interest was that virtually 100% of the audience indicated that the first clinical trial should be of pig kidney transplantation. This was based on such considerations as the ability to resort to chronic dialysis if the pig kidney graft failed, thus providing life-saving support if the experiment was not successful. For patients receiving a cardiac xenotransplant, despite increasing experience with mechanical circulatory assist devices, in the event that the pig graft fails, there is no realistic life-supporting alternative similar to dialysis.

The presentations given at the conference have now been edited and are presented in this volume as a record of the conference. To provide information on topics that were not fully addressed at the conference, additional chapters have been added. We hope that this book provides an outline of how the field might progress from the experimental laboratory to the clinic, and thus make available the many advantages of xenotransplantation to patients with end-stage organ failure. In other words, we hope the collected papers constitute a pathway to clinical xenotransplantation.

Birmingham, AL, USA David K. C. Cooper
London, UK Guerard Byrne

Contents

Contributors

Andrew B. Adams, MD Emory Transplant Center, Department of Surgery, Emory University, Atlanta, GA, USA

Kent Adams, DVM Revivicor, Inc., Blacksburg, VA, USA

Noriko Ammerman, MD, FASN, FAST Comprehensive Transplant Center, Transplant Immunotherapy Program, Cedars-Sinai Medical Center, Los Angeles, CA, USA

Douglas J. Anderson, MD Xenotransplantation Program, Department of Surgery, University of Alabama at Birmingham (UAB), Birmingham, AL, USA

Division of Transplantation, Department of Surgery, University of Alabama at Birmingham (UAB), Birmingham, AL, USA

Judith Arcidiacono, MS U.S. Food and Drug Administration, Silver Spring, MD, USA

David Ayares, PhD Revivicor, Inc., Blacksburg, VA, USA

Suyapa Ball, MS Revivicor, Inc., Blacksburg, VA, USA

John Bianchi, PhD Revivicor, Inc., Blacksburg, VA, USA

Rita Bottino, PhD Institute for Cellular Therapeutics, Allegheny-Singer Research Institute, Pittsburgh, PA, USA

Department of Biological Sciences, Carnegie-Mellon University, Pittsburgh, PA, USA

Michael E. Breimer, MD Department of Surgery, Institute of Clinical Sciences, Sahlgrenska Academy at University of Gothenburg, Gothenburg, Sweden

Lars Burdorf, MD Division of Cardiac Surgery and Center for Transplantation Sciences, Department of Surgery, Massachusetts General Hospital and Harvard University, Boston, MA, USA

Steve Butler, MS Revivicor, Inc., Blacksburg, VA, USA

Guerard Byrne, PhD Department of Surgery, University of Alabama at Birmingham, Birmingham, AL, USA

Institute of Cardiovascular Science, University College London, London, UK

Marilia Cascalho, MD Departments of Surgery and Microbiology and Immunology, University of Michigan, Ann Arbor, MI, USA

David C. Cleveland, MD, MBA Division of Cardiothoracic Surgery, Department of Surgery, University of Alabama at Birmingham, Children's Hospital of Alabama, Birmingham, AL, USA

David K. C. Cooper, MD, PhD, FRCS Xenotransplantation Program, Department of Surgery, University of Alabama at Birmingham (UAB), Birmingham, AL, USA

Formerly Thomas E. Starzl Transplantation Institute, Department of Surgery, University of Pittsburgh Medical Center, Pittsburgh, PA, USA

David A. D'Alessandro, MD Division of Cardiac Surgery, Department of Surgery, Massachusetts General Hospital and Harvard University, Boston, MA, USA

Amy Dandro, MS Revivicor, Inc., Blacksburg, VA, USA

Devin E. Eckhoff, MD Xenotransplantation Program, Department of Surgery, University of Alabama at Birmingham, Birmingham, AL, USA

Division of Transplantation, Department of Surgery, University of Alabama at Birmingham (UAB), Birmingham, AL, USA

Willard Eyestone, PhD Revivicor, Inc., Blacksburg, VA, USA

Mohamed B. Ezzelarab, MD Thomas E. Starzl Transplantation Institute, University of Pittsburgh, Pittsburgh, PA, USA

Alton B. Farris, MD Department of Pathology, Emory School of Medicine, Atlanta, GA, USA

Ana Laura Fazio, PhD Revivicor, Inc., Blacksburg, VA, USA

Jay A. Fishman, MD Transplantation Infectious Disease and Compromised Host Program, Infectious Disease Division and MGH Transplant Center, Massachusetts General Hospital and Harvard Medical School, Boston, MA, USA

Jon Friedman, MD, FAST Medical Benefit Management, Optum, Eden Prairie, MN, USA

Robert Gaston, MD Division of Nephrology, Department of Medicine, University of Alabama at Birmingham (UAB), Birmingham, AL, USA

Edward Gray Independent Consultant, Satellite Beach, FL, USA

Michael J. Hanaway, MD Division of Transplantation, Department of Surgery, University of Alabama at Birmingham (UAB), Birmingham, AL, USA

Hidetaka Hara, MD, PhD Xenotransplantation Program, Department of Surgery, University of Alabama at Birmingham (UAB), Birmingham, AL, USA

Hayato Iwase, MD, PhD Xenotransplantation Program, Department of Surgery, University of Alabama at Birmingham (UAB), Birmingham, AL, USA

Abhijit Jagdale, MD Xenotransplantation Program, Department of Surgery, University of Alabama at Birmingham (UAB), Birmingham, AL, USA

Stanley C. Jordan, MD Comprehensive Transplant Center, Transplant Immunotherapy Program, Cedars-Sinai Medical Center, Los Angeles, CA, USA

James K. Kirklin, MD Division of Cardiothoracic Surgery; Kirklin Institute for Research in Surgical Outcomes (KIRSO), Birmingham, AL, USA

Department of Surgery, University of Alabama at Birmingham (UAB), Birmingham, AL, USA

Stuart J. Knechtle, MD Duke Transplant Center, Department of Surgery, Duke University, Durham, NC, USA

Maria Kokkinaki, PhD Revivicor, Inc., Blacksburg, VA, USA

Karl Kraebber, MS Xenotransplantation Program, Department of Surgery, University of Alabama at Birmingham, Birmingham, AL, USA

Vineeta Kumar, MD Division of Nephrology, Department of Medicine, University of Alabama at Birmingham, Birmingham, AL, USA

Kasinath Kuravi, PhD Revivicor, Inc., Blacksburg, VA, USA

Jean Kwun, MD Duke Transplant Center, Department of Surgery, Duke University, Durham, NC, USA

Joseph M. Ladowski, MD, PhD Xenotransplantation Program, Department of Surgery, University of Alabama at Birmingham (UAB), Birmingham, AL, USA

Whayoung Lee, MD Thomas E. Starzl Transplantation Institute, University of Pittsburgh, Pittsburgh, PA, USA

Qi Li, MD Xenotransplantation Program, Department of Surgery, University of Alabama at Birmingham, Birmingham, AL, USA

Second Affiliated Hospital, University of South China, Hengyang City, Hunan, China

Alan Lipps, PhD School of Social Work, Abilene Christian University, Abilene, TX, USA

Jayme E. Locke, MD, MPH, FACS, FAST Division of Transplantation, Department of Surgery, University of Alabama at Birmingham, Birmingham, AL, USA

Brendon Lovasik, MD Emory Transplant Center, Department of Surgery, Emory University, Atlanta, GA, USA

Roslynn B. Mannon, MD Division of Nephrology, Department of Medicine, University of Alabama at Birmingham (UAB), Birmingham, AL, USA

Gregory R. Martens, MD, PhD Xenotransplantation Program, Department of Surgery, University of Alabama at Birmingham (UAB), Birmingham, AL, USA

Chace Mitchell, BS Xenotransplantation Program, Department of Surgery, University of Alabama at Birmingham (UAB), Birmingham, AL, USA

Jeff Monahan, BS Revivicor, Inc., Blacksburg, VA, USA

Robert A. Montgomery, MD, DPhil NYU Langone Transplant Institute, NYU Langone Health, New York, NY, USA

Ben Morrill, PhD Revivicor, Inc., Blacksburg, VA, USA

Luz A. Padilla, MD, MSPH Department of Epidemiology and Surgery, Pediatric and Congenital Heart Center of Alabama, Children's Hospital of Alabama, Birmingham, AL, USA

Wayne Paris, PhD, LCSW School of Social Work, Abilene Christian University, Abilene, TX, USA

Xenotransplantation Program, Department of Surgery, University of Alabama at Birmingham (UAB), Birmingham, AL, USA

Carol Phelps, MS Revivicor, Inc., Blacksburg, VA, USA

Richard N. Pierson III, MD Division of Cardiac Surgery and Center for Transplantation Sciences, Department of Surgery, Massachusetts General Hospital and Harvard University, Boston, MA, USA

Jeffrey L. Platt, MD Departments of Surgery and Microbiology & Immunology, University of Michigan, Ann Arbor, MI, USA

Farzana Rahman, BS Revivicor, Inc., Blacksburg, VA, USA

Jagdeece Ramsoondar, PhD Revivicor, Inc., Blacksburg, VA, USA

Lori Sorrells, BS Revivicor, Inc., Blacksburg, VA, USA

Winson W. Tang, MD U.S. Food and Drug Administration, Silver Spring, MD, USA

A. Joseph Tector, MD, PhD Department of Surgery, University of Miami, Miami, FL, USA

Todd Vaught, BS Revivicor, Inc., Blacksburg, VA, USA

Ashley Vo, MD, FASN, FAST Comprehensive Transplant Center, Transplant Immunotherapy Program, Cedars-Sinai Medical Center, Los Angeles, CA, USA

Anneke Walters, PhD Revivicor, Inc., Blacksburg, VA, USA

Yi Wang, MD Second Affiliated Hospital, University of South China, Hengyang City, Hunan, China

Zoie Werkheiser, BSc Xenotransplantation Program, Department of Surgery, University of Alabama at Birmingham (UAB), Birmingham, AL, USA

Martin Wijkstrom, MD Thomas E. Starzl Transplantation Institute, University of Pittsburgh, Pittsburgh, PA, USA

Takayuki Yamamoto, MD, PhD Xenotransplantation Program, Department of Surgery, University of Alabama at Birmingham (UAB), Birmingham, AL, USA

Zhongqiang Zhang, MD Thomas E. Starzl Transplantation Institute, University of Pittsburgh, Pittsburgh, PA, USA

Department of Organ Transplantation and General Surgery, Second Xiangya Hospital of Central South University, Changsha, Hunan, China

Part I

Clinical and Experimental Xenotransplantation: Background

A Brief History of Clinical Cross-Species Organ Xenotransplantation

1

David K. C. Cooper

Introduction

The increasing demand for organs, tissues, and cells for purposes of clinical transplantation, and the relative lack of improvement in the number of deceased human organs that become available each year have increased interest in the possibility of using organs and cells from an animal species [1, 2]. The concept of cross-species transplantation (or xenotransplantation) is not new, and there has been a surprisingly large number of clinical attempts during the past 300 years or more [1, 3, 4]. The barriers to xenotransplantation are considerable, but are steadily being overcome, largely by our ability to genetically engineer pigs to make their tissues more resistant to the human immune response.

Xenotransplantation in Mythology

A review of Greek mythology and of religious tracts, particularly, for example, from the Hindu religion, draws attention to the fact that humans have been interested in the possibility of merging physical features from various animal species for hundreds of years. For example, the chimera has been used to represent the *allo*transplantation of organs and cells (transplantation between members of the same

Based on a paper published in Proc (Bayl Univ Med Cent). 2012; 25:49-57, with permission of the editor.

D. K. C. Cooper (✉)
Thomas E. Starzl Transplantation Institute, Department of Surgery, University of Pittsburgh Medical Center, Pittsburgh, PA, USA

Xenotransplantation Program, Division of Transplantation, Department of Surgery, University of Alabama at Birmingham (UAB), Birmingham, AL, USA
e-mail: dkcooper@uabmc.edu

D. K. C. Cooper, G. Byrne (eds.), *Clinical Xenotransplantation*,
https://doi.org/10.1007/978-3-030-49127-7_1

Fig. 1.1 The lamassu – this mythological beast was adopted as the logo of the International Xenotransplantation Association and its official journal, *Xenotransplantation*

species), and the lamassu (Fig. 1.1) was selected as the mythological beast to represent the International Xenotransplantation Association and its official scientific journal, *Xenotransplantation*.

The late Keith Reemtsma pointed out that possibly one of the earliest examples of xenotransplantation was the attempt by Daedalus and his son, Icarus, to fly across the sea from Crete to mainland Greece with the help of bird's wings attached to their arms [5]. Icarus failed in the attempt, and Reemtsma (with tongue firmly in his cheek) put this forward as a possible case of hyperacute rejection (very rapid rejection of the graft), though he thought it was more likely to be related to failure of a thermo-labile adhesive. However, Daedalus successfully made the journey, which Reemtsma pointed out provided this pair with an enviable 50% success rate.

The first tissue xenograft was reputedly recorded in 1682, when a Russian nobleman, who had lost part of his scalp and skull in battle, had the defect in his skull "successfully repaired by a surgeon with a piece of bone from the skull of a dog" [6, 7]. The Russian church, however, believing that no man could be Christian if he had a dog bone in his head, threatened the nobleman with excommunication. Clearly, a God-fearing man, he chose to have the fragments of dog bone removed, thus presumably saving himself from a fate worse than death.

Blood Xenotransfusion

If we look beyond the realm of mythology and legend, we come to the seventeenth century, when Jean Baptiste Denis (Fig. 1.2) began the clinical practice of blood transfusion from animals to humans (Fig. 1.3) [8]. Perhaps not surprisingly, the results were mixed. As a result, xenotransfusion was banned in France for a number

Fig. 1.2 Jean-Baptiste Denis (c1635–1704)

Fig. 1.3 A blood transfusion being carried out from donkey to patient

of years. Today, with the increasing risk of transfer of infectious agents with human blood transfusions, a strong case could be made for using an animal, for example, the pig (housed under ideal "clean" conditions and monitored at intervals to ensure that no infectious agent would be transferred) as the source of blood cells and blood products in the future. In fact, this approach has recently been explored again by several groups [9].

Skin Xenotransplantation

According to skin graft pioneer, Thomas Gibson, the list of animal tissues that were transplanted into human subjects in the nineteenth century is extensive, with skin being the most common [6, 7]. But transplants of the urethra from a sheep and even the eye from a rabbit are recorded. Donors of skin included dogs, cats, rabbits, rats, pigs, chickens, cockerels, pigeons, and, most popular of all, frogs.

The fact that many of the species used as donors had hair, feathers, or fur growing from the skin did not appear to disconcert the surgeons involved, but the trend was to use animal species in which these accoutrements were not present. The ideal graft would appear to have been from frogs, which were sometimes "skinned alive." It is possible that some of these grafts were "successful" in that, when used to cover a skin ulcer, they provided protection, at least for a number of days, while the ulcer healed beneath the graft. However, it is inconceivable that any of the grafts actually became permanent.

The grafts were either free skin grafts or pedicle skin grafts. Pedicle grafts were complicated because they required the donor, for example, a sheep, to be strapped immobile to the patient for several days, during which period of time the graft would reputedly be vascularized by the recipient. If this occurred, the graft could be disconnected from the donor. It is almost certain that none of these grafts was in any way successful, although some "successes" were reported.

Corneal Xenotransplantation

Remarkably, in 1838 the first corneal xenotransplant (from a pig) was performed in a patient, whereas the first corneal allograft (human-to-human) was not carried out until more than 65 years later (in 1905). The field of corneal xenotransplantation has been reviewed elsewhere [10, 11].

Alexis Carrel and Blood Vessel Anastomosis

More scientific efforts had to wait until the twentieth century, when the French experimental surgeon, Alexis Carrel [12] (Fig. 1.4), working first in France and subsequently in North America, developed surgical techniques for anastomosing blood vessels, which enabled organ transplantation to be carried out successfully for

Fig. 1.4 Alexis Carrel (1873–1944)

the first time. It was for this work that he was awarded the Nobel Prize in 1912. He developed an interest in cross-species transplantation, at least from an experimental perspective, and in 1907 wrote these prophetic words:

> The ideal method would be to transplant in man organs of animals easy to secure and operate on, such as hogs, for instance. But it would in all probability be necessary to immunize organs of the hog against the human serum. The future of transplantation of organs for therapeutic purposes depends on the feasibility of hetero (xeno) transplantation.

It is remarkable that, more than 100 years ago, Carrel indicated what we are now trying to do, which is to genetically modify pigs to make their tissues resistant to the human immune response. Carrel was clearly a man of vision.

Serge Voronoff and "Rejuvenation" by Cell Xenotransplantation

A few years later, Serge Voronoff [13, 14] (Fig. 1.5), a Russian émigré working in Paris, developed the concept of transplanting cells that produced a hormone in which the recipient was deficient. This is another example of a visionary scientist who was ahead of his time. Today, we are doing exactly what he envisaged, namely transplanting human pancreatic islets that produce insulin in patients with severe type 1 diabetes. In view of the limited number of human pancreases that become

Fig. 1.5 Serge Voronoff (1866–1951)

available each year, there is a growing interest in using pig islets for this purpose (see below).

Voronoff's main interest, however, was in reversing the effects of aging in elderly men who had lost their "zest for life." He carried out a significant number of chimpanzee or baboon testicular transplants in male human recipients [13–15]. His technique was to slice up the animal testicle and insert the slices into the recipient's testicle. (It can be looked upon as the "*Viagra*" of the 1920s.) The procedure became popular on both sides of the Atlantic, and several hundreds of these operations were performed. It is inconceivable that any of them had any beneficial effect whatsoever except psychological, but there were reports of remarkable "rejuvenation" of men who reported much increased energy after the operation. The complications of the operations must have been significant because presumably on occasions the slices of donor testicle would have necrosed and set up inflammatory or infectious complications. Surprisingly, reports of such complications appear to have been uncommon.

Voronoff was certainly a man ahead of his time because he also applied to the authorities in Paris to carry out what would have been the first clinical kidney allotransplant, using the kidneys from a criminal who was to be guillotined. His request was refused, and this allowed Yurii Voronoy to become the first surgeon to perform kidney allotransplantation in 1933 [16].

The concept of transplanting glandular tissue to produce hormones that would benefit the recipient was continued in the USA by a much less scientific doctor, John Brinkley, whose work was carried out largely in Kansas and Texas [17]. His chosen donor was the goat, as he had been convinced by a local farmer of its sexual potency. It would appear that Brinkley was a charlatan rather than a serious transplant surgeon, and, although it made him a fortune, his work fell into serious disrepute, and he was eventually driven out of business by the American Medical Association.

Nevertheless, this concept of cell xenotransplantation has been sustained until the present time, with the establishment of several clinics, particularly in Europe, in which animal tissue or serum is injected into patients for a variety of conditions. The results have been controversial [18].

Several clinical organ xenotransplants were carried out in the early part of the twentieth century (Tables 1.1a, 1.1b, and 1.1c).

Table 1.1a World experience in clinical renal xenotransplantation

Year	Surgeon	Donor	*n*	Patient survival (days)
1905	Princeteau	Rabbit (kidney slices)	1	16
1906	Jaboulay	Pig,	1	3
		goat	1	3
1910	Unger	Monkey	1	2
1913	Schonstadt	Monkey	1	Not stated
1923	Neuhof	Sheep	1	9
1964	Reemtsma	Chimpanzee,	6	<9 months
		monkey	1	10
1964	Hitchcock	Baboon	1	5
1964	Starzl	Baboon	6	<60
1964	Hume	Chimpanzee	1	1
1964	Traeger	Chimpanzee	3	<49
1965	Goldsmith	Chimpanzee	2	<4 months
1966	Cortesini	Chimpanzee	1	31
1966	Kuss	Pig	1	2

Major source: Reference [3]

Table 1.1b World experience in clinical heart xenotransplantation

Year	Surgeon	Donor	Type	Patient survival (days)
1964	Hardy	Chimpanzee	O	<1
1968	Cooley	Sheep	O	<1
1968	Ross	Pig	H	<1
1968	Ross	Pig	Perfused with human blood but not transplanted	<1
1969	Marion	Chimpanzee	?O	<1
1977	Barnard	Baboon	H	<1
1977	Barnard	Chimpanzee	H	4
1984	Bailey	Baboon	O	20
1992	Religa	Pig	O	1

Major source: Reference [3]

H heterotopic (auxiliary) heart transplantation, *O* orthotopic heart transplantation

Table 1.1c World experience in clinical liver xenotransplantation

Year	Surgeon	Donor	Type	Patient survival (days)
1966	Starzl	Chimpanzee	H	<1
1969	Starzl	Chimpanzee	O	9
		Chimpanzee	O	<2
1969	Bertoye	Baboon	H	<1
1970	Leger	Baboon	H	3
1970	Marion	Baboon	H	<1
1971	Poyet	Baboon	H	<1
1971	Motin	Baboon	H	3
1974	Starzl	Chimpanzee	O	14
1992	Starzl	Baboon	O	70
1993	Starzl	Baboon	O	26
1993	Makowka	Pig	H	<2

Major source: Reference [3]
H heterotopic (auxiliary) liver transplantation, *O* orthotopic liver transplantation

Clinical Kidney Xenotransplantation

Keith Reemtsma and Chimpanzee Kidney Xenotransplantation

By the 1960s, Keith Reemtsma (Fig. 1.6) – then at Tulane University in Louisiana – hypothesized that nonhuman primate kidneys might function in human recipients and thus be a successful treatment for renal failure. At that time, the concept of kidney transplantation had been established largely by French and American surgeons, but the availability of deceased human kidneys was extremely limited and chronic dialysis was not yet being undertaken. In Reemtsma's opinion, therefore, there was little alternative to death for the patient unless organs could be made available from nonhuman species. He selected the chimpanzee as the source of organs because of its close evolutionary relationship to humans. He carried out six of these transplants, on each occasion transplanting both kidneys from the chimpanzee (that generally weigh significantly less than adult humans) into the recipient (Table 1.1a) [19].

The majority of these failed within 4–8 weeks, either from rejection (because of the limited immunosuppressive agents available at the time) or from an infectious complication (because of the over-administration of these agents). Nevertheless, one of Reemtsma's patients lived for 9 months, returning to work as a schoolteacher, and evidently remaining in good health until she collapsed and died. At autopsy, the chimpanzee kidneys appeared normal and showed no signs of acute or chronic rejection (Fig. 1.7). It was suggested that she had died from an acute electrolyte disturbance. This is quite likely since the transplantation of nonhuman primate kidneys into patients was frequently associated with an immense diuresis in the early post-transplant period, often exceeding 20 liters in 24 hours, and it is likely that there were also later electrolyte imbalances. On one occasion, however, Reemtsma demonstrated that acute cellular rejection of a chimpanzee kidney could be reversed by a course of increased steroid therapy.

Fig. 1.6 Keith Reemtsma (1925–2000)

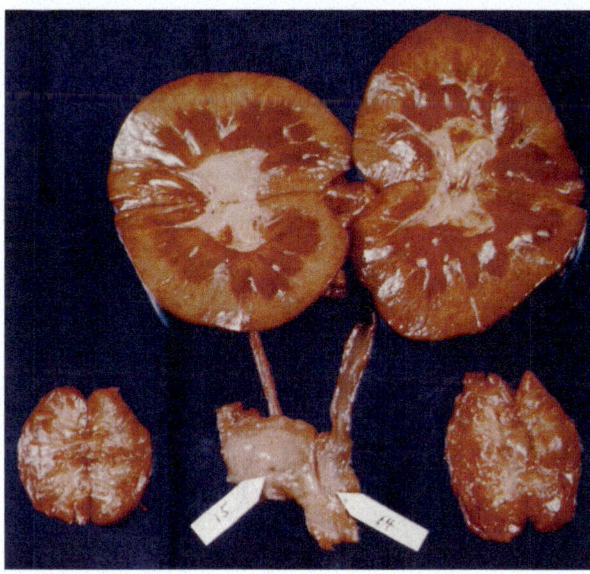

Fig. 1.7 Normal macroscopic appearance at autopsy of chimpanzee kidneys that had functioned well for a period of almost 9 months in a 23-year-old woman who had undergone renal xenotransplantation in 1963. This operation was one of a small series of kidney xenotransplants performed by Keith Reemtsma and his colleagues at Tulane University in New Orleans

Fig. 1.8 Tom Starzl (1926–2017)

Subsequent baboon kidney transplants, notably by Tom Starzl (Fig. 1.8) and his colleagues in Colorado [20], were rather less successful. Survival of patients with baboon kidneys ranged from 19 to 60 days. Subsequent chimpanzee kidney transplants also provided very mixed results, except for one patient who survived for almost 4 months (Table 1.1a). The chimpanzee donor kidneys were, in general, rejected more slowly and by a cellular mechanism, whereas the baboon donor kidneys were rejected more aggressively.

French surgeon, René Küss, a pioneer in kidney allotransplantation, attempted a clinical pig kidney transplant in 1966, with immediate graft loss [21].

Clinical Heart Xenotransplantation

James Hardy and the First Heart Xenotransplant

James Hardy (Fig. 1.9), who had carried out the first human lung *allo*transplant in 1963, visited Reemtsma and was impressed by the health of some of the patients with chimpanzee kidney transplants he examined. In 1964, he determined to carry out the first clinical heart transplantation and decided to acquire some chimpanzees as potential "donors" in case he could not identify a deceased human donor when needed (Table 1.1b). In the event, he had a less-than-ideal patient who would not be accepted for heart transplantation today, as he had widespread atheromatous vascular disease throughout his body – for which he had undergone amputations of both legs – and was in a semi-comatose state at the time the transplant was undertaken.

Fig. 1.9 James Hardy (1918–2003)

However, as the patient was rapidly dying, Hardy was stimulated to transplant a chimpanzee heart (Fig. 1.10) [22]. The chimpanzee heart was not large enough to support the circulation, and failed within a couple of hours. There was no histopathological evidence of antibody-mediated rejection [23].

In contrast to the response to the attempted lung *allo*transplantation, the public and medical professional response to the heart *xeno*transplantation was adverse, and dissuaded Hardy and his colleagues from carrying out any further attempts. The procedure of cardiac *allo*transplantation was later established by Barnard and his colleagues in 1967 [24], who later also carried out two cardiac *xeno*transplants [25].

It is of interest to note that the consent form for Hardy's operation (Fig. 1.11) – which, in view of the patient's semi-comatose condition, was signed by a close relative – stipulated that no heart transplant had ever been performed, but made no mention of the fact that an animal heart might be used for the procedure. Such was the medico-legal situation at that time that this "informed" consent was not considered in any way inadequate.

Two distinguished cardiac surgeons, Denton Cooley in the USA (Fig. 1.12) and Donald Ross in the UK (Fig. 1.13), carried out sheep and pig heart transplants, respectively, in patients that they could not wean from cardiopulmonary bypass after routine cardiac surgery (Table 1.1b). These attempts were rather misguided as

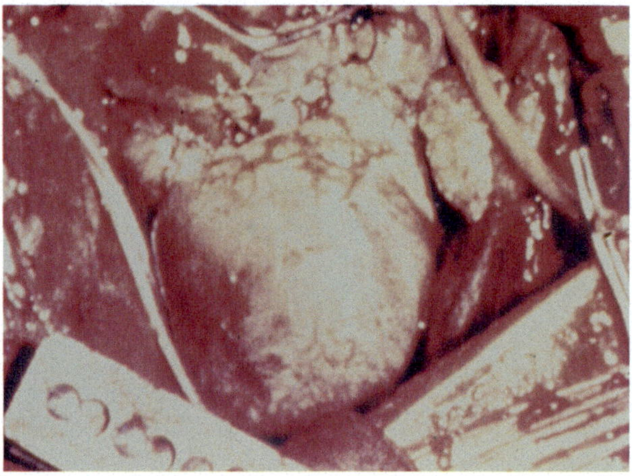

Fig. 1.10 The chimpanzee heart in the patient's chest after the world's first clinical heart transplant

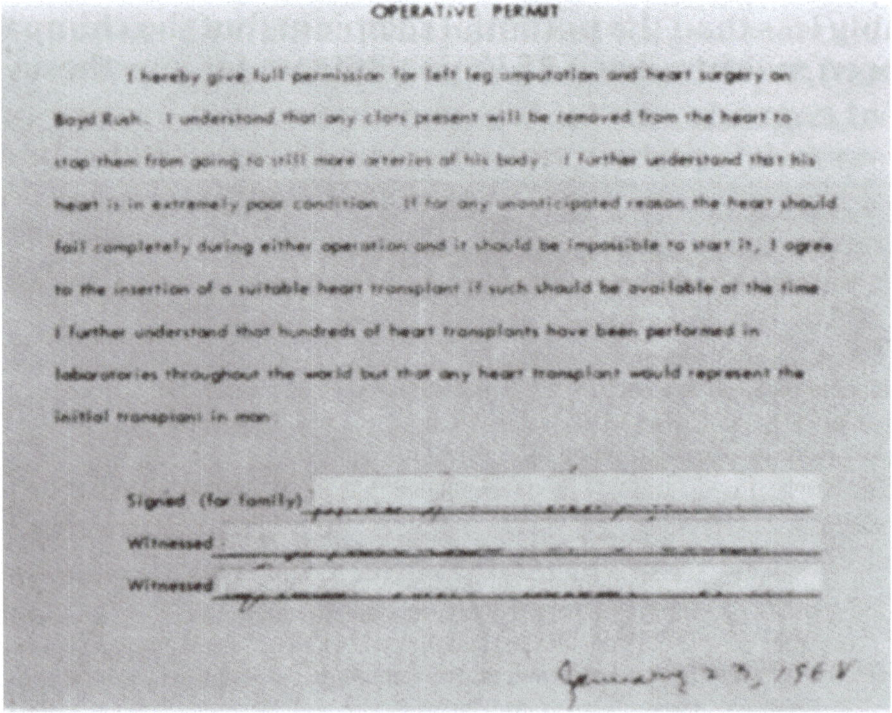

OPERATIVE PERMIT

I hereby give full permission for left leg amputation and heart surgery on Boyd Rush. I understand that any clots present will be removed from the heart to stop them from going to still more arteries of his body. I further understand that his heart is in extremely poor condition. If for any unanticipated reason the heart should fail completely during either operation and it should be impossible to start it, I agree to the insertion of a suitable heart transplant if such should be available at the time. I further understand that hundreds of heart transplants have been performed in laboratories throughout the world but that any heart transplant would represent the initial transplant in man.

Signed (for family)

Witnessed

Witnessed

January 23, 1964

Fig. 1.11 The consent form signed by the patient's family for the world's first heart transplant (using a chimpanzee heart)

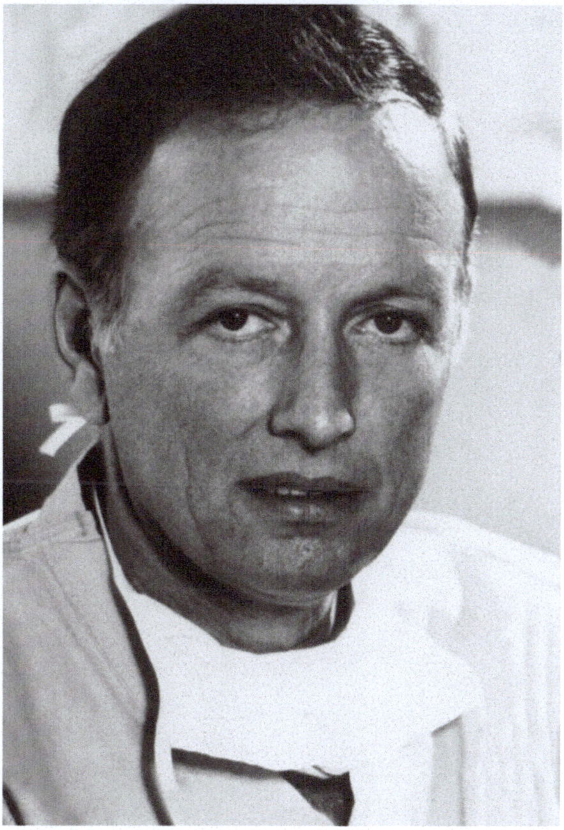

Fig. 1.12 Denton Cooley (1920–2016)

there was no evidence to suggest that either a sheep or pig heart would function for a prolonged period of time after transplantation into a primate. Both surgeons admitted to me that in retrospect, they were rather embarrassed by these attempts.

In 1977, Christiaan Barnard (Fig. 1.14) and his colleagues again attempted to support life of two patients who could not be weaned from cardiopulmonary bypass by attaching nonhuman primate hearts in the intra-thoracic heterotopic position [25] (Fig. 1.15). On the first occasion, they used a baboon heart which failed rapidly. On the second, they transplanted a chimpanzee heart, which functioned for 4 days before it failed. These attempts were made in an effort to support the patients until either the native heart recovered or a deceased human donor heart could become available for allotransplantation.

Leonard Bailey and "Baby Fae"

Perhaps the best known clinical cardiac xenotransplantation since Hardy's attempt was that by Leonard Bailey (Fig. 1.16) who transplanted a baboon heart into an infant girl, known as Baby Fae (Fig. 1.17), in 1983 [26]. At that time, it was almost impossible to obtain human organs from deceased infants, particularly those with anencephaly, for transplantation into infants with life-threatening congenital heart

Fig. 1.13 Donald Ross (1922–2014)

Fig. 1.14 Christiaan Barnard (1922–2001)

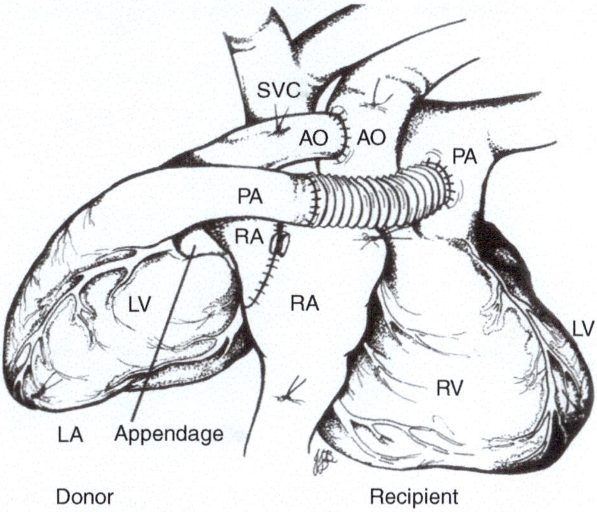

Fig. 1.15 The operation of heterotopic heart transplantation, developed by Christiaan Barnard and Jacques Losman, which Barnard used in the two patients in which he transplanted a nonhuman primate heart

Fig. 1.16 Leonard Bailey (1942–2019)

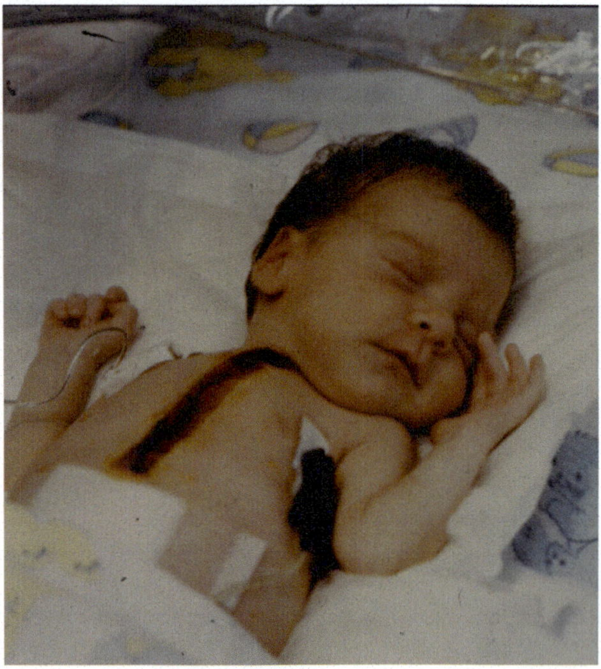

Fig. 1.17 Baby Fae

disease. The surgical procedure in Baby Fae was technically successful, but the graft underwent acute rejection and the patient died 20 days later. As the graft was necessarily taken from a baboon that was ABO incompatible with the recipient – as the O blood type is essentially not found in baboons – this might have added to the severity of rejection. Even though cyclosporine had become available by this time, the immunosuppressive therapy was not sufficient to prevent xenograft rejection.

This procedure did little to advance progress in xenotransplantation, but it did draw attention to the public and medical profession of the dearth of deceased human organs available for infants in need of a transplant. Following the procedure, particularly with the immense publicity associated with it, the situation with regard to donation of organs from infants became very much improved, and Bailey went on to develop an extremely successful cardiac *allo*transplantation program in infants and children at Loma Linda University (Fig. 1.18).

Clinical Liver Xenotransplantation

Thomas Starzl and Liver Xenotransplantation

Tom Starzl (Fig. 1.8), who was one of the greatest pioneers in the field of kidney and liver allotransplantation, performed a handful of liver transplants between nonhuman primates and young patients in Colorado in the 1960s, without lasting success (Table 1.1c) [27–30]. When the addition of tacrolimus had improved the

Fig. 1.18 Leonard Bailey with some of his first patients to undergo cardiac *allo*transplantation

immunosuppressive armamentarium, he and his team in Pittsburgh performed two liver transplants from baboons in adult patients in the 1990s, with one patient surviving for 70 days [31]. The first of these two cases can be considered a relative success in that there was little pathological evidence of rejection in the liver at any stage, but this was achieved probably at the expense of over-immunosuppression, the patient dying of overwhelming sepsis. The second case was less successful as the patient did not regain consciousness or renal function during the postoperative period, but again there was little histopathologic evidence of rejection in the transplanted liver. The results, however, were not successful enough to warrant continuing this experimental clinical trial.

Makowka et al. attempted to use a pig liver as a "bridge" to allotransplantation without success [32, 33].

In addition to the liver xenotransplants reviewed above, there have also been attempts to bridge patients with fulminant liver failure by perfusing their blood ex vivo through an animal liver until their native liver either recovered or a deceased human donor liver became available for transplantation. These were usually baboon or pig livers, and there is some evidence that this did prolong life, though the results were generally disappointing [34].

Comment

The data outlined above suggest that, if transplants were carried out today between nonhuman primates and humans, it is likely that, with the much-improved immunosuppressive drug therapy available to us now, relatively prolonged survival of these grafts would be obtained. However, it is extremely unlikely that nonhuman primates would be available in sufficient numbers to solve the shortage of donor organs today.

Table 1.2 The advantages and disadvantages of the pig as a potential source of organs and cells for humans, in contrast with those of the baboon in this role

	Pig	Baboon
Availability	Unlimited	Limited
Breeding potential	Good	Poor
Period to reproductive maturity	4–8 months	3–5 years
Length of pregnancy	114 ± 2 days	173–193 days
Number of offspring	5–12	1–2
Growth	Rapid (adult human size within 6 months)[a]	Slow (9 years to reach maximum size)
Size of adult organs	Adequate	Inadequate[b]
Cost of maintenance	Significantly lower	High
Anatomical similarity to humans	Moderately close	Close
Physiological similarity to humans	Moderately close	Close
Relationship of immune system to humans	Distant	Close
Knowledge of tissue typing	Considerable (in selected herds)	Limited
Necessity for blood type compatibility with humans	Probably unimportant	Important
Experience with genetic engineering	Considerable	None
Risk of transfer of infection (xenozoonosis)	Low	High
Availability of specific pathogen-free animals	Yes	No
Public opinion	More in favor	Mixed

[a]Breeds of miniature swine are approximately 50% of the weight of domestic pigs at birth and sexual maturity, and reach a maximum weight of approximately 30% of standard breeds
[b]The size of certain organs, for example, the heart, would be inadequate for transplantation into adult humans

Even with extensive breeding programs, the logistics would be such as to preclude the availability of these animals in large numbers. For example, baboons usually have only one or two offspring, and these take up to 9 years to grow to full size (Table 1.2). Furthermore, there is likely to be considerable ethical protest against the use of nonhuman primates for this purpose.

There is also the question of size (Table 1.2). Even chimpanzees do not grow to the same size as most adult humans, and therefore there may be restrictions on the function of organs, such as the heart, if transplanted into a large adult human. This problem would be exacerbated if the baboon or other Old World monkeys were utilized as the source of the organs. Even the largest baboon heart would not be of a size sufficient to support the circulation of a full-grown adult human. A pair of baboon kidneys or a baboon liver, however, may be sufficient to support life in adult humans, particularly with the well-known ability of the liver to hypertrophy rapidly under such circumstances, as was clearly demonstrated by the two Pittsburgh baboon-to-human liver transplants.

Most of the early attempts at clinical organ xenotransplantation used nonhuman primate species as sources of the organ, although there have been a few attempts using the pig and other non-primate mammals (Tables 1.1a, 1.1b, and 1.1c), but without significant success [3]. All of the early attempts at clinical xenotransplantation using widely disparate animal species, for example, pig and sheep, as sources of the organs were doomed to early failure. Histopathological examination of the rejected organ showed the typical features of hyperacute rejection, with massive capillary destruction with severe interstitial hemorrhage and edema.

Carl-Gustav Groth and the First Islet Xenotransplantation

There are an estimated 1–2 million patients with type 1 diabetes in the USA alone, and the number worldwide must be very large indeed. As pig insulin differs from human insulin by only one amino acid, and pig insulin was administered successfully for the treatment of diabetic patients for decades until recombinant human insulin became available, it is reasonable to anticipate that successful pig islet transplantation would result in normoglycemia. The Swedish group headed by Carl Groth (Fig. 1.19) was the first to attempt pig islet transplantation in patients with diabetes in 1993 [35]. Although porcine C-peptide was documented in the blood of some of the patients, indicating that some islets survived, no clinical benefit was obtained.

In recent years, there have been encouraging results from islet *allo*transplantation in patients with type 1 diabetes, but with such large numbers of patients suffering from the disease, the number of human pancreases that become available will never be sufficient to treat all of the potential patients, particularly as quite often two (or even three) human pancreases are required to provide enough islets to render a single recipient normoglycemic.

Xenotransplantation Using Pigs as Sources of Organs and Cells

The advantages of xenotransplantation, particularly if we could use a readily available animal source, such as the pig, would be numerous (Table 1.2) [36]. First, there would be an unlimited supply of "donor" organs, which would resolve the current increasing and severe shortage of human organs.

Second, these organs would be available electively whenever required, which is an equally important point. Currently, a patient may wait several months in an intensive care unit or supported by a left ventricular assist device while awaiting a heart transplant, or many years on chronic dialysis awaiting a kidney transplant. If transplantation could be carried out as soon as the patient is in irreversible organ failure, then immediate transplantation would almost certainly result in significantly improved survival.

Third – a point that is generally overlooked – is that brain death has numerous adverse effects on the donor organs, particularly the heart, that may lead to primary

Fig. 1.19 Carl-Gustav Groth (1933–2014)

graft non-function or other injury [37, 38]. In the case of the xenotransplantation of pig organs, this would be avoided, as the organs would be excised from a healthy pig under anesthesia.

Fourth, almost no year passes without a novel microorganism being transferred from a deceased human donor to the recipient with the organ graft. In recent years, West Nile virus, rabies, and other microorganisms have been transferred with fatal results. There has been some concern that a porcine microorganism might be transferred with a pig organ [39–41]. The pig herd will be housed under ideal (designated pathogen-free) conditions and be monitored at regular intervals for infectious agents, providing a much greater chance that the "donor" animal will be free of all known pathogenic organisms than the average deceased human donor.

Fifth, in several countries, there are cultural barriers to deceased organ donation, for example, Japan, and yet there are no barriers to xenotransplantation. The number of transplants performed in these countries would be vastly increased. The lack of deceased human donors, particularly with regard to kidney and liver transplantation, has popularized *living* donor transplants. When liver transplantation is undertaken in an adult recipient, this involves excision of almost two-thirds of the living donor liver which is then transferred to the adult recipient. This is a major surgical procedure, and not without some (small) risk to the living donor. There have been a small number of deaths of donors after these procedures, and the postoperative complication rate is significant. These tragedies would be avoided if pig livers could be used for the purpose.

The immunological and pathophysiological problems associated with pig xenotransplantation, however, are significant, and probably reflect the fact that it is 80 million years since the pig and human diverged on the evolutionary scale, and, therefore, in the words of Claus Hammer, what we are trying to do is to "outwit evolution."

Nevertheless, very significant progress has been made since we began to develop the ability to genetically modify large animals, particularly the pig. The "creation" of "Dolly" the sheep, the first cloned mammal, opened the gates to the possibility of rendering pig tissues resistant to the human immune response. It is only through this route that we have overcome many of the remaining barriers. Most of the barriers have now been identified and overcome (Chap. 2).

The experimental results of cell xenotransplantation, for example, islet or neuronal cells, were initially better than those of pig organ xenotransplantation. For example, pig islets have continued to function effectively in immunosuppressed nonhuman primates for periods in excess of a year [42–47]. Indeed, clinical trials of encapsulated pig islet transplantation have been carried out in diabetic patients in New Zealand [48] and Argentina.

There is also a great need for corneas, particularly in Asia and Africa. For example, it is estimated there are 4 million patients in need of corneal transplantation in China alone [11]. Experimental corneal xenotransplantation has made significant progress in recent years, with survival of pig lamellar grafts in monkeys surviving for periods in excess of 1 year, with the recipient receiving only corticosteroid drops to the eyes. Partly because the risk to the recipient would be small, it is likely that corneas will be among the first xenotransplants to be carried out in clinical trials, perhaps followed soon after by islet cell transplantation.

There are an estimated 8 million patients in the USA with a neurogenerative disease, such as Parkinson's disease. Human embryonic neural precursor cells can restore local motor activity in patients with Parkinson's disease, but the use of human embryos is largely precluded on ethical grounds or on logistic grounds as too few become available. Genetically engineered pig embryos might provide an alternative source. Indeed, a European group has reported encouraging improvement for >1 year in locomotor function in monkeys with a Parkinson-like condition after the transplantation of genetically modified pig dopamine-producing cells into the brain [49–51].

The words of George Orwell in *"Animal Farm"* will be apposite to organ transplantation in humans.

> The creatures outside looked from pig to man, and from man to pig, and from pig to man again; but already it was impossible to say which was which.

The same will 1 day be said about the doctor examining a patient with an organ transplant – the doctor will not be able to determine whether the organ is an allograft or a xenograft. Eventually, *allo*transplantation will be of historic interest only.

Acknowledgments The author thanks the editor of the Proceedings of the Baylor University Medical Center for permission to reproduce sections of his published paper. Work on xenotransplantation in the author's laboratory in the Department of Surgery at the University of Alabama at Birmingham has been supported in part by NIH grants U01 AI068642, R21 AI074844, and U19 AI090959, and by Sponsored Research Agreements between the UAB and United Therapeutics, Silver Spring, MD.

References

1. Cooper DKC, Lanza RP. Xeno – the promise of transplanting animal organs into humans. New York: Oxford University Press; 2000. p. 1–274.
2. Ekser B, Ezzelarab M, Hara H, et al. Clinical xenotransplantation – the next great medical revolution? Lancet. 2012;379:672–83.
3. Taniguchi S, Cooper DKC. Clinical xenotransplantation – past, present and future. Ann R Coll Surg Engl. 1997;79:13–9.
4. Deschamps JY, Roux FA, Sai P, Gouin E. History of xenotransplantation. Xenotransplantation. 2005;12:91–109.
5. Reemtsma K. Xenotransplantation – a brief history of clinical experience: 1900-1965. In: Cooper DKC, Kemp E, Reemtsma K, White DJG, editors. Xenotransplantation: the transplantation of organs and tissues between species. 1st ed. Heidelberg: Springer; 1991. p. 9–22.
6. Gibson T. Zoografting: a curious chapter in the history of plastic surgery. Br J Plast Surg. 1955;8:234–42.
7. Cooper DKC. Xenografting: the early, early years. Xeno. 1997;5:21–2.
8. Roux FA, Dai P, Deschamps J-Y. Xenotransfusions, past and present. Xenotransplantation. 2007;14:208–16.
9. Smood B, Bottino R, Cooper DKC. Xenotransplantation of the endocrine pancreas. In: Orlando G, Gruesner R, Piemonti L, Stratta R, Ricordi C, editors. Transplantation, bioengineering, and regeneration of the endocrine pancreas, vol. 1–2. New York: Elsevier/Academic Press; 2019.
10. Hara H, Cooper DKC. The immunology of corneal xenotransplantation: a review of the literature. Xenotransplantation. 2010;17:338–49.
11. Hara H, Cooper DKC. Xenotransplantation – the future of corneal transplantation? Cornea. 2011;30:371–8.
12. Hamilton D. The first transplant surgeon: the flawed genius of Nobel prize winner, Alexis Carrel. London: World Scientific Publishing; 1986.
13. Real J. Voronoff. (French). Paris: Editors Stock; 2001.
14. Voronoff S. Rejuvenation by grafting. New York: Adelphi; 1925.
15. Hamilton D. The monkey gland affair. London: Chatto and Windus; 1986.
16. Matevossian E, Kern H, Hüser N, et al. Surgeon Yurii Voronoy (1895-1961) – a pioneer in the history of clinical transplantation: in memoriam at the 75th anniversary of the first human kidney transplantation. Transpl Int. 2009;22:1132–9.
17. Lee RA. The bizarre careers of John R. Brinkley. Lexington: University Press of Kentucky; 2002.
18. Sgroi A, Buhler LH, Morel P, Sykes M, Noel L. International human xenotransplantation inventory. Transplantation. 2010;90:597–603.
19. Reemtsma K, McCracken BH, Schlegel JU, et al. Renal heterotransplantation in man. Ann Surg. 1964;160:384–410.
20. Starzl TE, Marchioro TL, Peters GN, et al. Renal heterotransplantation from baboon to man: experience with 6 cases. Transplantation. 1964;2:752–76.

21. Küss R. Human renal transplantation memories, 1951-1981. In: Terasaki PI, editor. History of transplantation: thirty-five recollections. Los Angeles: UCLA Tissue Typing Laboratory; 1991. p. 37–59.
22. Hardy JD, Chavez CM, Kurrus FD, et al. Heart transplantation in man: developmental studies and report of a case. JAMA. 1964;188:1132–40.
23. Cooper, D.K.C., Ye, Y. Experience with clinical heart xenotransplantation. In: Xenotransplantation: the transplantation of organs and tissues between species. 1st edition. Cooper, D.K.C., Kemp, E., Reemtsma, K., White, D.J.G. (eds). Springer; Heidelberg, 1991, pp. 541–557.
24. Barnard CN. The operation. A human cardiac transplant: an interim report of a successful operation performed at Groote Schuur Hospital, Cape Town. S Afr Med J. 1967;41:1271–4.
25. Barnard CN, Wolpowitz A, Losman JG. Heterotopic cardiac transplantation with a xenograft for assistance of the left heart in cardiogenic shock after cardiopulmonary bypass. S Afr Med J. 1977;52:1035–8.
26. Bailey LL, Nehlsen-Cannarella SL, Concepcion W, Jolley WB. Baboon-to-human cardiac xenotransplantation in a neonate. JAMA. 1985;254:3321–9.
27. Starzl TE, Marchioro TL, Faris TD, McArdle MJ, Iwasaki Y. Avenues of future research in homotransplantations of the liver. Am J Surg. 1966;112:391–400.
28. Starzl TE. Orthotopic heterotransplantation. In: Starzl TE, editor. Experience in hepatic transplantation. Philadelphia: W.B. Saunders; 1969. p. 408.
29. Starzl TE, Ishikawa M, Putnam CW, et al. Progress in and deterents to orthotopic liver transplantation, with special reference to survival, resistance to hyperacute rejection, and biliary duct reconstruction. Transplant Proc. 1974;6:129–39.
30. Giles GR, Boehmig HJ, Amemiya H, Halgrimson CG, Starzl TE. Clinical heterotransplantation of the liver. Transplant Proc. 1970;2:506–12.
31. Starzl TE, Fung JJ, Tzakis A, et al. Baboon-to-human liver transplantation. Lancet. 1993;341:65–71.
32. Makowka L, Cramer DV, Hoffman A, et al. The use of a pig liver xenograft for temporary support of a patient with fulminant hepatic failure. Transplantation. 1995;59:1654.
33. Makowka L, Wu GD, Hoffman A, et al. Immunohistopathologic lesions associated with the rejection of a pig-to-human liver xenograft. Transplant Proc. 1994;26:1074–5.
34. Abouna G. Extracorporeal xenogeneic liver perfusion for the treatment of hepatic failure. In: Cooper DKC, et al., editors. Xenotransplantation. 2nd ed. Heidelberg: Springer; 1997.
35. Groth CG, Korsgren O, Tibell A, et al. Transplantation of porcine fetal pancreas to diabetic patients. Lancet. 1994;344:1402–4.
36. Cooper DKC, Gollackner B, Sachs DH. Will the pig solve the transplantation backlog? Ann Rev Med. 2002;53:133–47.
37. Cooper DKC, Novitzky D, Wicomb WN. The pathophysiological effects brain death on potential donor organs, with particular reference to the heart. Ann R Coll Surg Engl. 1989;71:261–6.
38. Novitzky D, Wicomb WN, Cooper DKC. Pathophysiology of brain death and effects of hormonal therapy in large animal models. In: Novitzky D, Cooper DKC, editors. The brain-dead organ donor: pathophysiology and management. Springer: Heidelberg; 2012. p. 65–90.
39. Patience C, Takeuchi Y, Weiss RA. Infection of human cells by an endogenous retrovirus of pigs. Nat Med. 1997;3:282–6.
40. Onions D, Cooper DKC, Alexander TJ, et al. An assessment of the risk of xenozoonotic disease in pig-to-human xenotransplantation. Xenotransplantation. 2000;7:143–55.
41. Takeuchi Y, Fishman J. Long life with or without PERV. Xenotransplantation. 2010;17:429–30.
42. Hering BJ, Wijkstrom M, Graham ML, et al. Prolonged diabetes reversal after intraportal xenotransplantation of wild-type porcine islets in immunosuppressed nonhuman primates. Nat Med. 2006;12:301–3.
43. Cardona K, Korbutt GS, Milas Z, et al. Long-term survival of neonatal porcine islets in nonhuman primates by targeting costimulation pathways. Nat Med. 2006;12:304–6.

44. van der Windt DJ, Bottino R, Casu A, et al. Long-term controlled normoglycemia in diabetic non-human primates after transplantation with hCD46 transgenic porcine islets. Am J Transplant. 2009;9:2716–26.
45. Dufrane D, Goebbels RM, Gianello P. Alginate macroencapsulation of pig islets allows correction of streptozotocin-induced diabetes in primates up to 6 months without immunosuppression. Transplantation. 2010;90:1054–62.
46. Bottino R, Bottino R, Wijkstrom M, et al. Pig-to-monkey islet xenotransplantation using multi-transgenic pigs. Am J Transplant. 2014;14:2275–87.
47. Shin JS, Kim JM, Kim JS, et al. Long-term control of diabetes in immunosuppressed non-human primates (NHP) by the transplantation of adult porcine islets. Am J Transplant. 2015;15:2837–50. Erratum in: Am J Transplant. 2016; 16:1941.
48. Elliott RB. Towards xenotransplantation of pig islets in the clinic. Curr Opin Organ Transplant. 2011;16:195–200.
49. Martin C, Plat M, Nerriére-Daguin V, et al. Transgenic expression of CTLA4-Ig by fetal pig neurons for xenotransplantation. Transgenic Res. 2005;14:373–84.
50. Lévêque X, Cozzi E, Naveilhan P, Neveu I. Intracerebral xenotransplantation: recent findings and perspectives for local immunosuppression. Curr Opin Organ Transplant. 2011;16:190–4.
51. Vadori M, Aron Badin R, Hantraye P, Cozzi E. Current status of neuronal cell xenotransplantation. Int J Surg. 2015;23:267–72.

The Pathobiology of Pig-to-Primate Xeno.: A Historical Review

2

David K. C. Cooper, Mohamed B. Ezzelarab, Hidetaka Hara, Hayato Iwase, Whayoung Lee, Martin Wijkstrom, and Rita Bottino

Abbreviations

AHXR	Acute humoral xenograft rejection
EPCR	Endothelial protein C receptor
Gal	galactose-α1,3-galactose
GTKO	α1,3-galactosyltransferase gene-knockout
HAR	Hyperacute rejection
IBMIR	Instant blood-mediated inflammatory reaction
Neu5Gc	N-glycolylneuraminic acid
NHP	Nonhuman primate
TFPI	Tissue factor pathway inhibitor

Based on a paper published in *Xeno. 2016; 23:83–105*, with the editor's permission.

D. K. C. Cooper (✉)
Thomas E. Starzl Transplantation Institute, Department of Surgery, University of Pittsburgh Medical Center, Pittsburgh, PA, USA

Xeno. Program, Division of Transplantation, Department of Surgery, University of Alabama at Birmingham (UAB), Birmingham, AL, USA
e-mail: dkcooper@uabmc.edu

M. B. Ezzelarab · W. Lee · M. Wijkstrom
Thomas E. Starzl Transplantation Institute, University of Pittsburgh, Pittsburgh, PA, USA

H. Hara · H. Iwase
Thomas E. Starzl Transplantation Institute, University of Pittsburgh, Pittsburgh, PA, USA

Xeno. Program, Department of Surgery, University of Alabama at Birmingham (UAB), Birmingham, AL, USA

R. Bottino
Institute for Cellular Therapeutics, Allegheny-Singer Research Institute, Pittsburgh, PA, USA

Department of Biological Sciences, Carnegie-Mellon University, Pittsburgh, PA, USA

D. K. C. Cooper, G. Byrne (eds.), *Clinical Xenotransplantation*,
https://doi.org/10.1007/978-3-030-49127-7_2

Introduction

By the early 1960s, it was known from studies in pig-to-dog and dog-to-pig organ transplantation models that a xenograft would undergo an immediate immune response, known as hyperacute rejection (HAR) (reviewed in [1]). Subsequent studies in these models have added to our knowledge [2–7], but the pig-to-nonhuman primate (NHP) model is much more pertinent to the pig-to-human situation. Little was known about the pathobiology of pig-to-primate organ transplantation until studies were initiated in the clinically relevant wild-type (genetically unmodified) pig-to-baboon heart and kidney transplant models [8–11].

Organ Xenotransplantation

The Innate Immune Response

Following pig organ transplantation into a human [12, 13] or NHP [14, 15], HAR frequently occurs within minutes, although it has generally been defined as antibody-mediated complement activation leading to destruction of the graft within 24 hours. It is related to binding of primate natural (preformed) anti-pig antibodies to the vascular endothelial cells of the graft. Antibody deposition initiates complement-mediated injury of the endothelium, resulting in thrombosis, interstitial hemorrhage, and edema that disrupts graft function (Figs. 2.1 and 2.2), the histopathological features of which were initially described by Rose [16–18].

Before the nature of these natural preformed antibodies was known, an approach to prevent HAR was to administer an agent that depleted or inhibited complement, e.g., cobra venom factor or soluble complement receptor-1 (sCR-1), which extended graft survival significantly, but had only a temporary effect [19–23].

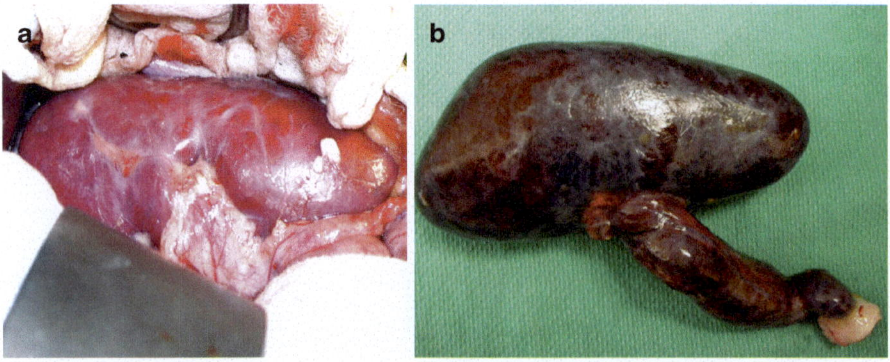

Fig. 2.1 Macroscopic appearance of a wild-type pig kidney immediately after transplantation and reperfusion in a baboon (**a**) and 10 minutes later when hyperacute rejection had occurred (**b**)

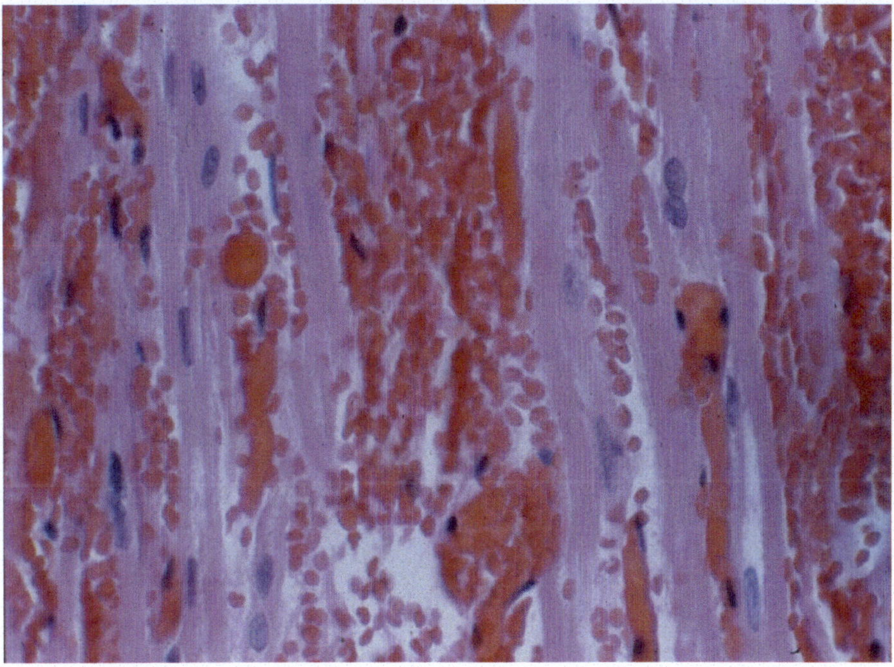

Fig. 2.2 Histopathology of hyperacute rejection in a wild-type pig heart graft. Complement-mediated injury associated with the binding of baboon natural preformed anti-pig antibodies to antigens expressed on the vascular endothelium of the pig organ results in intravascular thrombosis and interstitial hemorrhage. Acute humoral xenograft rejection (AHXR), a delayed antibody-mediated response, is often, but not always, associated with the production of elicited antibodies and has a similar histopathological appearance but possibly with the presence of rather more innate immune cells, e.g., macrophages, neutrophils. (Magnification ×400)

Pig Genetic Engineering

When genetic modification of the organ-source pig became possible, an approach to overcoming HAR was suggested by Dalmasso (in the USA) [24] and, independently, by White (in the UK) [25–27] and their respective colleagues, and by others [28]. The presence of complement-regulatory proteins on the surface of human vascular endothelial cells, e.g., decay accelerating factor (DAF, CD55), membrane cofactor protein (MCP, CD46), or membrane attack complex inhibitor protein (CD59) to some extent protects them from complement-mediated injury. Pig cells have equivalent complement-regulatory proteins, but these are less able to provide protection from the effects of *human* complement [29, 30]. In the mid-1990s, Dalmasso and White suggested introducing into the pig a transgene for a human complement-regulatory protein.

Several groups investigated this approach [25–27, 31–38]. As a consequence, the first genetically engineered pig directed toward xenotransplantation was produced by White and his colleagues in 1992 [26, 31, 39]. The expression of hCD55 provided considerable protection to the heart and kidney from HAR [15, 40, 41].

Anti-pig Antibodies

As the causative factors associated with HAR of a xenograft were seen to be similar to those of ABO-incompatible allograft rejection [42–45], a similar approach was taken to prevent rejection by depleting the potential recipient of anti-pig antibodies by plasmapheresis [11]. This prolonged graft survival beyond 24 hours and sometimes for a week or more, but the steady return of antibody resulted in graft failure, a phenomenon known variously as acute humoral xenograft rejection (AHXR), delayed xenograft rejection, or acute vascular rejection.

It was subsequently determined that the most important antibodies (IgM, IgG) directed against pig tissues bind to a carbohydrate epitope, galactose-α1,3-galactose (Gal) [46–52] reviewed in [53, 54]. This oligosaccharide is found on the surface of all pig vascular endothelial cells and some other tissues [55–58], and is also present in all other mammals, with the exception of humans and Old World NHPs (i.e., great apes, baboons, Old World monkeys) [55] reviewed in [59]. These primate species lost expression of Gal several million years ago, probably from a genetic mutation, and the absence of Gal resulted in primates making antibodies against this now "foreign" antigen [60].

These and other "natural" antibodies almost certainly develop as a response to Gal-expressing microorganisms and viruses that colonize the primate's gastrointestinal tract during neonatal life [61–65]. Of interest, baboons bred and housed in a specific pathogen-free facility have lower anti-pig antibody levels [66, 67]. Antibodies that develop in this way are known as "natural" or "preformed" antibodies. They were initially believed to be T cell-independent (as opposed to elicited antibodies that are T cell-dependent and develop after direct exposure to an antigen, e.g., antibodies that develop after an organ transplant), but some studies suggest that natural antibodies may also be T cell-dependent [63, 68].

The identification of anti-Gal antibodies enabled them to be specifically depleted by one of two methods (reviewed in [69]). Again based on experience with ABO-incompatibility studies [70, 71], the intravenous infusion of natural or synthetic Gal oligosaccharides was tested, which were bound by anti-Gal antibodies followed by elimination of the antibody-antigen complexes [72–82]. Alternatively, perfusing the potential recipient's blood or plasma through Gal immunoaffinity columns successfully depleted anti-Gal antibody [83–90]. However, even when combined with conventional immunosuppressive therapy (e.g., cyclosporine-based), these approaches were only modestly successful; they delayed antibody-mediated rejection, but the graft was lost when antibody levels recovered (AHXR) (Fig. 2.3).

α1, 3-Galactosyltransferase Gene-Knockout Pigs

When the importance of Gal antigens had first been established in 1992, it was suggested that the gene responsible for the enzyme that attached Gal terminally on oligosaccharide chains, α1,3-galactosyltransferase, should be deleted or knocked-out [91]. At this time, this was possible in mice [92–94], but not in pigs. Insights into the molecular basis of evolutionary inactivation of Gal, as well as the

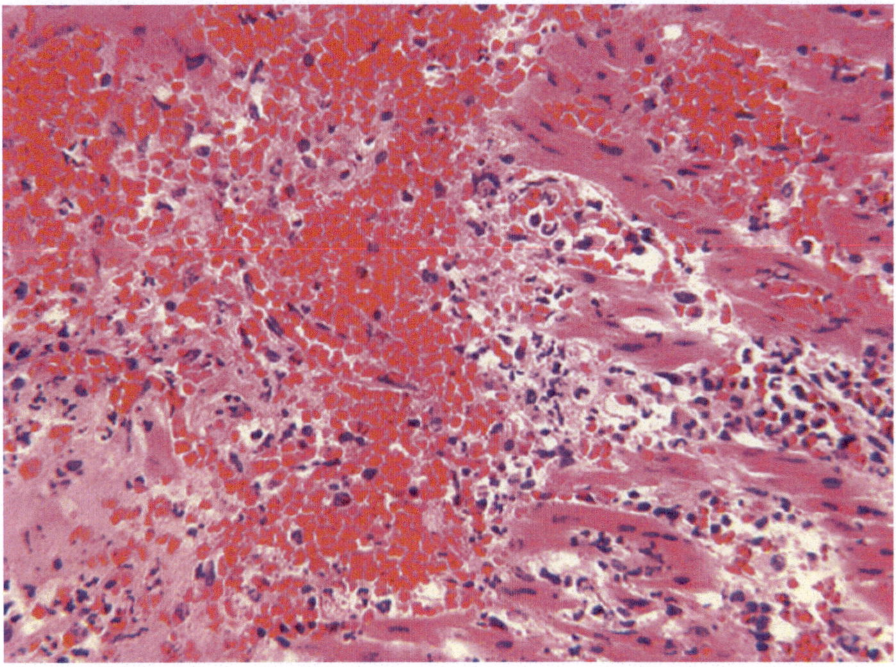

Fig. 2.3 Histopathology of acute humoral xenograft rejection (AHXR) in a GTKO heart transplanted into a baboon 12 days previously. Features of humoral rejection similar to those seen in hyperacute rejection (interstitial hemorrhage, edema) are present, but there is also a significant cellular infiltrate, mainly of polymorphonuclear neutrophils. (Magnification ×200). (Reprinted with permission from Ezzelarab et al. [375])

introduction of nuclear transfer/embryo transfer (cloning) technology in mammals [95–97], enabled α1,3-galactosyltransferase gene-knockout (GTKO) in pigs. The first GTKO pigs were reported in 2003 [98, 99].

Initial studies of the transplantation of organs from GTKO pigs in NHPs showed protection from HAR, with prolonged survival of pig heart and kidney grafts in some cases for weeks or months, though it should be noted that these studies were carried out in baboons selected for their low levels of anti-pig antibodies (i.e., in these cases, of anti-nonGal antibodies) [100–104]. However, in nonselected recipients [105] or when the immunosuppressive regimen was inadequate to prevent an adaptive immune response [106, 107], survival was limited.

When a GTKO pig organ is transplanted into a NHP, the natural antibodies involved in rejection of the graft are directed against nonGal antigens [90, 108–111], the exact nature of which remained uncertain [112–114] until recently, although one had been identified [115] (see below). HAR has only rarely been described after the transplantation of organs from pigs in which the Gal antigen is *not* expressed (GTKO pigs), but its occasional occurrence indicates that it can be initiated by binding of high levels of anti-nonGal antibodies [116].

The combination of GTKO and expression of a human complement-regulatory protein was even more successful in preventing early graft failure of a transplanted pig organ [117–120], as predicted by early studies in mice [121].

Nevertheless, even when HAR was prevented, AHXR could develop within a few days or weeks, almost certainly related to the binding of antibody and deposition of complement, and to the effect of innate immune cells (e.g., polymorphonuclear leukocytes, monocyte/macrophages, natural killer [NK] cells) and/or platelets that together activate the endothelium [105, 122, 123], resulting in graft injury.

The Adaptive Immune Response

If the innate immune response is prevented or delayed by the approaches outlined above, but immunosuppressive therapy is inadequate, a T cell-dependent elicited antibody response develops, resulting in high levels of anti-pig IgG [107, 124, 125]. Binding of these antibodies (whether directed to Gal or nonGal antigens on the vascular endothelium) initiates histopathological changes indistinguishable from AHXR. Surprisingly, histopathological features of pure acute cellular rejection (massive lymphocyte infiltration), as seen in the majority of inadequately immunosuppressed recipients of allografts, have virtually never been recorded after pig-to-NHP organ xenotransplantation, most probably because the innate response develops more rapidly and overwhelms the cellular response, though small numbers of T and B cells may be seen in the graft.

Activation of the adaptive response has been found to be more complex than after allotransplantation. For example, thrombin can induce porcine vascular endothelial cell activation without upregulation of swine leukocyte antigens (SLA), resulting in an increased primate T-cell response [126].

Immunosuppressive Therapy

Graft failure can be delayed by very high doses of conventional immunosuppressive therapy, e.g., tacrolimus, mycophenolate mofetil, and corticosteroids [127–129], but has been associated with a relatively high incidence of infectious complications [130]. Costimulation blockade-based immunosuppressive therapy, first introduced into xenotransplantation by Buhler in 2000 [124], has been more successful [66, 105, 131–138].

Anti-CD154 monoclonal antibody (mAb) was the first costimulation-blockade agent that was found to be effective in xenotransplantation [100, 101, 105, 124, 139, 140]. However, its administration is associated with thrombotic complications in both allogeneic [141, 142] and xenogeneic [143] organ and artery patch transplant models, and is currently not in clinical development. Its administration may well have contributed to the early development of thrombotic microangiopathy (see below), though it is clearly not the sole causative factor [135, 136].

Although blockade of the CD28/B7 pathway with CTLA4-Ig appears to potently suppress a T-cell response to pig cells in vitro [144, 145], it has been found to be inadequate in inhibiting an anti-pig antibody response in vivo [131, 135, 137, 138]. This suggests that other mechanisms of inducing sensitization (other than the T cell-dependent adaptive immune response) may be involved [126, 146]. For example, evidence has been provided that indicates that soluble CD154 plays a significant role [147]. In this respect, anti-CD154 mAbs may reduce class switching of anti-pig antibodies by binding both T-cell surface CD154 and circulating soluble CD154, thus preventing subsequent stimulation of B cells and activation of lymphoid follicles in secondary lymphoid tissues.

In contrast, blockade of the CD40/CD154 pathway with an anti-CD40mAb-based regimen (or a combined anti-CD40/CTLA4-Ig-based regimen) is effective in vivo in pig heart, kidney, or artery patch transplant models [133, 135, 136]. Although in some studies B-cell depletion has been associated with increased infectious complications [135], and although it does not immediately reduce anti-pig antibody levels [148], B-cell depletion using an anti-CD20mAb appears to be advantageous [129, 134].

Although not the original intention, genetic engineering of the organ-source pigs directed toward protection from the innate response in some cases also reduces the adaptive immune response [149, 150]. Deletion of Gal antigens [151] or expression of a human complement-regulatory protein [152–154] has been demonstrated to reduce the T-cell response to pig cells.

Pig Genetic Engineering

Pigs transgenic for the T-cell costimulation blockade agent, CTLA4-Ig, have been produced successfully [155]. In vitro assays demonstrated the potent immunosuppressive effect on human T-cell activation [156]. However, blood levels of soluble CTLA4-Ig in these pigs significantly exceeded the therapeutic level in patients receiving the agent after organ allotransplantation. The high levels of CTLA4-Ig rendered the pigs immunocompromised, prohibiting their long-term survival and therefore reproduction. Local expression of CTLA4-Ig in selected tissues, e.g., the islet beta cells (using an insulin promoter), is not associated with health problems in the pigs. When the islets are isolated and transplanted into a recipient NHP, it is hoped they will contribute a local immunosuppressive effect, though it is difficult to detect this in in vivo models, especially when associated with multiple other genetic modifications [157].

A second approach has been to express a mutant human Major Histocompatibility Complex (MHC) class II transactivator transgene in the pig [158]. This mutation results in downregulation of swine leukocyte antigen (SLA) class II expression and inhibits upregulation of expression after activation of the pig endothelial cells, thus reducing the human T-cell response in vitro [158]. In vivo studies in a pig artery patch transplantation model have demonstrated a modest immunomodulatory effect [137].

MHC class I-knockout pigs are also now available, but their effect on graft survival in NHPs has not yet been assessed [159]. Several other genetic approaches to reduce the T-cell response have been suggested.

Natural Killer (NK) Cells
NK cells have been demonstrated to participate in the cellular response to a pig graft [160–164] (although this has been difficult to detect in vivo), and transgenic expression of human leukocyte antigen (HLA)-G and/or E and/or Cw3 may inhibit their response [165–171]. Pigs expressing HLA-E of HLA-G have been produced, but not yet fully tested when on the preferable GTKO/human complement-regulatory protein background [172–174]. Studies by Miyagawa's group indicate that expression of these transgenes may also inhibit macrophage activity, which is likely to be important after organ, islet, and cell transplantation [173, 174].

Coagulation Dysfunction

Coagulation/thrombosis within the vessels of the graft has been observed even when HAR, AHXR, and the T-cell response are successfully controlled and has played a significant role in graft failure. This was predicted by early work in vitro and in rodent models by the groups of Robson/Bach and Platt [175–181].

In the pig-to-NHP model, coagulation dysregulation was first definitively described by Ierino, Kozlowski, and their respective colleagues in the late 1990s [83, 88, 182–184]. It can result in the development of a thrombotic microangiopathy, in which the vasculature of the organ is steadily occluded by thrombus, resulting in ischemic necrosis of the tissues (Fig. 2.4) [185]. The rapidity at which these pathologic changes develop varies, but they can occur within a few days or may be delayed for several weeks or months [100, 135, 136, 139, 186].

Small vessels in the graft become occluded by fibrin and platelet aggregation, leading to ischemic injury [103–105, 185, 187–191]. When thrombotic microangiopathy becomes advanced, the loss of platelets and clotting factors in the graft may result in the development of a consumptive coagulopathy in the recipient primate. The initiating cause of the thrombotic microangiopathy is almost certainly in part immune-related and probably results from activation of the graft vascular endothelial cells by antibody, complement, platelets, and innate immune cells, changing the phenotype of the vascular endothelial cells from anticoagulant to procoagulant. For example, the inducible procoagulant protein, fibrinogen-like protein 2 (fgl2), has been demonstrated to be upregulated on activated pig vascular endothelial cells in a pig-to-baboon kidney transplantation model, resulting in thrombin generation [192, 193].

However, other mechanisms play a role. For example, Lin and colleagues demonstrated in vitro that, in the *absence* of antibody and complement, procoagulant tissue factor could be induced on human platelets and monocytes by contact with porcine vascular endothelial cells [194]. In vivo, after pig kidney transplantation, activation of baboon platelets to express tissue factor was associated with the initiation of a consumptive coagulopathy in the relative absence of features of AHXR [186]. Furthermore, this upregulation of tissue factor on primate platelets occurs equally when exposed to GTKO, GTKO/CD46, or wild-type pig vascular endothelial cells [195]. This is associated with concomitant expression of P-selectin and

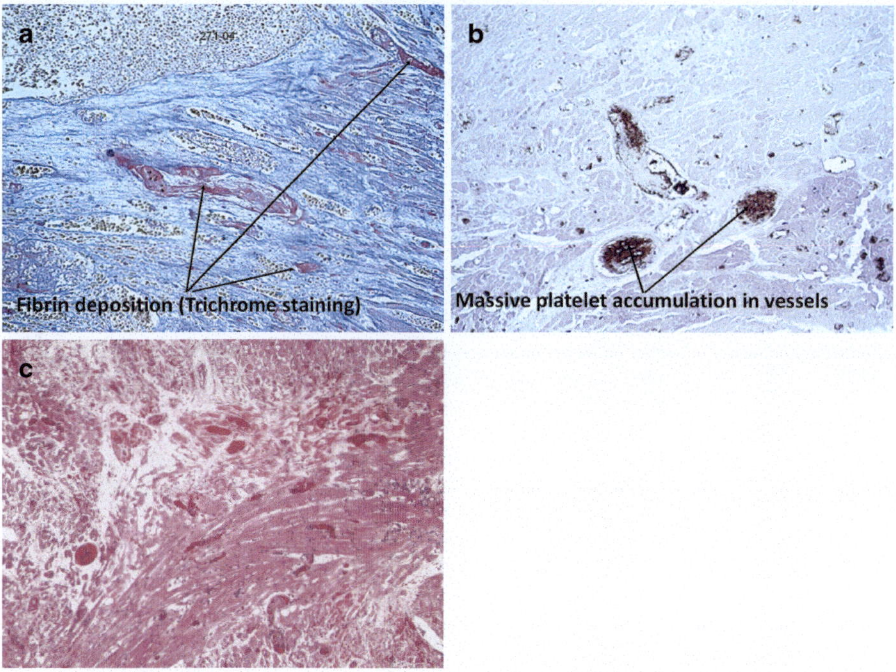

Fig. 2.4 Thrombotic microangiopathy in a GTKO pig heart graft. Fibrin deposition (**a**) and platelet aggregation (**b**) result in thrombosis within the vessels of the graft. Occlusion of small vessels results in ischemic injury with replacement fibrosis (**c**). (Magnification ×100)

P-selectin glycoprotein ligand-1 (PSGL-1), forming an auto-augmented loop of endothelial cells and platelet activation.

Almost certainly contributing to the coagulation dysregulation between graft and host are several known molecular incompatibilities in the coagulation/anticoagulation factors between pig and primate (Table 2.1) [106, 196–204]. For example, pig tissue factor pathway inhibitor (TFPI) does not successfully inhibit primate factor Xa, pig thrombomodulin does not catalyze primate protein C, and pig von Willebrand factor is associated with excessive primate platelet aggregation. The mechanism is complicated in that not only do the activated porcine endothelial cells express high levels of tissue factor (a procoagulant molecule) and increased tissue factor activity, but direct exposure of primate platelets and monocytes to porcine endothelial cells results in increased tissue factor activity on these primate structures also [186, 194].

Spontaneous bleeding in the NHP recipient can be the end result of the consumption of clotting factors [182, 183, 186]. If the organ graft is excised at the first sign of consumptive coagulopathy, then the condition can be rapidly reversed, confirming its association with the presence of the graft [124]. A consumptive coagulopathy appears to develop more rapidly in NHPs with pig *kidney* grafts, whereas a thrombotic microangiopathy predominates after pig *heart* transplantation (followed

Table 2.1 Major molecular incompatibilities relating to the coagulation system between pigs and primates

"Anticoagulant" genes	Molecular incompatibilities	Solution
von Willebrand factor (vWF)	Pig vWF spontaneously aggregates human platelets in the absence of shear stress, due to an aberrant interaction between the O-glycosylated A1 domain of pig vWF and human GPIb.	Replace pig vWF with human vWF in the pig vascular endothelial cells (EC) (by knockout-knockin technology)
Tissue factor pathway inhibitor (TFPI)	Pig TFPI does not effectively neutralize human factor Xa, although the recombinant forms expressed in primate cells have full anticoagulant activity.	Expression of human TFPI on pig EC
Thrombomodulin (TBM)	Pig TBM binds human thrombin, but is a poor cofactor for the activation of human protein C, with only 1–10% of the activity of human TBM.	Expression of human TBM on pig EC
Endothelial protein C receptor (EPCR)	Pig EPCR promotes activation of human protein C by human TBM, albeit less efficiently than human EPCR	Expression of human EPCR on pig EC
Ectonucleoside triphosphate diphosphohydrolase-1 (CD39)	The loss of CD39 activity following EC activation and injury results in the generation of procoagulant	Expression of human CD39 on pig EC
Ecto-5′-nucleotidases (CD73)	The loss of CD73 activity following EC activation and injury results in the generation of procoagulant	Expression of human CD73 on pig EC

EC endothelial cells, *EPCR* endothelial protein C receptor, *TBM* thrombomodulin, *TFPI* tissue factor pathway inhibitor, *vWF* von Willebrand Factor

terminally by a consumptive coagulopathy) [100, 186]. This observation has been investigated at the molecular level by Knosalla et al. [205].

The administration of anticoagulants and/or anti-thrombotic agents to the recipient NHP met with only partial success in preventing this problem [100, 102, 103, 105, 124, 187, 206, 207]. Heparin seemed beneficial, possibly because it also has anti-inflammatory properties. However, evidence has been presented indicating that increased immunosuppression may be more effective than anticoagulant agents, probably by suppressing the vascular endothelial-activating effect of the innate immune system [206–208]. The encouraging results achieved after pig heart [100, 101] or kidney [102, 138] transplantation in NHPs with low anti-pig antibody levels is likely related to the absence of endothelial cell activation.

Pig Genetic Engineering

This is one more pathobiological barrier that is being overcome by further genetic manipulation of the organ source pig, e.g., by the insertion of a human coagulation-regulatory gene (an "anticoagulant" or "anti-thrombotic" gene), such as thrombomodulin, endothelial protein C receptor (EPCR), TFPI, CD39, or CD73 [197, 209–225] or by knockdown of tissue factor expression [226]. For example, the expression of thrombomodulin and/or EPCR in the pig graft appears to inhibit or delay the development of thrombotic microangiopathy [132] even in the absence of heparin therapy [135, 136], though the beneficial effect of a potent immunosuppressive regimen cannot be excluded.

Most encouragingly, Mohiuddin et al. have reported >24 months survival of a heterotopic heart graft in a baboon in the absence of the development of thrombotic microangiopathy when a GTKO/hCD46 graft expressed thrombomodulin [132, 134]. (There may be added protection if other transgenes are also expressed [135]). By suppressing tissue factor activity on the porcine endothelial cells, TFPI may prevent the change to a procoagulant phenotype and may also prevent activation of the recipient cells and platelets. CD39 is another critical thromboregulatory molecule on endothelial cells that may limit platelet activation [227]. With the expression of several of these human transgenes, additional drug therapy to enhance an anti-thrombotic state may not be required, although the administration of aspirin or other oral antiplatelet agent may prove advantageous [228].

Recent studies indicate that expression of human thrombomodulin is also partially protective against complement activation, illustrating the complexity of the immune and coagulation responses [229].

One other factor must not be ignored, and that is that the presence of certain microorganisms in the organ-source pig, e.g., cytomegalovirus, may be associated with greater coagulation dysfunction (±rejection) [230–237]. Early weaning of pigs results in an absence of cytomegalovirus and reduces the development of consumptive coagulopathy after organ transplantation in NHPs. Indeed, this may have been a factor in the studies reported by Schirmer and Byrne [206–208] (see above) as their organ-source pigs were bred and housed under "clean" designated pathogen-free isolation conditions, whereas those used in most other studies were not.

Inflammatory Response

Recently, attention has been directed to the inflammatory response that develops after pig organ transplantation in NHPs, a topic that was again previously investigated in vitro and in rodent models. Evidence has been presented that inflammation precedes the development of coagulation dysfunction [146, 238], illustrating the inter-relationships between the immune, coagulation, and inflammatory responses, for which there is increasing evidence [239–242]. In particular, the potential role of interleukin-6 (IL-6) has been highlighted [146, 238], and the beneficial effect of an IL-6 receptor antagonist is being explored [136, 238]. Statin therapy may also prove beneficial by reducing both the T-cell response [243] and tissue factor activity on pig vascular endothelial cells [244].

In addition to pharmacological anti-inflammatory approaches, the infusion of mesenchymal stromal cells, derived from either primates of possibly isolated from the same genetically-engineered pigs that provide the organ or cells, is likely to have a beneficial anti-inflammatory effect (in addition to an immunomodulatory effect) [152, 245–247].

Pig Genetic Engineering

An alternate approach, of course, is to produce genetically engineered pigs that have some protection against the inflammatory response, e.g., pigs expressing human hemeoxygenase-1 (HO-1) or A20 [248–251]. Expression of these transgenes would likely have the added advantage of reducing the apoptosis of pig cells. These pigs are available, but not necessarily on a background of GTKO ± a human complement-regulatory protein ± a human coagulation-regulatory protein, and therefore the specific effect of the anti-inflammatory manipulation has not been determined in vivo. Nevertheless, initial ex vivo perfusion experiments using human blood show encouraging results [252].

Graft Vasculopathy (Chronic Rejection)

Graft vasculopathy (chronic rejection) has been documented in some pig cardiac xenografts that have functioned for >3 months in baboons and has a similar histopathological appearance to that seen after allotransplantation (Fig. 2.5) [139]. Its pathogenesis in both allotransplantation and xenotransplantation remains poorly understood, but is almost certainly associated with a prolonged and continuing low-grade immune response that induces chronic, low-grade activation of the graft

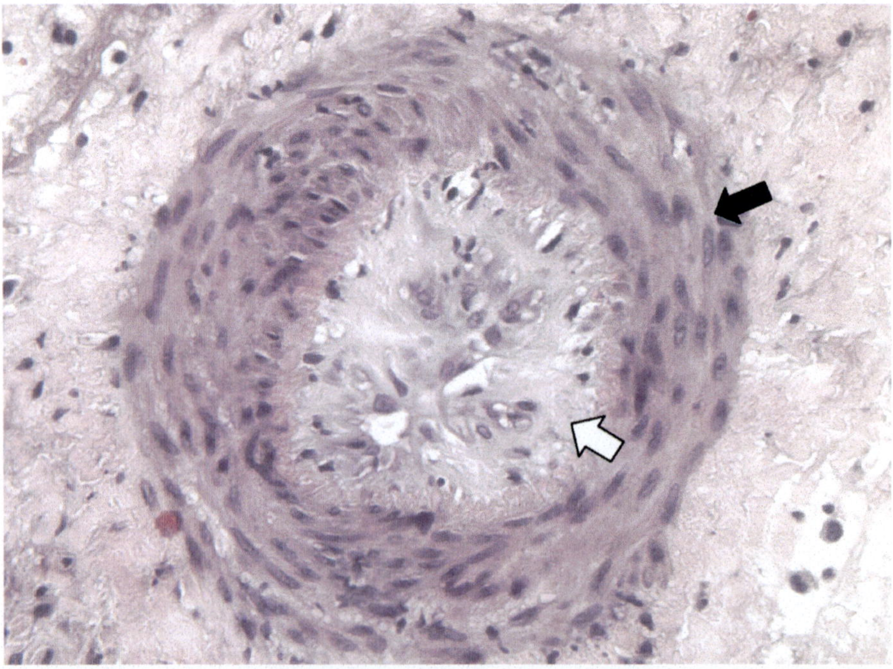

Fig. 2.5 Graft vasculopathy (chronic rejection) in a hCD55 pig heart graft 3 months after transplantation into an immunosuppressed baboon. Smooth muscle proliferation (black arrow) and neo-intimal proliferation (white arrow) are indicated. (Magnification ×200)

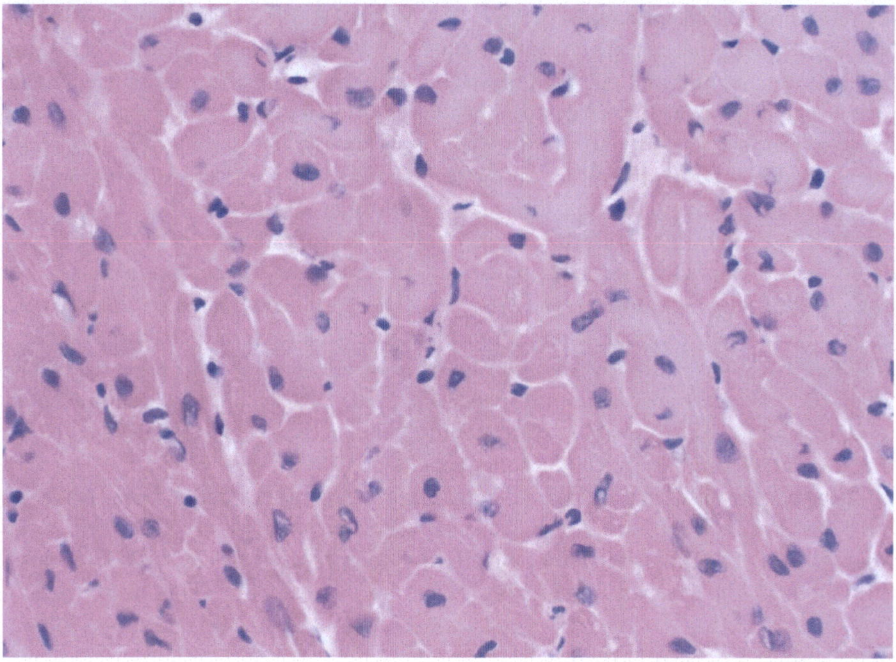

Fig. 2.6 Microscopic appearance of the myocardium of a GTKO/CD46/TBM pig heart transplanted heterotopically into an immunosuppressed baboon 1 year previously. No significant histopathological changes are seen. (Magnification ×200). (Courtesy of Muhammad Mohiuddin MD, NHLBI, NIH, Bethesda, MD, USA)

vascular endothelium. Some encouragement, however, is provided by the ongoing studies of Mohiuddin et al. [132] in which graft vasculopathy has not been reported in heart grafts from GTKO/hCD46 pigs expressing human thrombomodulin, with follow-up in one case for >2 years (Fig. 2.6).

The Special Cases of Pig Liver or Lung Transplantation

The pathobiological problems associated with transplantation of pig livers and lungs are more complex and to date have limited graft survival to days rather than weeks, months, or years. A major problem, particularly with pig liver transplantation, is phagocytosis of recipient platelets and possibly erythrocytes by pig macrophages [253–255]. Overcoming these problems may require pigs with multiple genetic manipulations, as has been discussed elsewhere [14, 209, 256–270].

Remaining Barriers to Pig Organ Xenotransplantation

Humans are known to have natural antibodies to at least one other antigen on pig cells, namely N-glycolylneuraminic acid (Neu5Gc) [271–275] reviewed in [276].

As most nonhuman mammals express this oligosaccharide, including Old World NHPs (even chimpanzees), it is not possible to investigate its effect in the standard in vivo experimental models, e.g., pig-to-baboon organ transplantation. The recent report that New World monkeys do *not* express Neu5Gc [277], and therefore are likely to produce anti-Neu5Gc antibodies, suggests that these monkeys might provide a model in which the effect on the immune system of the transplantation of a Neu5Gc-expressing pig organ (or tissues or cells) can be studied.

Most information to date has been obtained from staining of pig tissues for expression of Neu5Gc (Fig. 2.7), and laboratory assays, such as antibody binding by flow cytometry using human sera [278–280]. These indicate that Neu5Gc may be a significant barrier to clinical xenotransplantation There is increasing evidence, therefore, that this antibody–antigen interaction will need to be prevented if pig organs or cells are transplanted into human patients. The recent production of pigs in which the gene responsible for producing Neu5Gc has been deleted (as well as that for Gal) is a major step toward this goal [145, 279, 281].

A further pig antigen that may have importance in graft rejection, though this has not yet been confirmed in in vivo experiments, is porcine β1,4 N-acetylgalactosaminyltransferase [115]. Baboons have preformed antibodies to this glycan [115], as do most humans [280].

Physiologic Considerations

If problems related to the immune response can be completely resolved, then attention can be directed to whether pig organs will function normally in primate hosts. The very few studies to date suggest that some problems that were initially perceived to be related to physiological differences between pig and primate were in fact related to the effects of the immune response [136, 282, 283]. The high levels of proteinuria seen in the early baboons with pig kidney grafts [102, 282] have not been seen in recent experiments when protection from the immune response was increased by the genetic engineering of the pig [136] or when the NHP recipient was selected for low levels of anti-pig antibody [138]. Any true physiological differences that are identified are likely to be resolved by genetic manipulation of the pig, e.g., to replace a porcine metabolic product with a human one. Rapid growth of a pig organ may also be problematic [136, 284], and the rate of growth of a pig organ after transplantation into a NHP is as yet uncertain.

Cell and Tissue Xenotransplantation

Pigs could also act as a source of other tissues, such as pancreatic islets (for the treatment of diabetes) [285, 286], neuronal cells (for the treatment of neurodegenerative conditions, such as Parkinson's disease) [287, 288], corneas (for patients with corneal blindness) [289–294], and red blood cells (for clinical transfusion) [295–299]. Indeed, clinical trials of encapsulated pig islet [300, 301] and partial-thickness corneal [302] transplantation have already been undertaken.

The pathobiology of graft injury is similar to that seen in pig organs, though the absence of the vascular endothelium in some cases modifies the histopathological

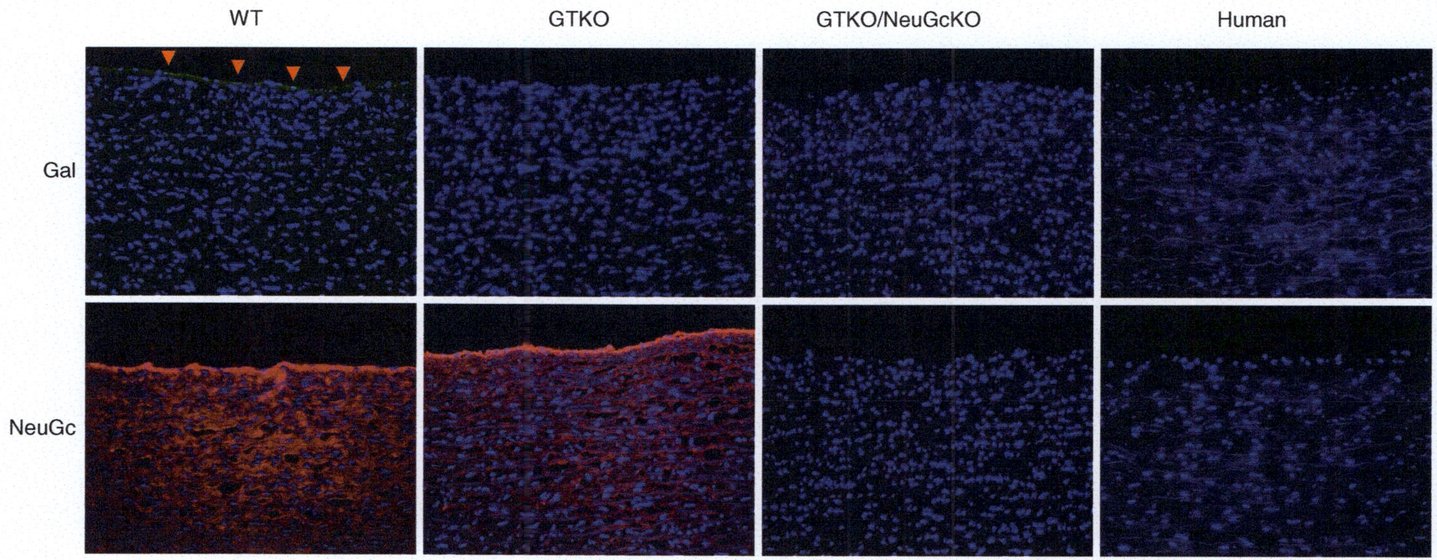

Fig. 2.7 *(Left to right)* Expression of Gal and Neu5Gc on aortas from wild-type, GTKO, and GTKO/Neu5GcKO pigs, and also on a human aorta. Expression of Gal is determined by staining with the isolectin B4 from *Bandeiraea simplicifolia*, and expression of NeuGc by staining with a chicken-derived anti-Neu5Gc immunohistochemistry set. Therefore, it is not possible to make a direct quantitative comparison of the level of expression between the two oligosaccharides. However, Gal (green) is expressed mainly on the vascular endothelium, whereas Neu5Gc (red) is much more widely expressed in all layers, including the vascular endothelium. (Cell nuclei – blue; Gal – green; NeuGc – red. Magnification ×200)

appearances. A T-cell-induced response appears to play a more important role, as in allotransplantation [303, 304].

Islet Xenotransplantation

Considerable attention has been paid to the transplantation of porcine islets of Langerhans as a potential means of providing a source of insulin in patients with Type 1 diabetes [286]. Islets from *adult* wild-type (unmodified) pigs, if carefully isolated, express very low levels of the Gal epitope (though expression can be upregulated by an inflammatory response), and therefore the problem of anti-Gal antibodies is reduced [305]. In contrast, fetal and neonatal islets (that may prove preferable sources of islets for clinical transplantation [306–309]), express significant levels of Gal [310], and therefore GTKO pigs will likely be essential, or at least preferable [311].

Similarly, Neu5Gc will require deletion because humans have antibodies targeting Neu5Gc. However, preliminary data from in vitro studies show that human antibody binding to GTKO/CD46/CMAHKO (i.e., Neu5GcKO) pig islets is no less than to GTKO/CD46 islets [281], suggesting that additional antigens may need to be knocked-out to sustain islet function in this species combination [312–314].

Techniques are available, e.g., by the use of an insulin promoter, that enable a desired transgene to be expressed in the islet beta cells alone [315], thus negating any potential detrimental effects (e.g., systemic anticoagulation) that might be associated with widespread expression of a human coagulation-regulatory protein, e.g., TFPI [157, 316]. However, despite promising results in the wild-type pig-to-NHP model with islet graft function exceeding 600 days [317], experimental islet xenotransplantation has not yet been consistently successful [157], and further study is required.

As with pig organ grafts, the cellular immune response to porcine cells can be inhibited by the currently available immunosuppressive agents, particularly by those that result in blockade of T-cell costimulation [318]. Both genetically engineered and wild-type *adult* porcine islets have been demonstrated to maintain normoglycemia in diabetic monkeys for >6 months [319, 320] and even for >1 year (Fig. 2.8) [157, 317, 321]. Wild-type *neonatal* pig islets have maintained normoglycemia in diabetic monkeys for >6 months [311, 322, 323].

However, the immunosuppressive regimens utilized in most of these studies would currently not be clinically applicable (mainly because they included an anti-CD154mAb), and therefore present studies are aimed toward reducing the intensity of the regimen and ensuring that all agents administered are approved for clinical use. Genetic manipulation of the pig to increase the resistance of the islets to the primate T-cell response is a potential option, as is the cotransplantation of islets with mesenchymal stromal cells [152, 245–247, 324–326] or Sertoli cells [327–332].

The Instant Blood-Mediated Inflammatory Reaction (IBMIR)

The current site of transplantation of both allogeneic islets and pig xenogeneic islets is into the portal vein, where the islets lodge in the liver. Direct contact between

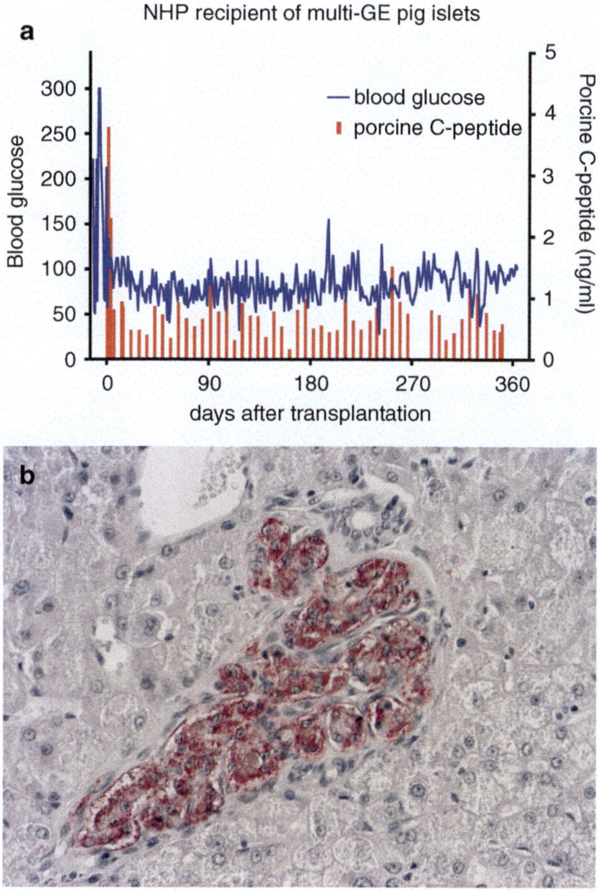

Fig. 2.8 (**a**) Blood glucose and pig C-peptide levels in a streptozotocin-induced diabetic cynomolgus monkey before and after intraportal transplantation of islets from a pig expressing the human complement-regulatory protein, CD46. No exogenous insulin was administered after the transplant. The normoglycemic monkey was electively euthanized after 12 months. Tx = day of islet transplantation. (**b**) Insulin immunostaining (in red) of a liver section in a monkey recipient of islets from a pig transgenic for human CD46, showing a healthy pig islet 12 months after transplantation. (Magnification ×200)

islets (particularly xenogeneic islets) and blood leads to an immediate inflammatory response that destroys a large percentage of the infused islets; this is known as the instant blood-mediated inflammatory reaction (IBMIR) [333–338]; reviewed in [339]. It is a response to cells that are not normally present in the blood – in this case, the islets. Until relatively recently, it was thought to be primarily a nonspecific inflammatory response that involved activation of the coagulation and complement systems. Recent evidence suggests that the anti-pig immune response is playing a significant role and that IBMIR may even be a form of HAR [309, 340, 341] (Fig. 2.9). It can be reduced, though not completely prevented, by treatment of the recipient with anticoagulant and anticomplement agents, but genetic engineering of the pig will probably prove more beneficial [157, 315].

An alternative approach to avoid IBMIR is to transplant the islets in a site where they are not immediately exposed to blood, e.g., the gastric submucosal space [342, 343], skeletal muscle [344], omental pouch [345], etc.

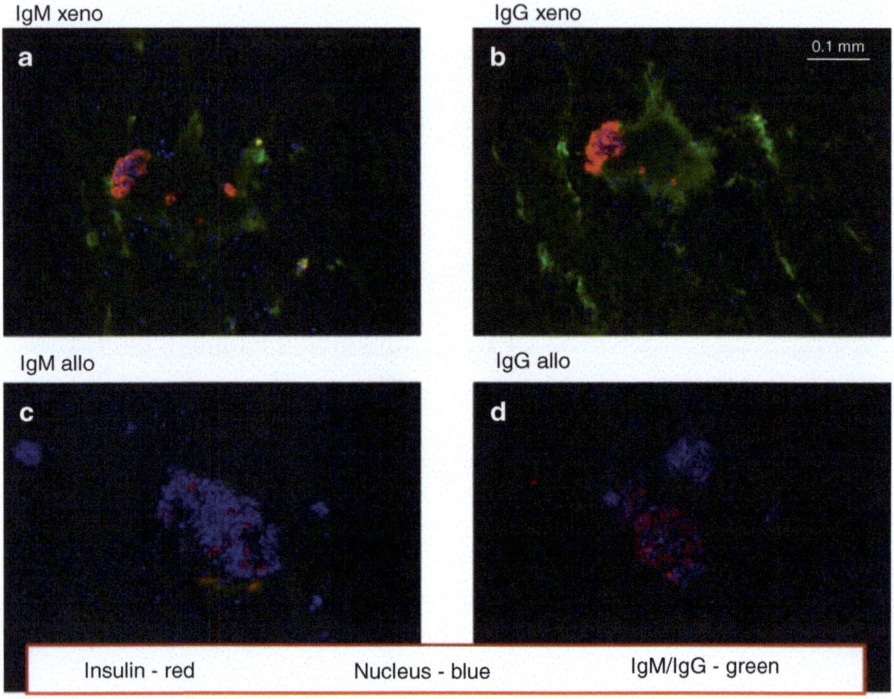

Fig. 2.9 Binding of human IgM and IgG antibody to pig islets (xenogeneic) (**a**, **b**) and to human islets (allogeneic) (**c**, **d**). IgM (green, **a**, **c**), IgG (green, **b**, **d**), insulin (red), nucleus (DAPI/blue). Yellow indicates co-localization of insulin and IgM/IgG. There is much greater antibody binding (both of IgM and IgG) to the pig (xenogeneic) islets than to human (allogeneic) islets

There have been some encouraging reports of prolonged islet graft survival in nonhuman primates in the absence of exogenous immunosuppressive therapy when the islets have been encapsulated [346], but it is unlikely this approach will be entirely successful. Any microcapsule that allows insulin to escape is almost certain to allow cytokines and chemokines to enter, with the risk of injury to the islets. Furthermore, it remains uncertain whether microcapsules allow sufficient nutrients and oxygen to access the islets, and so graft failure may result.

Physiologically, as pig insulin was administered to diabetic patients for many years, there is every evidence that pig insulin will function well in primates [347, 348], and they may even have some advantages [349].

Induction of Tolerance to Pig Organs and Cells

The ultimate goal in both allotransplantation and xenotransplantation is the induction of immunological "tolerance," in which the immune system of the recipient is manipulated so that the transplanted organ or cells are accepted as "self" with no effort made to reject them, thus allowing immunosuppressive therapy to be tapered and discontinued. In xenotransplantation, the identification and availability of the "donor" prior to the transplant allows the immunologic manipulation to be accurately "timed" and may facilitate the induction of tolerance to the transplanted organs or cells.

Efforts in this respect have been made through attempting to induce hematopoietic cell chimerism [350–352] or by pig thymic transplantation [102, 353–356], but without complete success to date. A major problem has been the inability to prevent recipient macrophages from phagocytosing pig hematopoietic cells. Expression of human CD47 on the pig cells may reduce this problem [357]. It is likely that tolerance will only be achieved when the other barriers (discussed above) have been overcome.

Conclusions

The technology of genetic engineering is steadily improving and new techniques have been introduced, such as zinc finger nucleases [358–363] and transcription activator-like effector nucleases (TALENS) [364], which may lead to greater efficiency [365]. In particular, genome editing by RNA-guided endonucleases (also known as clustered regularly interspaced short palindromic repeat, CRISPR) significantly increases gene-editing efficiency. The CRISPR/Cas9 system technology allows the rapid production of pigs with multiple genetic modifications [366–371], which is having an impact on the development of new pigs for xenotransplantation [280, 372].

Worldwide, there are currently at least 25 different genetic modifications expressed in pigs, with some pigs expressing six of these (Table 2.2) [157, 315, 373].

Table 2.2 Selected genetically modified pigs currently available for Xeno. research

Complement regulation by human complement-regulatory gene expression
CD46 (membrane cofactor protein)
CD55 (decay-accelerating factor)
CD59 (protectin or membrane inhibitor of reactive lysis)
Gal or nonGal antigen "masking"' or deletion
Human H-transferase gene expression (expression of blood type O antigen)
Endo-beta-galactosidase C (reduction of Gal antigen expression)
α1,3-galactosyltransferase gene-knockout (GTKO)
Cytidine monophosphate-N-acetylneuraminic acid hydroxylase (CMAH) gene-knockout (NeuGcKO)
β4GalNT2 (β1,4 N-acetylgalactosaminyltransferase) gene-knockout (β4GalNT2KO)
Suppression of cellular immune response by gene expression or downregulation
CIITA-DN (MHC class II transactivator knockdown, resulting in swine leukocyte antigen class II knockdown)
Class I MHC-knockout (MHC-IKO)
HLA-E/human β2-microglobulin (inhibits human natural killer cell cytotoxicity)
Human FAS ligand (CD95L)
Human GnT-III (*N*-acetylglucosaminyltransferase III) gene
Porcine CTLA4-Ig (Cytotoxic T-Lymphocyte Antigen 4 or CD152)
Human TRAIL (tumor necrosis factor-alpha-related apoptosis-inducing ligand)

(continued)

Table 2.2 (continued)

Anticoagulation and anti-inflammatory gene expression or deletion
von Willebrand factor (vWF)-deficient (natural mutant)
Human tissue factor pathway inhibitor (TFPI)
Human thrombomodulin
Human endothelial protein C receptor (EPCR)
Human CD39 (ectonucleoside triphosphate diphosphohydrolase-1)
Anticoagulation, anti-inflammatory, and antiapoptotic gene expression
Human A20 (tumor necrosis factor-alpha-induced protein 3)
Human heme oxygenase-1 (HO-1)
Human CD47 (species-specific interaction with SIRP-α inhibits phagocytosis)
Porcine asialoglycoprotein receptor 1 gene-knockout (ASGR1-KO) (decreases platelet phagocytosis)
Human signal regulatory protein α (SIRPα) (decreases platelet phagocytosis by "self" recognition)
Prevention of porcine endogenous retrovirus (PERV) activation
PERV siRNA

Table courtesy of Burcin Ekser MD, PhD

There is a Native American proverb that states, "Timing has a lot to do with the success of a rain dance." The availability of these multi-transgenic pigs, together with the novel immunosuppressive and anti-inflammatory agents now becoming available, is likely to overcome any remaining pathobiological problems that currently prevent xenotransplantation from being introduced successfully into clinical practice. For xenotransplantation, the timing appears to be right [374].

Acknowledgments We thank Burcin Ekser MD, PhD, for permission to include Table 2.1. Work on Xenotransplantation in the Thomas E. Starzl Transplantation Institute at the University of Pittsburgh was supported in part by NIH grants U01 AI068642, R21 AI074844, and U19 AI090959, and by Sponsored Research Agreements between the University of Pittsburgh and Revivicor, Inc., Blacksburg, VA.

Conflict of Interest The authors report no conflict of interest.

References

1. Najarian JS. Experimental Xenotransplantation: a personal history. Xenotransplantation 2003;10:10–5.
2. Tector AJ, Chen X, Soderland C, Tchervenkov JI. Complement activation in discordant hepatic xenotransplantation. Xenotransplantation. 1998;5:257–61.
3. Tector AJ, Fridell JA, Watanabe T, et al. Pulmonary injury in recipients of discordant hepatic and renal xenografts in the dog-to-pig model. Xenotransplantation 1998;5:44–9.
4. Tector AJ, Fridell JA, Watanabe T, et al. Aberrations in hemostasis and coagulation in untreated discordant hepatic Xenotransplantation: studies in the dog-to-pig model. Liver Transpl. 2002;8:153–9.
5. Tector AJ, Fridell JA, Ruiz P, et al. Experimental discordant hepatic Xenotransplantation in the recipient with liver failure: implications for clinical bridging trials. J Am Coll Surg. 2000;191:54–64.

6. Baker J, Martin M, Ye Y, Oleinick S, Zuhdi N, Cooper DKC. Presence of a complement-dependent "cytotoxic factor" in dog serum: relevance to experimental discordant Xenotransplantation Transplant Proc. 1992;24:490–1.
7. Ye Y, Cooper DKC, Niekrasz M, et al. Removal of dog antipig antibody by adsorption with pig red blood cell stroma columns. Transplant Proc. 1992;24:563–5.
8. Lexer G, Cooper DKC, Rose AG, et al. Hyperacute rejection in a discordant (pig to baboon) cardiac xenograft model. J Heart Transplant. 1986;5:411–8.
9. Cooper DKC, Human PA, Lexer G, et al. Effects of cyclosporine and antibody adsorption on pig cardiac xenograft survival in the baboon. J Heart Transplant. 1988;7:238–46.
10. Cooper DKC, Rose AG. Experience with experimental xenografting in primates. In: Hardy MA, editor. Xenograft 25. Amsterdam/New York/Oxford: Excerpta Medica; 1989. p. 95–105.
11. Alexandre GPJ, Gianello P, Latinne D, Carlier M, Dewaele A, van Obbergh L. Plasmapheresis and splenectomy in experimental renal Xenotransplantation In: Hardy M, editor. Xenograft 25. Amsterdam/New York/Oxford: Excerpta Medica; 1989. p. 259–66.
12. Taniguchi S, Cooper DKC. Clinical Xenotransplantation: past, present and future. Ann R Coll Surg Engl. 1997;79:13–9.
13. Cooper DKC. A brief history of cross-species organ transplantation. Proc (Bayl Univ Med Cent). 2012;25:49–57.
14. Cooper DKC, Satyananda V, Ekser B, et al. Progress in pig-to-nonhuman primate transplantation models (1998-2013): a comprehensive review of the literature. Xenotransplantation 2014;21:397–419.
15. Lambrigts D, Sachs DH, Cooper DKC. Discordant organ Xenotransplantation in primates: world experience and current status. Transplantation. 1998;66:547–61.
16. Rose AG, Cooper DKC, Human PA, Reichenspurner H, Reichart B. Histopathology of hyperacute rejection of the heart: experimental and clinical observations in allografts and xenografts. J Heart Lung Transplant. 1991;10:223–34.
17. Rose AG, Cooper DKC. A histopathologic grading system of hyperacute (humoral, antibody-mediated) cardiac xenograft and allograft rejection. J Heart Lung Transplant. 1996;15:804–17.
18. Rose AG, Cooper DKC. Venular thrombosis is the key event in the pathogenesis of antibody-mediated cardiac rejection. Xenotransplantation 2000;7:31–41.
19. Leventhal JR, Dalmasso AP, Cromwell JW, et al. Prolongation of cardiac xenograft survival by depletion of complement. Transplantation. 1993;55:857–65; discussion 865-856.
20. Leventhal JR, Sakiyalak P, Witson J, et al. The synergistic effect of combined antibody and complement depletion on discordant cardiac xenograft survival in nonhuman primates. Transplantation. 1994;57:974–8.
21. Kobayashi T, Neethling FA, Taniguchi S, et al. Investigation of the anti-complement agnets, fut-175 and k76cooh, in discordant xenotransplantation. Xenotransplantation. 1996;3:237–45.
22. Kobayashi T, Taniguchi S, Neethling FA, et al. Delayed xenograft rejection of pig-to-baboon cardiac transplants after cobra venom factor therapy. Transplantation. 1997;64:1255–61.
23. Pruitt SK, Kirk AD, Bollinger RR, et al. The effect of soluble complement receptor type 1 on hyperacute rejection of porcine xenografts. Transplantation. 1994;57:363–70.
24. Dalmasso AP, Vercellotti GM, Platt JL, Bach FH. Inhibition of complement-mediated endothelial cell cytotoxicity by decay-accelerating factor. Potential for prevention of xenograft hyperacute rejection. Transplantation. 1991;52:530–3.
25. White DJ, Oglesby T, Liszewski MK, et al. Expression of human decay accelerating factor or membrane cofactor protein genes on mouse cells inhibits lysis by human complement. Transpl Int. 1992;5(Suppl 1):S648–50.
26. White DJG, Langford GA, Cozzi E, Young VK. Production of pigs transgenic for human daf: a strategy for xenotransplantation. Xenotransplantation. 1995;2:213–7.
27. Oglesby TJ, Allen CJ, Liszewski MK, White DJ, Atkinson JP. Membrane cofactor protein (cd46) protects cells from complement-mediated attack by an intrinsic mechanism. J Exp Med. 1992;175:1547–51.
28. Walsh LA, Tone M, Waldmann H. Transfection of human cd59 complementary DNA into rat cells confers resistance to human complement. Eur J Immunol. 1991;21:847–50.

29. Atkinson JP, Oglesby TJ, White D, Adams EA, Liszewski MK. Separation of self from non-self in the complement system: a role for membrane cofactor protein and decay accelerating factor. Clin Exp Immunol. 1991;86(Suppl 1):27–30.
30. Morgan BP, Berg CW, Harris CL. "Homologous restriction" in complement lysis: roles of membrane complement regulators. Xenotransplantation 2005;12:258–65.
31. Langford GA, Yannoutsos N, Cozzi E, et al. Production of pigs transgenic for human decay accelerating factor. Transplant Proc. 1994;26:1400–1.
32. Cozzi E, White DJ. The generation of transgenic pigs as potential organ donors for humans. Nat Med. 1995;1:964–6.
33. Fodor WL, Williams BL, Matis LA, et al. Expression of a functional human complement inhibitor in a transgenic pig as a model for the prevention of xenogeneic hyperacute organ rejection. Proc Natl Acad Sci U S A. 1994;91:11153–7.
34. Rosengard AM, Cary NR, Langford GA, Tucker AW, Wallwork J, White DJ. Tissue expression of human complement inhibitor, decay-accelerating factor, in transgenic pigs. A potential approach for preventing xenograft rejection. Transplantation. 1995;59:1325–33.
35. McCurry KR, Kooyman DL, Alvarado CG, et al. Human complement regulatory proteins protect swine-to-primate cardiac xenografts from humoral injury. Nat Med. 1995;1:423–7.
36. Diamond LE, McCurry KR, Martin MJ, et al. Characterization of transgenic pigs expressing functionally active human cd59 on cardiac endothelium. Transplantation. 1996;61:1241–9.
37. Loveland BE, Milland J, Kyriakou P, et al. Characterization of a cd46 transgenic pig and protection of transgenic kidneys against hyperacute rejection in non-immunosuppressed baboons. Xenotransplantation 2004;11:171–83.
38. Cozzi E, Yannoutsos N, Langford GA, Pino-Chavez G, Wallwork J, White DJG. Effects of transgenic expression of human decay accelerating factor on the inhibition of hyperacute rejection of pig organs. In: Cooper D, Kemp E, Platt JL, White DJG, editors. Xenotransplantation – the transplantation of organs and tissues between species. Berlin: Springer; 1997. p. 665–82.
39. Cozzi E, Langford GA, Richards A, et al. Expression of human decay accelerating factor in transgenic pigs. Transplant Proc. 1994;26:1402–3.
40. Schuurman HJ, Pino-Chavez G, Phillips MJ, Thomas L, White DJ, Cozzi E. Incidence of hyperacute rejection in pig-to-primate transplantation using organs from HDAF-transgenic donors. Transplantation. 2002;73:1146–51.
41. Schuurman HJ, Cheng J, Lam T. Pathology of xenograft rejection: a commentary. Xenotransplantation 2003;10:293–9.
42. Alexandre GP, De Bruyere M, Squifflet JP, Moriau M, Latinne D, Pirson Y. Human abo-incompatible living donor renal homografts. Neth J Med. 1985;28:231–4.
43. Alexandre GP, Squifflet JP, De Bruyere M, et al. Present experiences in a series of 26 abo-incompatible living donor renal allografts. Transplant Proc. 1987;19:4538–42.
44. Alexandre GP. From abo-incompatible human kidney transplantation to xenotransplantation. Xenotransplantation. 2004;11:233–6.
45. Stussi G, West L, Cooper DKC, Seebach JD. Abo-incompatible allotransplantation as a basis for clinical xenotransplantation. Xenotransplantation. 2006;13:390–9.
46. Good AH, Cooper DKC, Malcolm AJ, et al. Identification of carbohydrate structures that bind human antiporcine antibodies: implications for discordant xenografting in humans. Transplant Proc. 1992;24:559–62.
47. Cooper DKC. Depletion of natural antibodies in non-human primates--a step towards successful discordant xenografting in humans. Clin Transpl. 1992;6:178–83.
48. Cooper DKC, Good AH, Koren E, et al. Identification of alpha-galactosyl and other carbohydrate epitopes that are bound by human anti-pig antibodies: relevance to discordant xenografting in man. Transpl Immunol. 1993;1:198–205.
49. Koren E, Neethling FA, Richards S, et al. Binding and specificity of major immunoglobulin classes of preformed human anti-pig heart antibodies. Transpl Int. 1993;6:351–3.
50. Sandrin MS, Vaughan HA, Dabkowski PL, McKenzie IF. Anti-pig igm antibodies in human serum react predominantly with gal(alpha 1-3)gal epitopes. Proc Natl Acad Sci U S A. 1993;90:11391–5.

51. Cooper DKC. Identification of alpha gal as the major target for human anti-pig antibodies. Xenotransplantation 2009;16:47–9.
52. Kujundzic M, Koren E, Neethling FA, et al. Variability of anti-alphagal antibodies in human serum and their relation to serum cytotoxicity against pig cells. Xenotransplantation 1994;1:58–65.
53. Kobayashi T, Cooper DKC. Anti-gal, alpha-gal epitopes, and Xenotransplantation Subcell Biochem. 1999;32:229–57.
54. Cooper DKC, Koren E, Oriol R. Oligosaccharides and discordant Xenotransplantation Immunol Rev. 1994;141:31–58.
55. Galili U, Shohet SB, Kobrin E, Stults CL, Macher BA. Man, apes, and old world monkeys differ from other mammals in the expression of alpha-galactosyl epitopes on nucleated cells. J Biol Chem. 1988;263:17755–62.
56. Oriol R, Ye Y, Koren E, Cooper DKC. Carbohydrate antigens of pig tissues reacting with human natural antibodies as potential targets for hyperacute vascular rejection in pig-to-man organ Xenotransplantation Transplantation. 1993;56:1433–42.
57. Oriol R, Barthod F, Bergemer AM, Ye Y, Koren E, Cooper DKC. Monomorphic and polymorphic carbohydrate antigens on pig tissues: implications for organ Xenotransplantation in the pig-to-human model. Transpl Int. 1994;7:405–13.
58. McKenzie IF, Xing PX, Vaughan HA, Prenzoska J, Dabkowski PL, Sandrin MS. Distribution of the major xenoantigen (gal (alpha 1-3) gal) for pig to human xenografts. Transpl Immunol. 1994;2:81–6.
59. Galili U. Evolution of alpha 1,3galactosyltransferase and of the alpha-gal epitope. Subcell Biochem. 1999;32:1–23.
60. Koike C, Uddin M, Wildman DE, et al. Functionally important glycosyltransferase gain and loss during catarrhine primate emergence. Proc Natl Acad Sci U S A. 2007;104:559–64.
61. Galili U, Mandrell RE, Hamadeh RM, Shohet SB, Griffiss JM. Interaction between human natural anti-alpha-galactosyl immunoglobulin g and bacteria of the human flora. Infect Immun. 1988;56:1730–7.
62. Rood PP, Tai HC, Hara H, et al. Late onset of development of natural anti-nongal antibodies in infant humans and baboons: implications for Xenotransplantation in infants. Transpl Int. 2007;20:1050–8.
63. Dons EM, Montoya C, Long CE, et al. T-cell-based immunosuppressive therapy inhibits the development of natural antibodies in infant baboons. Transplantation. 2012;93:769–76.
64. Neethling F, Cooper DKC, Xu H, Michler RE. Newborn baboon serum anti-alpha galactosyl antibody levels and cytotoxicity to cultured pig kidney (pk15) cells. Transplantation. 1995;60:520–1.
65. Minanov OP, Itescu S, Neethling FA, et al. Anti-gal igg antibodies in sera of newborn humans and baboons and its significance in pig Xenotransplantation Transplantation. 1997;63:182–6.
66. Mohiuddin MM, Corcoran PC, Singh AK, et al. B-cell depletion extends the survival of gtko. Hcd46tg pig heart xenografts in baboons for up to 8 months. Am J Transplant. 2012;12:763–71.
67. Zhou H, Iwase H, Wolf RF, et al. Are there advantages in the use of specific pathogen-free baboons in pig organ Xenotransplantation models? Xenotransplantation 2014;21:287–90.
68. Cretin N, Bracy J, Hanson K, Iacomini J. The role of t cell help in the production of antibodies specific for gal alpha 1-3gal. J Immunol (Baltimore, MD: 1950). 2002;168:1479–83.
69. Alwayn IP, Basker M, Buhler L, Cooper DKC. The problem of anti-pig antibodies in pig-to-primate xenografting: current and novel methods of depletion and/or suppression of production of anti-pig antibodies. Xenotransplantation 1999;6:157–68.
70. Cooper DKC, Ye Y, Niekrasz M, et al. Specific intravenous carbohydrate therapy. A new concept in inhibiting antibody-mediated rejection--experience with abo-incompatible cardiac allografting in the baboon. Transplantation. 1993;56:769–77.
71. Ye Y, Niekrasz M, Kehoe M, et al. Cardiac allotransplantation across the abo-blood group barrier by the neutralization of preformed antibodies: the baboon as a model for the human. Lab Anim Sci. 1994;44:121–4.

72. Ye Y, Neethling FA, Niekrasz M, et al. Evidence that intravenously administered alpha-galactosyl carbohydrates reduce baboon serum cytotoxicity to pig kidney cells (pk15) and transplanted pig hearts. Transplantation. 1994;58:330–7.
73. Simon PM, Neethling FA, Taniguchi S, et al. Intravenous infusion of galalpha1-3gal oligosaccharides in baboons delays hyperacute rejection of porcine heart xenografts. Transplantation. 1998;65:346–53.
74. Romano E, Neethling FA, Nilsson K, et al. Intravenous synthetic alphagal saccharides delay hyperacute rejection following pig-to-baboon heart transplantation. Xenotransplantation 1999;6:36–42.
75. Teranishi K, Alwayn IP, Buhler L, et al. Depletion of anti-gal antibodies by the intravenous infusion of gal type 2 and 6 glycoconjugates in baboons. Xenotransplantation 2003;10:357–67.
76. Teranishi K, Gollackner B, Buhler L, et al. Depletion of anti-gal antibodies in baboons by intravenous therapy with bovine serum albumin conjugated to gal oligosaccharides. Transplantation. 2002;73:129–39.
77. Kuwaki K, Knosalla C, Moran K, et al. Reduction of anti-galalpha1,3gal antibodies by infusion of types 2 and 6 gal trisaccharides conjugated to poly-l-lysine. Xenotransplantation 2004;11:210–5.
78. Rieben R, Von Allmen E, Korchagina EY, et al. Detection, immunoabsorption, and inhibition of cytotoxic activity of anti-alphagal antibodies using newly developed substances with synthetic gal alpha1-3gal disaccharide epitopes. Xenotransplantation 1995;2:98–106.
79. Neethling FA, Koren E, Ye Y, et al. Protection of pig kidney (pk15) cells from the cytotoxic effect of anti-pig antibodies by alpha-galactosyl oligosaccharides. Transplantation. 1994;57:959–63.
80. Neethling FA, Joziasse D, Bovin N, Cooper DKC. Oriol. R. the reducing end of alpha gal oligosaccharides contributes to their efficiency in blocking natural antibodies of human and baboon sera. Transpl Int. 1996;9:98–101.
81. Neethling FA, Cooper DKC. Serum cytotoxicity to pig cells and anti-alphagal antibody level and specificity in humans and baboons. Transplantation. 1999;67:658–65.
82. Gollackner B, Knosalla C, Houser S, et al. Pig kidney transplantation in baboons treated intravenously with a bovine serum albumin-galalpha1-3gal conjugate. Xenotransplantation 2003;10:606–14.
83. Ierino FL, Kozlowski T, Siegel JB, et al. Disseminated intravascular coagulation in association with the delayed rejection of pig-to-baboon renal xenografts. Transplantation. 1998;66:1439–50.
84. Kozlowski T, Ierino FL, Lambrigts D, et al. Depletion of anti-gal(alpha)1-3gal antibody in baboons by specific alpha-gal immunoaffinity columns. Xenotransplantation 1998;5:122–31.
85. Xu Y, Lorf T, Sablinski T, et al. Removal of anti-porcine natural antibodies from human and nonhuman primate plasma in vitro and in vivo by a galalpha1-3galbeta1-4betaglc-x immunoaffinity column. Transplantation. 1998;65:172–9.
86. Taniguchi S, Neethling FA, Korchagina EY, et al. In vivo immunoadsorption of antipig antibodies in baboons using a specific gal(alpha)1-3gal column. Transplantation. 1996;62:1379–84.
87. Lambrigts D, Van Calster P, Xu Y, et al. Pharmacologic immunosuppressive therapy and extracorporeal immunoadsorption in the suppression of anti-alphagal antibody in the baboon. Xenotransplantation 1998;5:274–83.
88. Kozlowski T, Shimizu A, Lambrigts D, et al. Porcine kidney and heart transplantation in baboons undergoing a tolerance induction regimen and antibody adsorption. Transplantation. 1999;67:18–30.
89. Cooper DKC, Cairns TD, Taube DH. Extracorporeal immunoadsorption of alphagal antibodies. Xeno. 1996;4:27–9.
90. Watts A, Foley A, Awwad M, et al. Plasma perfusion by apheresis through a gal immunoaffinity column successfully depletes anti-gal antibody: experience with 320 aphereses in baboons. Xenotransplantation 2000;7:181–5.
91. Cooper DKC, Koren E, Oriol R. Genetically engineered pigs. Lancet. 1993;342:682–3.

92. Thall AD, Maly P, Lowe JB. Oocyte gal alpha 1,3gal epitopes implicated in sperm adhesion to the zona pellucida glycoprotein zp3 are not required for fertilization in the mouse. J Biol Chem. 1995;270:21437–40.
93. Tearle RG, Tange MJ, Zannettino ZL, et al. The alpha-1,3-galactosyltransferase knockout mouse. Implications for Xenotransplantation Transplantation. 1996;61:13–9.
94. McKenzie IF, Li YQ, Patton K, Thall AD, Sandrin MS. A murine model of antibody-mediated hyperacute rejection by galactose-alpha(1,3)galactose antibodies in gal o/o mice. Transplantation. 1998;66:754–63.
95. Campbell KH, McWhir J, Ritchie WA, Wilmut I. Sheep cloned by nuclear transfer from a cultured cell line. Nature. 1996;380:64–6.
96. Polejaeva IA, Chen SH, Vaught TD, et al. Cloned pigs produced by nuclear transfer from adult somatic cells. Nature. 2000;407:86–90.
97. Koike C, Fung JJ, Geller DA, et al. Molecular basis of evolutionary loss of the alpha 1,3-galactosyltransferase gene in higher primates. J Biol Chem. 2002;277:10114–20.
98. Phelps CJ, Koike C, Vaught TD, et al. Production of alpha 1,3-galactosyltransferase-deficient pigs. Science (New York, NY). 2003;299:411–4.
99. Kolber-Simonds D, Lai L, Watt SR, et al. Production of alpha-1,3-galactosyltransferase null pigs by means of nuclear transfer with fibroblasts bearing loss of heterozygosity mutations. Proc Natl Acad Sci U S A. 2004;101:7335–40.
100. Kuwaki K, Tseng YL, Dor FJ, et al. Heart transplantation in baboons using alpha1,3-galactosyltransferase gene-knockout pigs as donors: initial experience. Nat Med. 2005;11:29–31.
101. Tseng YL, Kuwaki K, Dor FJ, et al. Alpha1,3-galactosyltransferase gene-knockout pig heart transplantation in baboons with survival approaching 6 months. Transplantation. 2005;80:1493–500.
102. Yamada K, Yazawa K, Shimizu A, et al. Marked prolongation of porcine renal xenograft survival in baboons through the use of alpha1,3-galactosyltransferase gene-knockout donors and the cotransplantation of vascularized thymic tissue. Nat Med. 2005;11:32–4.
103. Shimizu A, Hisashi Y, Kuwaki K, et al. Thrombotic microangiopathy associated with humoral rejection of cardiac xenografts from alpha1,3-galactosyltransferase gene-knockout pigs in baboons. Am J Pathol. 2008;172:1471–81.
104. Hisashi Y, Yamada K, Kuwaki K, et al. Rejection of cardiac xenografts transplanted from alpha1,3-galactosyltransferase gene-knockout (galt-ko) pigs to baboons. Am J Transplant. 2008;8:2516–26.
105. Ezzelarab M, Garcia B, Azimzadeh A, et al. The innate immune response and activation of coagulation in alpha1,3-galactosyltransferase gene-knockout xenograft recipients. Transplantation. 2009;87:805–12.
106. Chen D, Dorling A. Microcoagulation processes after Xenotransplantation Curr Opin Organ Transplant. 2005;10:240–5.
107. Chen G, Sun H, Yang H, et al. The role of anti-non-gal antibodies in the development of acute humoral xenograft rejection of hdaf transgenic porcine kidneys in baboons receiving anti-gal antibody neutralization therapy. Transplantation. 2006;81:273–83.
108. Rood PP, Hara H, Busch JL, et al. Incidence and cytotoxicity of antibodies in cynomolgus monkeys directed to nongal antigens, and their relevance for experimental models. Transpl Int. 2006;19:158–65.
109. Rood PP, Hara H, Ezzelarab M, et al. Preformed antibodies to alpha1,3-galactosyltransferase gene-knockout (gt-ko) pig cells in humans, baboons, and monkeys: implications for Xenotransplantation Transplant Proc. 2005;37:3514–5.
110. Ezzelarab M, Hara H, Busch J, et al. Antibodies directed to pig non-gal antigens in naive and sensitized baboons. Xenotransplantation 2006;13:400–7.
111. Tseng YL, Moran K, Dor FJ, et al. Elicited antibodies in baboons exposed to tissues from alpha1,3-galactosyltransferase gene-knockout pigs. Transplantation. 2006;81:1058–62.
112. Cooper DKC. Xenoantigens and xenoantibodies. Xenotransplantation 1998;5:6–17.

113. Ezzelarab M, Ayares D, Cooper DKC. Carbohydrates in Xenotransplantation Immunol Cell Biol. 2005;83:396–404.
114. Yeh P, Ezzelarab M, Bovin N, et al. Investigation of potential carbohydrate antigen targets for human and baboon antibodies. Xenotransplantation 2010;17:197–206.
115. Byrne GW, Du Z, Stalboerger P, Kogelberg H, McGregor CG. Cloning and expression of porcine beta1,4 n-acetylgalactosaminyl transferase encoding a new xenoreactive antigen. Xenotransplantation 2014;21:543–54.
116. Byrne GW, Azimzadeh AM, Ezzelarab M, et al. Histopathologic insights into the mechanism of anti-non-gal antibody-mediated pig cardiac xenograft rejection. Xenotransplantation 2013;20:292–307.
117. Hara H, Long C, Lin YJ, et al. In vitro investigation of pig cells for resistance to human antibody-mediated rejection. Transpl Int. 2008;21:1163–74.
118. Azimzadeh A, Kelishadi S, Ezzelarab M, et al. Early graft failure of gtko pig organs in baboons is reduced in hcprp expression. Xenotransplantation 2009;16:356; (Abstract).
119. Azimzadeh A, Kelishadi S, Ezzelarab MB, et al. Early graft failure of galtko pig organs in baboons is reduced by expression of a human complement-regulatory protein. Xenotransplantation 2015;22:310.
120. McGregor CG, Ricci D, Miyagi N, et al. Human cd55 expression blocks hyperacute rejection and restricts complement activation in gal knockout cardiac xenografts. Transplantation. 2012;93:686–92.
121. van Denderen BJ, Salvaris E, Romanella M, et al. Combination of decay-accelerating factor expression and alpha1,3-galactosyltransferase knockout affords added protection from human complement-mediated injury. Transplantation. 1997;64:882–8.
122. Gollackner B, Goh SK, Qawi I, et al. Acute vascular rejection of xenografts: roles of natural and elicited xenoreactive antibodies in activation of vascular endothelial cells and induction of procoagulant activity. Transplantation. 2004;77:1735–41.
123. Lin SS, Weidner BC, Byrne GW, et al. The role of antibodies in acute vascular rejection of pig-to-baboon cardiac transplants. J Clin Invest. 1998;101:1745–56.
124. Buhler L, Awwad M, Basker M, et al. High-dose porcine hematopoietic cell transplantation combined with cd40 ligand blockade in baboons prevents an induced anti-pig humoral response. Transplantation. 2000;69:2296–304.
125. Chen G, Qian H, Starzl T, et al. Acute rejection is associated with antibodies to non-gal antigens in baboons using gal-knockout pig kidneys. Nat Med. 2005;11:1295–8.
126. Ezzelarab C, Ayares D, Cooper DKC, Ezzelarab MB. Human t-cell proliferation in response to thrombin-activated gtko pig endothelial cells. Xenotransplantation 2012;19:311–6.
127. Cozz IE, Vial C, Ostlie D, et al. Maintenance triple immunosuppression with cyclosporin a, mycophenolate sodium and steroids allows prolonged survival of primate recipients of HDAF porcine renal xenografts. Xenotransplantation 2003;10:300–10.
128. McGregor CG, Davies WR, Oi K, et al. Cardiac Xenotransplantation: recent preclinical progress with 3-month median survival. J Thorac Cardiovasc Surg. 2005;130:844–51.
129. McGregor CG, Teotia SS, Byrne GW, et al. Cardiac Xenotransplantation: progress toward the clinic. Transplantation. 2004;78:1569–75.
130. Teotia SS, Walker RC, Schirmer JM, et al. Prevention, detection, and management of early bacterial and fungal infections in a preclinical cardiac Xenotransplantation model that achieves prolonged survival. Xenotransplantation 2005;12:127–33.
131. Ezzelarab MB, Ekser B, Echeverri G, et al. Costimulation blockade in pig artery patch Xenotransplantation – a simple model to monitor the adaptive immune response in nonhuman primates. Xenotransplantation 2012;19:221–32.
132. Mohiuddin MM, Singh AK, Corcoran PC, et al. Genetically engineered pigs and target-specific immunomodulation provide significant graft survival and hope for clinical cardiac Xenotransplantation J Thorac Cardiovasc Surg. 2014;148:1106–13; discussion 1113-1114.
133. Mohiuddin MM, Singh AK, Corcoran PC, et al. Role of anti-cd40 antibody-mediated costimulation blockade on non-gal antibody production and heterotopic cardiac xenograft survival in a gtko.Hcd46tg pig-to-baboon model. Xenotransplantation 2014;21:35–45.

134. Mohiuddin MM, Singh AK, Corcoran PC, et al. One-year heterotopic cardiac xenograft survival in a pig to baboon model. Am J Transplant. 2014;14:488–9.
135. Iwase H, Ekser B, Satyananda V, et al. Pig-to-baboon heart transplantation - first experience with pigs transgenic for human thrombomodulin and comparison of three costimulation blockade-based regimens. Xenotransplantation 2015;22:211–20.
136. Iwase H, Liu H, Wijkstrom M, et al. Pig kidney graft survival in a baboon for 136 days: longest life-supporting organ graft survival to date. Xenotransplantation 2015;22:302–9.
137. Iwase H, Satyananda V, Zhou H, et al. Initial in vivo experience of pig artery patch transplantation in baboons using mutant mhc (ciita-dn) pigs. Transpl Immunol. 2015;32:99–108.
138. Higginbotham L, Mathews D, Breeden CA, et al. Pre-transplant antibody screening and anti-cd154 costimulation blockade promote long-term xenograft survival in a pig-to-primate kidney transplant model. Xenotransplantation 2015;22:221–30.
139. Kuwaki K, Knosalla C, Dor FJ, et al. Suppression of natural and elicited antibodies in pig-to-baboon heart transplantation using a human anti-human cd154 mab-based regimen. Am J Transplant. 2004;4:363–72.
140. Knosalla C, Ryan DJ, Moran K, et al. Initial experience with the human anti-human cd154 monoclonal antibody, abi793, in pig-to-baboon xenotransplantation. Xenotransplantation. 2004;11:353–60.
141. Kawai T, Andrews D, Colvin RB, Sachs DH, Cosimi AB. Thromboembolic complications after treatment with monoclonal antibody against cd40 ligand. Nat Med. 2000;6:114.
142. Kirk AD, Knechtle SJ, Sollinger HW. Preliminary results of the use of humanized anti-cd154 in human renal allotransplantation. Am J Transplant. 2001;1(Suppl 1):S191; (Abstract).
143. Knosalla C, Gollackner B, Cooper DKC. Anti-cd154 monoclonal antibody and thromboembolism revisited. Transplantation. 2002;74:416–7.
144. Lee RS, Yamada K, Womer KL, et al. Blockade of cd28-b7, but not cd40-cd154, prevents costimulation of allogeneic porcine and xenogeneic human anti-porcine t cell responses. J Immunol (Baltimore, MD: 1950). 2000;164:3434–44.
145. Lee W, Satyananda V, Iwase H, et al. In vitro testing of an anti-cdd40 monoclonal antibody, clone 2c10, in primates and pigs. Transpl Immunol. 2015;33:185.
146. Ezzelarab MB, Ekser B, Azimzadeh A, et al. Systemic inflammation in xenograft recipients precedes activation of coagulation. Xenotransplantation 2015;22:32–47.
147. Ezzelarab MB, Ekser B, Isse K, et al. Increased soluble CD154 (CD40 ligand) levels in xenograft recipients correlate with the development of de novo anti-pig igg antibodies. Transplantation. 2014;97:502–8.
148. Alwayn IP, Xu Y, Basker M, et al. Effects of specific anti-b and/or anti-plasma cell immunotherapy on antibody production in baboons: depletion of cd20- and cd22-positive b cells does not result in significantly decreased production of anti-alphagal antibody. Xenotransplantation 2001;8:157–71.
149. Satyananda V, Hara H, Ezzelarab MB, Phelps C, Ayares D, Cooper DKC. New concepts of immune modulation in Xenotransplantation Transplantation. 2013;96:937–45.
150. Cowan PJ, Cooper DKC, d'Apice AJ. Kidney Xenotransplantation Kidney Int. 2014;85:265–75.
151. Wilhite T, Ezzelarab C, Hara H, et al. The effect of gal expression on pig cells on the human t-cell xenoresponse. Xenotransplantation 2012;19:56–63.
152. Ezzelarab M, Ezzelarab C, Wilhite T, et al. Genetically-modified pig mesenchymal stromal cells: xenoantigenicity and effect on human t-cell xenoresponses. Xenotransplantation 2011;18:183–95.
153. Schubert T, Poilvache H, Galli C, Gianello P, Dufrane D. Galactosyl-knock-out engineered pig as a xenogenic donor source of adipose mscs for bone regeneration. Biomaterials. 2013;34:3279–89.
154. Ezzelarab MB, Ayares D, Cooper DKC. Transgenic expression of human cd46: does it reduce the primate t cell response to pig xenografts? Xenotransplantation 2015;22:487.
155. Phelps CJ, Ball SF, Vaught TD, et al. Production and characterization of transgenic pigs expressing porcine ctla4-ig. Xenotransplantation 2009;16:477–85.

156. Koshika T, Phelps C, Fang J, et al. Relative efficiency of porcine and human cytotoxic t-lymphocyte antigen 4 immunoglobulin in inhibiting human cd4+ t-cell responses co-stimulated by porcine and human b7 molecules. Immunology. 2011;134:386–97.
157. Bottino R, Wijkstrom M, van der Windt DJ, et al. Pig-to-monkey islet Xenotransplantation using multi-transgenic pigs. Am J Transplant. 2014;14:2275–87.
158. Hara H, Witt W, Crossley T, et al. Human dominant-negative class ii transactivator transgenic pigs – effect on the human anti-pig t-cell immune response and immune status. Immunology. 2013;140:39–46.
159. Reyes LM, Estrada JL, Wang ZY, et al. Creating class i mhc-null pigs using guide rna and the cas9 endonuclease. J Immunol (Baltimore, MD: 1950). 2014;193:5751–7.
160. Inverardi L, Clissi B, Stolzer AL, Bender JR, Sandrin MS, Pardi R. Human natural killer lymphocytes directly recognize evolutionarily conserved oligosaccharide ligands expressed by xenogeneic tissues. Transplantation. 1997;63:1318–30.
161. Baumann BC, Forte P, Hawley RJ, Rieben R, Schneider MK, Seebach JD. Lack of galactose-alpha-1,3-galactose expression on porcine endothelial cells prevents complement-induced lysis but not direct xenogeneic nk cytotoxicity. J Immunol (Baltimore, MD: 1950). 2004;172:6460–7.
162. Rieben R, Seebach JD. Xenograft rejection: Igg1, complement and nk cells team up to activate and destroy the endothelium. Trends Immunol. 2005;26:2–5.
163. Horvath-Arcidiacono JA, Porter CM, Bloom ET. Human nk cells can lyse porcine endothelial cells independent of their expression of galalpha(1,3)-gal and killing is enhanced by activation of either effector or target cells. Xenotransplantation 2006;13:318–27.
164. Kennett SB, Porter CM, Horvath-Arcidiacono JA, Bloom ET. Characterization of baboon nk cells and their xenogeneic activity. Xenotransplantation 2010;17:288–99.
165. Sasaki H, Xu XC, Smith DM, Howard T, Mohanakumar T. Hla-g expression protects porcine endothelial cells against natural killer cell-mediated xenogeneic cytotoxicity. Transplantation. 1999;67:31–7.
166. Dorling A, Monk N, Lechler R. Hla-g inhibits the transendothelial cell migration of human nk cells: a strategy for inhibiting xenograft rejection. Transplant Proc. 2000;32:938.
167. Matsunami K, Miyagawa S, Nakai R, Murase A, Shirakura R. The possible use of hla-g1 and g3 in the inhibition of nk cell-mediated swine endothelial cell lysis. Clin Exp Immunol. 2001;126:165–72.
168. Forte P, Pazmany L, Matter-Reissmann UB, Stussi G, Schneider MK, Seebach JD. Hla-g inhibits rolling adhesion of activated human nk cells on porcine endothelial cells. J Immunol (Baltimore, MD: 1950). 2001;167:6002–8.
169. Forte P, Baumann BC, Weiss EH, Seebach JD. Hla-e expression on porcine cells: protection from human nk cytotoxicity depends on peptide loading. Am J Transplant. 2005;5:2085–93.
170. Crew MD. Play it in e or g: utilization of hla-e and -g in xenotransplantation. Xenotransplantation. 2007;14:198–207.
171. Seebach JD, Comrack C, Germana S, LeGuern C, Sachs DH, DerSimonian H. Hla-cw3 expression on porcine endothelial cells protects against xenogeneic cytotoxicity mediated by a subset of human nk cells. J Immunol (Baltimore, MD: 1950). 1997;159:3655–61.
172. Weiss EH, Lilienfeld BG, Muller S, et al. Hla-e/human beta2-microglobulin transgenic pigs: protection against xenogeneic human anti-pig natural killer cell cytotoxicity. Transplantation. 2009;87:35–43.
173. Maeda A, Kawamura T, Ueno T, Usui N, Eguchi H, Miyagawa S. The suppression of inflammatory macrophage-mediated cytotoxicity and proinflammatory cytokine production by transgenic expression of hla-e. Transpl Immunol. 2013;29:76–81.
174. Esquivel EL, Maeda A, Eguchi H, et al. Suppression of human macrophage-mediated cytotoxicity by transgenic swine endothelial cell expression of hla-g. Transpl Immunol. 2015;32:109–15.
175. Platt JL, Vercellotti GM, Lindman BJ, Oegema TR Jr, Bach FH, Dalmasso AP. Release of heparan sulfate from endothelial cells. Implications for pathogenesis of hyperacute rejection. J Exp Med. 1990;171:1363–8.

176. Bach FH, Robson SC, Ferran C, et al. Endothelial cell activation and thromboregulation during xenograft rejection. Immunol Rev. 1994;141:5–30.
177. Kopp CW, Siegel JB, Hancock WW, et al. Effect of porcine endothelial tissue factor pathway inhibitor on human coagulation factors. Transplantation. 1997;63:749–58.
178. Kopp CW, Grey ST, Siegel JB, et al. Expression of human thrombomodulin cofactor activity in porcine endothelial cells. Transplantation. 1998;66:244–51.
179. Lawson JH, Daniels LJ, Platt JL. The evaluation of thrombomodulin activity in porcine to human Xenotransplantation Transplant Proc. 1997;29:884–5.
180. Siegel JB, Grey ST, Lesnikoski BA, et al. Xenogeneic endothelial cells activate human prothrombin. Transplantation. 1997;64:888–96.
181. Saadi S, Holzknecht RA, Patte CP, Stern DM, Platt JL. Complement-mediated regulation of tissue factor activity in endothelium. J Exp Med. 1995;182:1807–14.
182. Buhler L, Basker M, Alwayn IP, et al. Coagulation and thrombotic disorders associated with pig organ and hematopoietic cell transplantation in nonhuman primates. Transplantation. 2000;70:1323–31.
183. Cowan PJ, Aminian A, Barlow H, et al. Renal xenografts from triple-transgenic pigs are not hyperacutely rejected but cause coagulopathy in non-immunosuppressed baboons. Transplantation. 2000;69:2504–15.
184. Buhler L, Yamada K, Kitamura H, et al. Pig kidney transplantation in baboons: anti-gal(alpha)1-3gal igm alone is associated with acute humoral xenograft rejection and disseminated intravascular coagulation. Transplantation. 2001;72:1743–52.
185. Houser SL, Kuwaki K, Knosalla C, et al. Thrombotic microangiopathy and graft arteriopathy in pig hearts following transplantation into baboons. Xenotransplantation 2004;11:416–25.
186. Lin CC, Ezzelarab M, Shapiro R, et al. Recipient tissue factor expression is associated with consumptive coagulopathy in pig-to-primate kidney Xenotransplantation Am J Transplant. 2010;10:1556–68.
187. Shimizu A, Yamada K, Yamamoto S, et al. Thrombotic microangiopathic glomerulopathy in human decay accelerating factor-transgenic swine-to-baboon kidney xenografts. J Am Soc Nephrol. 2005;16:2732–45.
188. Shimizu A, Yamada K, Robson SC, Sachs DH, Colvin RB. Pathologic characteristics of transplanted kidney xenografts. J Am Soc Nephrol. 2012;23:225–35.
189. Iwase H, Ekser B, Zhou H, Dons EM, Cooper DKC, Ezzelarab MB. Platelet aggregation in humans and nonhuman primates: relevance to xenotransplantation. Xenotransplantation. 2012;19:233–43.
190. Schulte em Esch J 2nd, Cruz MA, Siegel JB, Anrather J, Robson SC. Activation of human platelets by the membrane-expressed a1 domain of von Willebrand factor. Blood. 1997;90:4425–37.
191. Robson SC, Young VK, Cook NS, et al. Thrombin inhibition in an ex vivo model of porcine heart xenograft hyperacute rejection. Transplantation. 1996;61:862–8.
192. Ghanekar A, Mendicino M, Liu H, et al. Endothelial induction of fgl2 contributes to thrombosis during acute vascular xenograft rejection. J Immunol (Baltimore, MD: 1950). 2004;172:5693–701.
193. Mendicino M, Liu M, Ghanekar A, et al. Targeted deletion of fgl-2/fibroleukin in the donor modulates immunologic response and acute vascular rejection in cardiac xenografts. Circulation. 2005;112:248–56.
194. Lin CC, Chen D, McVey JH, Cooper DKC, Dorling A. Expression of tissue factor and initiation of clotting by human platelets and monocytes after incubation with porcine endothelial cells. Transplantation. 2008;86:702–9.
195. Ezzelarab M, Liu YW, Lin CC, et al. Role of p-selectin and p-selectin glycoprotein ligand-1 interaction in the induction of tissue factor expression on human platelets after incubation with porcine aortic endothelial cells. Xenotransplantation 2014;21:15–24.
196. Robson SC, Cooper DKC, d'Apice AJ. Disordered regulation of coagulation and platelet activation in xenotransplantation. Xenotransplantation. 2000;7:166–76.

197. Cowan PJ, Robson SC, d'Apice AJ. Controlling coagulation dysregulation in Xenotransplantation Curr Opin Organ Transplant. 2011;16:214–21.
198. Cowan PJ, d'Apice AJ. The coagulation barrier in Xenotransplantation: incompatibilities and strategies to overcome them. Curr Opin Organ Transplant. 2008;13:178–83.
199. Cowan PJ, Roussel JC, d'Apice AJ. The vascular and coagulation issues in Xenotransplantation Curr Opin Organ Transplant. 2009;14:161–7.
200. Iwase H, Ezzelarab MB, Ekser B, Cooper DKC. The role of platelets in coagulation dysfunction in Xenotransplantation, and therapeutic options. Xenotransplantation 2014;21:201–20.
201. Schulte em Esch J 2nd, Rogiers X, Robson SC. Molecular incompatibilities in hemostasis between swine and men--impact on xenografting. Ann Transplant. 2001;6:12–6.
202. Dorling A, Lechler R. Disordered thromboregulation after xenografting. Curr Opin Organ Transplant. 2001;6:36–41.
203. Roussel JC, Moran CJ, Salvaris EJ, Nandurkar HH, d'Apice AJ, Cowan PJ. Pig thrombomodulin binds human thrombin but is a poor cofactor for activation of human protein c and tafi. Am J Transplant. 2008;8:1101–12.
204. Lee KF, Salvaris EJ, Roussel JC, Robson SC, d'Apice AJ, Cowan PJ. Recombinant pig tfpi efficiently regulates human tissue factor pathways. Xenotransplantation 2008;15:191–7.
205. Knosalla C, Yazawa K, Behdad A, et al. Renal and cardiac endothelial heterogeneity impact acute vascular rejection in pig-to-baboon Xenotransplantation Am J Transplant. 2009;9:1006–16.
206. Schirmer JM, Fass DN, Byrne GW, Tazelaar HD, Logan JS, McGregor CG. Effective antiplatelet therapy does not prolong transgenic pig to baboon cardiac xenograft survival. Xenotransplantation 2004;11:436–43.
207. Byrne GW, Schirmer JM, Fass DN, et al. Warfarin or low-molecular-weight heparin therapy does not prolong pig-to-primate cardiac xenograft function. Am J Transplant. 2005;5:1011–20.
208. Byrne GW, Davies WR, Oi K, et al. Increased immunosuppression, not anticoagulation, extends cardiac xenograft survival. Transplantation. 2006;82:1787–91.
209. Cooper DKC, Ekser B, Burlak C, et al. Clinical lung Xenotransplantation--what donor genetic modifications may be necessary? Xenotransplantation 2012;19:144–58.
210. Petersen B, Ramackers W, Tiede A, et al. Pigs transgenic for human thrombomodulin have elevated production of activated protein c. Xenotransplantation 2009;16:486–95.
211. Miwa Y, Yamamoto K, Onishi A, et al. Potential value of human thrombomodulin and daf expression for coagulation control in pig-to-human xenotransplantation. Xenotransplantation. 2010;17:26–37.
212. Klymiuk N, Wuensch A, Kurome M, et al. Galt-ko/cd46/htm triple-transgenic donor animals for pig-to-baboon heart transplantation. Xenotransplantation 2011;18:271; (Abstract #126).
213. Yazaki S, Iwamoto M, Onishi A, et al. Production of cloned pigs expressing human thrombomodulin in endothelial cells. Xenotransplantation 2012;19:82–91.
214. Wuensch A, Baehr A, Bongoni AK, et al. Regulatory sequences of the porcine thbd gene facilitate endothelial-specific expression of bioactive human thrombomodulin in single- and multitransgenic pigs. Transplantation. 2014;97:138–47.
215. Wrighton CJ, Kopp CW, McShea A, Vetr H, Bach FH. High-level expression of functional human thrombomodulin in cultured porcine aortic endothelial cells. Transplant Proc. 1995;27:288–9.
216. Kaczmarek E, Koziak K, Sevigny J, et al. Identification and characterization of cd39/vascular atp diphosphohydrolase. J Biol Chem. 1996;271:33116–22.
217. Smolenski R, Kochan Z, Karbowska J, et al. Low expression of ecto-5′-nucleotidase in the pig heart – a potential barrier to Xenotransplantation J Heart Lung Transplant. 2001;20:214–5.
218. Smolenski RT, Khalpey Z, Yuen AC, et al. Purine metabolism in pigs and humans and its implications for Xenotransplantation Nucleosides Nucleotides Nucleic Acids. 2005;24:263–6.
219. Smolenski RT, Khalpey Z, Osborne FN, et al. Species differences of endothelial extracellular nucleotide metabolism and its implications for Xenotransplantation Pharmacol Rep. 2006;58(Suppl):118–25.
220. Deaglio S, Dwyer KM, Gao W, et al. Adenosine generation catalyzed by cd39 and cd73 expressed on regulatory t cells mediates immune suppression. J Exp Med. 2007;204:1257–65.

221. Dwyer KM, Robson SC, Nandurkar HH, et al. Thromboregulatory manifestations in human cd39 transgenic mice and the implications for thrombotic disease and transplantation. J Clin Invest. 2004;113:1440–6.
222. Dwyer KM, Mysore TB, Crikis S, et al. The transgenic expression of human cd39 on murine islets inhibits clotting of human blood. Transplantation. 2006;82:428–32.
223. Dwyer KM, Deaglio S, Gao W, Friedman D, Strom TB, Robson SC. Cd39 and control of cellular immune responses. Purinergic Signal. 2007;3:171–80.
224. DiVirgilio F, Boeynaems JM, Robson SC. Extracellular nucleotides as negative modulators of immunity. Curr Opin Pharmacol. 2009;9:507–13.
225. Crikis S, Lu B, Murray-Segal LM, et al. Transgenic overexpression of cd39 protects against renal ischemia-reperfusion and transplant vascular injury. Am J Transplant. 2010;10:2586–95.
226. Ahrens HE, Petersen B, Herrmann D, et al. Sirna mediated knockdown of tissue factor expression in pigs for Xenotransplantation Am J Transplant. 2015;15:1407–14.
227. Robson SC, Candinas D, Siegel JB, et al. Potential mechanism of abnormal thromboregulation in xenograft rejection: loss of ecto-atpases upon endothelial cell activation. Transplant Proc. 1996;28:536.
228. Dor FJ, Kuwaki K, Tseng YL, et al. Potential of aspirin to inhibit thrombotic microangiopathy in alpha1,3-galactosyltransferase gene-knockout pig hearts after transplantation in baboons. Transplant Proc. 2005;37:489–90.
229. Kim H, Hawthorne WJ, Kang HJ, et al. Human thrombomodulin regulates complement activation as well as the coagulation cascade in xeno-immune response. Xenotransplantation 2015;22:260–72.
230. Mueller NJ, Barth RN, Yamamoto S, et al. Activation of cytomegalovirus in pig-to-primate organ Xenotransplantation J Virol. 2002;76:4734–40.
231. Mueller NJ, Kuwaki K, Dor FJ, et al. Reduction of consumptive coagulopathy using porcine cytomegalovirus-free cardiac porcine grafts in pig-to-primate Xenotransplantation Transplantation. 2004;78:1449–53.
232. Mueller NJ, Kuwaki K, Knosalla C, et al. Early weaning of piglets fails to exclude porcine lymphotropic herpesvirus. Xenotransplantation 2005;12:59–62.
233. Gollackner B, Mueller NJ, Houser S, et al. Porcine cytomegalovirus and coagulopathy in pig-to-primate Xenotransplantation Transplantation. 2003;75:1841–7.
234. Yamada K, Tasaki M, Sekijima M, et al. Porcine cytomegalovirus infection is associated with early rejection of kidney grafts in a pig to baboon Xenotransplantation model. Transplantation. 2014;98:411–8.
235. Mueller NJ, Sulling K, Gollackner B, et al. Reduced efficacy of ganciclovir against porcine and baboon cytomegalovirus in pig-to-baboon Xenotransplantation Am J Transplant. 2003;3:1057–64.
236. Mueller NJ, Livingston C, Knosalla C, et al. Activation of porcine cytomegalovirus, but not porcine lymphotropic herpesvirus, in pig-to-baboon Xenotransplantation J Infect Dis. 2004;189:1628–33.
237. Mueller NJ, Ezzelarab M, Buhler L, Haeberli L, Ayares D, Cooper DKC. Monitoring of porcine and baboon cytomegalovirus infection in xenotransplantation. Xenotransplantation. 2009;16:535–6.
238. Iwase H, Ekser B, Zhou H, et al. Further evidence for a sustained systemic inflammatory response in xenograft recipients (sixr). Xenotransplantation 2015;22:399.
239. Esmon CT. Inflammation and the activated protein c anticoagulant pathway. Semin Thromb Hemost. 2006;32(Suppl 1):49–60.
240. Esmon CT. The discovery of the endothelial cell protein c receptor. J Thromb Haemost. 2010;8:2–5.
241. Esmon CT, Xu J, Lupu F. Innate immunity and coagulation. J Thromb Haemost. 2011;9(Suppl 1):182–8.
242. Esmon CT. Protein c anticoagulant system--anti-inflammatory effects. Semin Immunopathol. 2012;34:127–32.
243. Ezzelarab M, Welchons D, Torres C, et al. Atorvastatin down-regulates the primate cellular response to porcine aortic endothelial cells in vitro. Transplantation. 2008;86:733–7.

244. Lin CC, Ezzelarab M, Hara H, et al. Atorvastatin or transgenic expression of tfpi inhibits coagulation initiated by anti-nongal igg binding to porcine aortic endothelial cells. J Thromb Haemost. 2010;8:2001–10.
245. Ezzelarab M, Ayares D, Cooper DKC. The potential of genetically-modified pig mesenchymal stromal cells in xenotransplantation. Xenotransplantation. 2010;17:3–5.
246. Li J, Ezzelarab MB, Ayares D, Cooper DKC. The potential role of genetically-modified pig mesenchymal stromal cells in Xenotransplantation Stem Cell Rev. 2014;10:79–85.
247. Kumar G, Hara H, Long C, et al. Adipose-derived mesenchymal stromal cells from genetically modified pigs: immunogenicity and immune modulatory properties. Cytotherapy. 2012;14:494–504.
248. Oropeza M, Petersen B, Carnwath JW, et al. Transgenic expression of the human a20 gene in cloned pigs provides protection against apoptotic and inflammatory stimuli. Xenotransplantation 2009;16:522–34.
249. Petersen B, Lucas-Hahn A, Lemme E, et al. Generation and characterization of pigs transgenic for human hemeoxygenase-1 (hho-1). Xenotransplantation 2010;17:102–3.
250. Petersen B, Ramackers W, Lucas-Hahn A, et al. Transgenic expression of human heme oxygenase-1 in pigs confers resistance against xenograft rejection during ex vivo perfusion of porcine kidneys. Xenotransplantation 2011;18:355–68.
251. Schwinzer R, Baars W, Borns K, Petersen B, Kues W, Niemann H. Over-expression of human ho-1 and a20 in porcine cells: effects on susceptibility to cell-mediated lysis and inflammatory cytokines. Xenotransplantation 2011;18:269; (Abstract #122).
252. Ahrens HE, Petersen B, Ramackers W, et al. Kidneys from alpha1,3-galactosyltransferase knockout/human heme oxygenase-1/human a20 transgenic pigs are protected from rejection during ex vivo perfusion with human blood. Transplant Direct. 2015; Online 28 July 2015;1:E23. https://doi.org/10.1097/txd.0000000000000533.
253. Ide K, Ohdan H, Kobayashi T, Hara H, Ishiyama K, Asahara T. Antibody- and complement-independent phagocytotic and cytolytic activities of human macrophages toward porcine cells. Xenotransplantation 2005;12:181–8.
254. Burlak C, Twining LM, Rees MA. Terminal sialic acid residues on human glycophorin a are recognized by porcine kupffer cells. Transplantation. 2005;80:344–52.
255. Rees MA, Butler AJ, Brons IG, Negus MC, Skepper JN, Friend PJ. Evidence of macrophage receptors capable of direct recognition of xenogeneic epitopes without opsonization. Xenotransplantation 2005;12:13–9.
256. Ekser B, Echeverri GJ, Hassett AC, et al. Hepatic function after genetically engineered pig liver transplantation in baboons. Transplantation. 2010;90:483–93.
257. Ekser B, Long C, Echeverri GJ, et al. Impact of thrombocytopenia on survival of baboons with genetically modified pig liver transplants: clinical relevance. Am J Transplant. 2010;10:273–85.
258. Ekser B, Klein E, He J, et al. Genetically-engineered pig-to-baboon liver Xenotransplantation: histopathology of xenografts and native organs. PLoS One. 2012;7:e29720.
259. Ekser B, Burlak C, Waldman JP, et al. Immunobiology of liver Xenotransplantation Expert Rev Clin Immunol. 2012;8:621–34.
260. Ezzelarab M, Ekser B, Gridelli B, Iwase H, Ayares D, Cooper DKC. Thrombocytopenia after pig-to-baboon liver Xenotransplantation: where do platelets go? Xenotransplantation 2011;18:320–7.
261. Bush EL, Barbas AS, Holzknecht ZE, et al. Coagulopathy in alpha-galactosyl transferase knockout pulmonary xenotransplants. Xenotransplantation 2011;18:6–13.
262. LaMattina JC, Burdorf L, Zhang T, et al. Pig-to-baboon liver xenoperfusion utilizing galtko. Hcd46 pigs and glycoprotein ib blockade. Xenotransplantation 2014;21:274–86.
263. Burlak C, Paris LL, Chihara RK, et al. The fate of human platelets perfused through the pig liver: implications for xenotransplantation. Xenotransplantation. 2010;17:350–61.
264. Burdorf L, Stoddard T, Zhang T, et al. Expression of human cd46 modulates inflammation associated with galtko lung xenograft injury. Am J Transplant. 2014;14:1084–95.
265. Yeh H, Machaidze Z, Wamala I, et al. Increased transfusion-free survival following auxiliary pig liver xenotransplantation. Xenotransplantation. 2014;21:454–64.

266. Harris DG, Quinn KJ, French BM, et al. Meta-analysis of the independent and cumulative effects of multiple genetic modifications on pig lung xenograft performance during ex vivo perfusion with human blood. Xenotransplantation 2015;22:102–11.
267. Paris LL, Estrada JL, Li P, et al. Reduced human platelet uptake by pig livers deficient in the asialoglycoprotein receptor 1 protein. Xenotransplantation 2015;22:203–10.
268. Cantu E, Balsara KR, Li B, et al. Prolonged function of macrophage, von Willebrand factor-deficient porcine pulmonary xenografts. Am J Transplant. 2007;7:66–75.
269. Hara H, Campanile N, Tai HC, et al. An in vitro model of pig liver Xenotransplantation--pig complement is associated with reduced lysis of wild-type and genetically modified pig cells. Xenotransplantation 2010;17:370–8.
270. Hara H, Gridelli B, Lin YJ, Marcos A, Cooper DKC. Liver xenografts for the treatment of acute liver failure: clinical and experimental experience and remaining immunologic barriers. Liver Transpl. 2008;14:425–34.
271. Bouhours D, Pourcel C, Bouhours JE. Simultaneous expression by porcine aorta endothelial cells of glycosphingolipids bearing the major epitope for human xenoreactive antibodies (gal alpha 1-3gal), blood group h determinant and n-glycolylneuraminic acid. Glycoconj J. 1996;13:947–53.
272. Zhu A, Hurst R. Anti-n-glycolylneuraminic acid antibodies identified in healthy human serum. Xenotransplantation 2002;9:376–81.
273. Salama A, Evanno G, Harb J, Soulillou JP. Potential deleterious role of anti-Neu5Gc antibodies in xenotransplantation. Xenotransplantation. 2015;22:85–94.
274. Miwa Y, Kobayashi T, Nagasaka T, et al. Are n-glycolylneuraminic acid (hanganutziu-deicher) antigens important in pig-to-human Xenotransplantation? Xenotransplantation 2004;11:247–53.
275. Bardor M, Nguyen DH, Diaz S, Varki A. Mechanism of uptake and incorporation of the non-human sialic acid n-glycolylneuraminic acid into human cells. J Biol Chem. 2005;280:4228–37.
276. Padler-Karavani V, Varki A. Potential impact of the non-human sialic acid n-glycolylneuraminic acid on transplant rejection risk. Xenotransplantation 2011;18:1–5.
277. Springer SA, Diaz SL, Gagneux P. Parallel evolution of a self-signal: humans and new world monkeys independently lost the cell surface sugar neu5gc. Immunogenetics. 2014;66:671–4.
278. Burlak C, Paris LL, Lutz AJ, et al. Reduced binding of human antibodies to cells from ggta1/cmah ko pigs. Am J Transplant. 2014;14:1895–900.
279. Lutz AJ, Li P, Estrada JL, et al. Double knockout pigs deficient in n-glycolylneuraminic acid and galactose alpha-1,3-galactose reduce the humoral barrier to xenotransplantation. Xenotransplantation. 2013;20:27–35.
280. Estrada JL, Martens G, Li P, et al. Evaluation of human and non-human primate antibody binding to pig cells lacking ggta1/cmah/beta4galnt2 genes. Xenotransplantation 2015;22:194–202.
281. Lee W, Hara H, Ezzelarab MB, et al. Initial in vitro studies on tissues and cells from gtko/hcd46/neugcko pigs. Xenotransplantation 2016;23:137.
282. Soin B, Smith KG, Zaidi A, et al. Physiological aspects of pig-to-primate renal Xenotransplantation Kidney Int. 2001;60:1592–7.
283. Ibrahim Z, Busch J, Awwad M, Wagner R, Wells K, Cooper DKC. Selected physiologic compatibilities and incompatibilities between human and porcine organ systems. Xenotransplantation 2006;13:488–99.
284. Soin B, Ostliem D, Cozzi E, et al. Growth of porcine kidneys in their native and xenograft environment. Xenotransplantation 2000;7:96–100.
285. Groth CG, Korsgren O, Tibell A, et al. Transplantation of porcine fetal pancreas to diabetic patients. Lancet. 1994;344:1402–4.
286. van der Windt DJ, Bottino R, Kumar G, et al. Clinical islet Xenotransplantation: how close are we? Diabetes. 2012;61:3046–55.
287. Martin C, Plat M, Nerriere-Daguin V, et al. Transgenic expression of ctla4-ig by fetal pig neurons for Xenotransplantation Transgenic Res. 2005;14:373–84.

288. Leveque X, Cozzi E, Naveilhan P, Neveu I. Intracerebral Xenotransplantation: recent findings and perspectives for local immunosuppression. Curr Opin Organ Transplant. 2011;16:190–4.
289. Hara H, Cooper DKC. The immunology of corneal Xenotransplantation: a review of the literature. Xenotransplantation 2010;17:338–49.
290. Hara H, Cooper DKC. Xenotransplantation--the future of corneal transplantation? Cornea. 2011;30:371–8.
291. Kim MK, Wee WR, Park CG, Kim SJ. Xenocorneal transplantation. Curr Opin Organ Transplant. 2011;16:231–6.
292. Choi HJ, Kim MK, Lee HJ, et al. Efficacy of pig-to-rhesus lamellar corneal Xenotransplantation Invest Ophthalmol Vis Sci. 2011;52:6643–50.
293. Choi HJ, Lee JJ, Kim MK, et al. Cross-reactivity between decellularized porcine corneal lamellae for corneal xenobridging and subsequent corneal allotransplants. Xenotransplantation 2014;21:115–23.
294. Lamm V, Hara H, Mammen A, Dhaliwal D, Cooper DKC. Corneal blindness and xenotransplantation. Xenotransplantation. 2014;21:99–114.
295. Long C, Hara H, Pawlikowski Z, et al. Genetically engineered pig red blood cells for clinical transfusion: initial in vitro studies. Transfusion. 2009;49:2418–29.
296. Cooper DKC, Hara H, Yazer M. Genetically engineered pigs as a source for clinical red blood cell transfusion. Clin Lab Med. 2010;30:365–80.
297. Wang ZY, Burlak C, Estrada JL, Li P, Tector MF, Tector AJ. Erythrocytes from ggta1/cmah knockout pigs: implications for xenotransfusion and testing in non-human primates. Xenotransplantation 2014;21:376–84.
298. Eckermann JM, Buhler LH, Zhu A, Dor FJ, Awwad M, Cooper DKC. Initial investigation of the potential of modified porcine erythrocytes for transfusion in primates. Xenotransplantation 2004;11:18–26.
299. Rouhani FJ, Dor FJ, Cooper DKC. Investigation of red blood cells from alpha1,3-galactosyltransferase-knockout pigs for human blood transfusion. Transfusion. 2004;44:1004–12.
300. Matsumoto S, Tan P, Baker J, et al. Clinical porcine islet Xenotransplantation under comprehensive regulation. Transplant Proc. 2014;46:1992–5.
301. Wynyard S, Nathu D, Garkavenko O, Denner J, Elliott R. Microbiological safety of the first clinical pig islet Xenotransplantation trial in New Zealand. Xenotransplantation 2014;21:309. Epub ahead of print. https://doi.org/10.1111/xen.12102.
302. Zhang MC, Liu X, Jin Y, Jiang DL, Wei XS, Xie HT. Lamellar keratoplasty treatment of fungal corneal ulcers with acellular porcine corneal stroma. Am J Transplant. 2015;15:1068–75.
303. Kirchhof N, Shibata S, Wijkstrom M, et al. Reversal of diabetes in non-immunosuppressed rhesus macaques by intraportal porcine islet xenografts precedes acute cellular rejection. Xenotransplantation 2004;11:396–407.
304. Soderlund J, Wennberg L, Castanos-Velez E, et al. Fetal porcine islet-like cell clusters transplanted to cynomolgus monkeys: an immunohistochemical study. Transplantation. 1999;67:784–91.
305. Dor FJ, Cheng J, Alt A, Cooper DKC, Schuurman HJ. Gal alpha 1,3gal expression on porcine pancreatic islets, testis, spleen, and thymus. Xenotransplantation 2004;11:101–6.
306. Korbutt GS, Elliott JF, Ao Z, Smith DK, Warnock GL, Rajotte RV. Large scale isolation, growth, and function of porcine neonatal islet cells. J Clin Invest. 1996;97:2119–29.
307. Emamaullee JA, Shapiro AM, Rajotte RV, Korbutt G, Elliott JF. Neonatal porcine islets exhibit natural resistance to hypoxia-induced apoptosis. Transplantation. 2006;82:945–52.
308. Murray AG, Nelson RC, Rayat GR, Elliott JF, Korbutt GS. Neonatal porcine islet cells induce human cd4+, but not cd8+, lymphocyte proliferation and resist cell-mediated cytolytic injury in vitro. Diabetes. 1999;48:1713–9.
309. Nagaraju S, Bottino R, Wijkstrom M, Trucco M, Cooper DKC. Islet Xenotransplantation: what is the optimal age of the islet-source pig? Xenotransplantation 2015;22:7–19.
310. Rayat GR, Rajotte RV, Hering BJ, Binette TM, Korbutt GS. In vitro and in vivo expression of galalpha-(1,3)gal on porcine islet cells is age dependent. J Endocrinol. 2003;177:127–35.

311. Thompson P, Badell IR, Lowe M, et al. Islet Xenotransplantation using gal-deficient neonatal donors improves engraftment and function. Am J Transplant. 2011;11:2593–602.
312. McKenzie IF, Koulmanda M, Mandel TE, Sandrin MS. Pig islet xenografts are susceptible to "anti-pig" but not gal alpha(1,3)gal antibody plus complement in gal o/o mice. J Immunol (Baltimore, MD: 1950). 1998;161:5116–9.
313. Komoda H, Miyagawa S, Kubo T, et al. A study of the xenoantigenicity of adult pig islets cells. Xenotransplantation 2004;11:237–46.
314. Komoda H, Miyagawa S, Omori T, et al. Survival of adult islet grafts from transgenic pigs with n-acetylglucosaminyltransferase-iii (gnt-iii) in cynomolgus monkeys. Xenotransplantation 2005;12:209–16.
315. Wijkstrom M, Bottino R, Iwase H, et al. Glucose metabolism in pigs expressing human genes under an insulin promoter. Xenotransplantation 2015;22:70–9.
316. Nagaraju S, Bottino R, Wijkstrom M, Hara H, Trucco M, Cooper DKC. Islet Xenotransplantation from genetically engineered pigs. Curr Opin Organ Transplant. 2013;18:695–702.
317. Shin JS, Kim JM, Kim JS, et al. Long-term control of diabetes in immunosuppressed non-human primates (NHP) by the transplantation of adult porcine islets. Am J Transplant. 2015;15:2837. Epub ahead of print. https://doi.org/10.1111/ajt.13345.
318. Buhler L, Deng S, O'Neil J, et al. Adult porcine islet transplantation in baboons treated with conventional immunosuppression or a non-myeloablative regimen and cd154 blockade. Xenotransplantation 2002;9:3–13.
319. Hering BJ, Wijkstrom M, Graham ML, et al. Prolonged diabetes reversal after intraportal Xenotransplantation of wild-type porcine islets in immunosuppressed nonhuman primates. Nat Med. 2006;12:301–3.
320. Cardona K, Milas Z, Strobert E, et al. Engraftment of adult porcine islet xenografts in diabetic nonhuman primates through targeting of costimulation pathways. Am J Transplant. 2007;7:2260–8.
321. van der Windt DJ, Bottino R, Casu A, et al. Long-term controlled normoglycemia in diabetic non-human primates after transplantation with hcd46 transgenic porcine islets. Am J Transplant. 2009;9:2716–26.
322. Cardona K, Korbutt GS, Milas Z, et al. Long-term survival of neonatal porcine islets in non-human primates by targeting costimulation pathways. Nat Med. 2006;12:304–6.
323. Thompson P, Cardona K, Russell M, et al. Cd40-specific costimulation blockade enhances neonatal porcine islet survival in nonhuman primates. Am J Transplant. 2011;11:947–57.
324. Li J, Ezzelarab MB, Cooper DKC. Do mesenchymal stem cells function across species barriers? Relevance for xenotransplantation. Xenotransplantation. 2012;19:273–85.
325. Li J, Andreyev O, Chen M, et al. Human t cells upregulate cd69 after coculture with xenogeneic genetically-modified pig mesenchymal stromal cells. Cell Immunol. 2013;285:23–30.
326. Perez-Basterrechea M, Obaya AJ, Meana A, Otero J, Esteban MM. Cooperation by fibroblasts and bone marrow-mesenchymal stem cells to improve pancreatic rat-to-mouse islet Xenotransplantation PLoS One. 2013;8:e73526.
327. Dufour JM, Rajotte RV, Korbutt GS, Emerich DF. Harnessing the immunomodulatory properties of sertoli cells to enable Xenotransplantation in type i diabetes. Immunol Investig. 2003;32:275–97.
328. Dufour JM, Rajotte RV, Kin T, Korbutt GS. Immunoprotection of rat islet xenografts by cotransplantation with sertoli cells and a single injection of antilymphocyte serum. Transplantation. 2003;75:1594–6.
329. Gores PF, Hayes DH, Copeland MJ, et al. Long-term survival of intratesticular porcine islets in nonimmunosuppressed beagles. Transplantation. 2003;75:613–8.
330. Halberstadt C, Emerich DF, Gores P. Use of sertoli cell transplants to provide local immunoprotection for tissue grafts. Expert Opin Biol Ther. 2004;4:813–25.
331. Wang DZ, Skinner S, Elliot R, et al. Xenotransplantation of neonatal porcine islets and sertoli cells into nonimmunosuppressed streptozotocin-induced diabetic rats. Transplant Proc. 2005;37:470–1.

332. Ramji QA, Bayrack K, Arefanian H, et al. Protection of porcine islet xenografts in mice using sertoli cells and monoclonal antibodies. Transplantation. 2011;92:1309–15.
333. Bennet W, Sundberg B, Lundgren T, et al. Damage to porcine islets of langerhans after exposure to human blood in vitro, or after intraportal transplantation to cynomologus monkeys: protective effects of scr1 and heparin. Transplantation. 2000;69:711–9.
334. Moberg L, Johansson H, Lukinius A, et al. Production of tissue factor by pancreatic islet cells as a trigger of detrimental thrombotic reactions in clinical islet transplantation. Lancet. 2002;360:2039–45.
335. Ozmen L, Ekdahl KN, Elgue G, Larsson R, Korsgren O, Nilsson B. Inhibition of thrombin abrogates the instant blood-mediated inflammatory reaction triggered by isolated human islets: possible application of the thrombin inhibitor melagatran in clinical islet transplantation. Diabetes. 2002;51:1779–84.
336. Johansson H, Lukinius A, Moberg L, et al. Tissue factor produced by the endocrine cells of the islets of langerhans is associated with a negative outcome of clinical islet transplantation. Diabetes. 2005;54:1755–62.
337. Goto M, Tjernberg J, Dufrane D, et al. Dissecting the instant blood-mediated inflammatory reaction in islet xenotransplantation. Xenotransplantation. 2008;15:225–34.
338. Ji M, Yi S, Smith-Hurst H, et al. The importance of tissue factor expression by porcine nicc in triggering ibmir in the xenograft setting. Transplantation. 2011;91:841–6.
339. van der Windt DJ, Bottino R, Casu A, Campanile N, Cooper DKC. Rapid loss of intraportally transplanted islets: an overview of pathophysiology and preventive strategies. Xenotransplantation 2007;14:288–97.
340. van der Windt DJ, Marigliano M, He J, et al. Early islet damage after direct exposure of pig islets to blood: has humoral immunity been underestimated? Cell Transplant. 2012;21:1791–802.
341. Nagaraju S, Bertera S, Tanaka T, et al. In vitro exposure of pig neonatal islet-like cell clusters to human blood. Xenotransplantation 2015;22:317.
342. Echeverri GJ, McGrath K, Bottino R, et al. Endoscopic gastric submucosal transplantation of islets (endo-sti): technique and initial results in diabetic pigs. Am J Transplant. 2009;9:2485–96.
343. Fujita M, McGrath KM, Bottino R, et al. Technique of endoscopic biopsy of islet allografts transplanted into the gastric submucosal space in pigs. Cell Transplant. 2013;22:2335–44.
344. Christoffersson G, Henriksnas J, Johansson L, et al. Clinical and experimental pancreatic islet transplantation to striated muscle: establishment of a vascular system similar to that in native islets. Diabetes. 2010;59:2569–78.
345. Berman DM, O'Neil JJ, Coffey LC, et al. Long-term survival of nonhuman primate islets implanted in an omental pouch on a biodegradable scaffold. Am J Transplant. 2009;9:91–104.
346. Dufrane D, Goebbels RM, Gianello P. Alginate macroencapsulation of pig islets allows correction of streptozotocin-induced diabetes in primates up to 6 months without immunosuppression. Transplantation. 2010;90:1054–62.
347. Casu A, Bottino R, Balamurugan AN, et al. Metabolic aspects of pig-to-monkey (macaca fascicularis) islet transplantation: implications for translation into clinical practice. Diabetologia. 2008;51:120–9.
348. Casu A, Echeverri GJ, Bottino R, et al. Insulin secretion and glucose metabolism in alpha 1,3-galactosyltransferase knock-out pigs compared to wild-type pigs. Xenotransplantation 2010;17:131–9.
349. Potter KJ, Abedini A, Marek P, et al. Islet amyloid deposition limits the viability of human islet grafts but not porcine islet grafts. Proc Natl Acad Sci U S A. 2010;107:4305–10.
350. Tseng YL, Sachs DH, Cooper DKC. Porcine hematopoietic progenitor cell transplantation in nonhuman primates: a review of progress. Transplantation. 2005;79:1–9.
351. Tseng YL, Dor FJ, Kuwaki K, et al. Bone marrow transplantation from alpha1,3-galactosyltransferase gene-knockout pigs in baboons. Xenotransplantation 2004;11:361–70.
352. Tasaki M, Wamala I, Tena A, et al. High incidence of xenogenic bone marrow engraftment in pig-to-baboon intra-bone bone marrow transplantation. Am J Transplant. 2015;15:974–83.

353. Barth RN, Yamamoto S, LaMattina JC, et al. Xenogeneic thymokidney and thymic tissue transplantation in a pig-to-baboon model: I. evidence for pig-specific t-cell unresponsiveness. Transplantation. 2003;75:1615–24.
354. Yamamoto S, Lavelle JM, Vagefi PA, et al. Vascularized thymic lobe transplantation in a pig-to-baboon model: a novel strategy for xenogeneic tolerance induction and t-cell reconstitution. Transplantation. 2005;80:1783–90.
355. Griesemer AD, Hirakata A, Shimizu A, et al. Results of gal-knockout porcine thymokidney xenografts. Am J Transplant. 2009;9:2669–78.
356. Tena A, Vallabhajosyula P, Hawley RJ, Griesemer A, Yamada K, Sachs DH. Quantification of baboon thymopoiesis in porcine thymokidney xenografts by the signal-joining t-cell receptor excision circle assay. Transplantation. 2011;91:639–44.
357. Tena A, Kurtz J, Leonard DA, et al. Transgenic expression of human cd47 markedly increases engraftment in a murine model of pig-to-human hematopoietic cell transplantation. Am J Transplant. 2014;14:2713–22.
358. Bibikova M, Golic M, Golic KG, Carroll D. Targeted chromosomal cleavage and mutagenesis in drosophila using zinc-finger nucleases. Genetics. 2002;161:1169–75.
359. Carroll D. Progress and prospects: zinc-finger nucleases as gene therapy agents. Gene Ther. 2008;15:1463–8.
360. Klug A. The discovery of zinc fingers and their applications in gene regulation and genome manipulation. Annu Rev Biochem. 2010;79:213–31.
361. Urnov FD, Rebar EJ, Holmes MC, Zhang HS, Gregory PD. Genome editing with engineered zinc finger nucleases. Nat Rev Genet. 2010;11:636–46.
362. Watanabe M, Umeyama K, Matsunari H, et al. Knockout of exogenous egfp gene in porcine somatic cells using zinc-finger nucleases. Biochem Biophys Res Commun. 2010;402:14–8.
363. Hauschild J, Petersen B, Santiago Y, et al. Efficient generation of a biallelic knockout in pigs using zinc-finger nucleases. Proc Natl Acad Sci U S A. 2011;108:12013–7.
364. Tesson L, Usal C, Menoret S, et al. Knockout rats generated by embryo microinjection of talens. Nat Biotechnol. 2011;29:695–6.
365. Galli C, Perota A, Brunetti D, Lagutina I, Lazzari G, Lucchini F. Genetic engineering including superseding microinjection: new ways to make gm pigs. Xenotransplantation 2010;17:397–410.
366. Hsu PD, Lander ES, Zhang F. Development and applications of crispr-cas9 for genome engineering. Cell. 2014;157:1262–78.
367. Doudna JA, Charpentier E. Genome editing. The new frontier of genome engineering with crispr-cas9. Science (New York, NY). 2014;346:1258096.
368. Ni W, Qiao J, Hu S, et al. Efficient gene knockout in goats using crispr/cas9 system. PLoS One. 2014;9:e106718.
369. Mali P, Esvelt KM, Church GM. Cas9 as a versatile tool for engineering biology. Nat Methods. 2013;10:957–63.
370. Seruggia D, Montoliu L. The new crispr-cas system: Rna-guided genome engineering to efficiently produce any desired genetic alteration in animals. Transgenic Res. 2014;23:707–16.
371. Kim H, Kim JS. A guide to genome engineering with programmable nucleases. Nat Rev Genet. 2014;15:321–34.
372. Li P, Estrada JL, Burlak C, et al. Efficient generation of genetically distinct pigs in a single pregnancy using multiplexed single-guide rna and carbohydrate selection. Xenotransplantation 2015;22:20–31.
373. Ayares D, Vaught T, Ball S, et al. Genetic engineering of source pigs for Xenotransplantation: progress and prospects. Xenotransplantation 2013;20:361; (Abstract 408).
374. Cooper DKC. The case for Xenotransplantation Clin Transpl. 2015;29:288–93.
375. Ezzelarab M, Cortese-Hassett A, Cooper DKC, Yazer MH. Extended coagulation profiles of healthy baboons and of baboons rejecting GT-KO pig heart grafts. Xenotransplantation 2006;13:522–528.

Is Sensitization to Pig Antigens Detrimental to Subsequent Allotransplantation?

3

Qi Li, Hidetaka Hara, Zhongqiang Zhang, Michael E. Breimer, Yi Wang, and David K. C. Cooper

Abbreviations

BAL	Bioartificial liver
Gal	Galactose-α1,3-galactose
GTKO	α1,3-galactosyltransferase gene-knockout
HLA	Human leukocyte antigens

Based on a paper published in Xenotransplantation 2018; 25:e12393. doi: 10.1111/xen.12393, with permission of the editor.

Q. Li
Xenotransplantation Program, Department of Surgery, University of Alabama at Birmingham (UAB), Birmingham, AL, USA

Second Affiliated Hospital, University of South China, Hengyang City, Hunan, China

H. Hara
Xenotransplantation Program, Department of Surgery, University of Alabama at Birmingham (UAB), Birmingham, AL, USA

Z. Zhang
Thomas E. Starzl Transplantation Institute, University of Pittsburgh, Pittsburgh, PA, USA

Department of Organ Transplantation and General Surgery, Second Xiangya Hospital of Central South University, Changsha, Hunan, China

M. E. Breimer
Department of Surgery, Institute of Clinical Sciences, Sahlgrenska Academy at University of Gothenburg, Gothenburg, Sweden

Y. Wang
Second Affiliated Hospital, University of South China, Hengyang City, Hunan, China

D. K. C. Cooper (✉)
Xenotransplantation Program, Division of Transplantation, Department of Surgery, University of Alabama at Birmingham (UAB), Birmingham, AL, USA
e-mail: dkcooper@uabmc.edu

D. K. C. Cooper, G. Byrne (eds.), *Clinical Xenotransplantation*,
https://doi.org/10.1007/978-3-030-49127-7_3

NHP	Nonhuman primate
SLA	Swine leukocyte antigens
WT	Wild type

Introduction

There is a continuing shortage of organs from deceased human donors for the purpose of transplantation into patients with end-stage organ failure [1]. Xenotransplantation could provide an alternative source of organs. Recently, there has been substantial progress in overcoming the barriers to xenotransplantation, especially through the transplantation of organs from genetically engineered pigs combined with effective immunosuppressive therapy [2].

Initial patients might receive a pig graft as a "bridge" to maintain life until a suitable allograft becomes available. Others might require allotransplantation in the event that a xenograft fails. If a pig xenograft were to fail, and a suitable allograft became available, could the patient undergo allotransplantation without a detrimental effect from the previous xenotransplant? In other words, do elicited xenoreactive antibodies cross-react with alloantigens? This important question has not yet been definitively answered. We have searched the literature, and identified a small number of studies of relevance.

We primarily limited our search to studies directly relevant to pig organ or cell transplantation in humans, which largely related to studies in pig-to-nonhuman primate (NHP) (discordant) models. However, some other experimental studies, including those relating to xenotransplantation between closely related NHP (concordant) models, will be briefly summarized. We also searched for experience in clinical xenotransplantation.

Experimental Secondary Allotransplantation After Xenosensitization in Discordant Models

"Discordant" relates to models in which transplantation is carried out between widely disparate species where hyperacute rejection usually results, for example, wild-type (WT, i.e., genetically unmodified) pig-to-NHP [3]. We identified only five relevant reports (Table 3.1). Within these reports were three in which in vitro studies were carried out after in vivo exposure to pig antigens, and three in which both in vitro studies and additional allotransplantation were carried out. In addition, however, we have carried out a new in vitro study which we report below.

Ye et al. [4] were the first (in 1995) to investigate this topic. In immunosuppressed baboons that were sensitized by WT pig heart transplants ($n = 2$, with the grafts undergoing hyperacute rejection) or by pig erythrocytes ($n = 1$), subsequent baboon liver allografts survived without evidence of antibody-mediated or accelerated cellular rejection. No baboon developed antibodies that cross-reacted with alloantigens.

Table 3.1 Secondary allotransplantation after xenosensitization in discordant NHP models

Author	Year	Recipient (*n*)	Primary xenograft donor	Primary organ/tissue/cell	Graft survival	Secondary allograft donor (*n*)	Secondary organ/tissue/cell	Graft survival	Reference
Ye et al.	1995	Baboon ($n = 3$)	Pig	Heart, Blood	20 minutes	Baboon ($n = 3$)	Liver	6 > 62 days	[4]
Ye et al.	1995	Baboon ($n = 3$)	Pig	Liver, Heart	1 hour–6 days	–	–	–	[4]
Baertschiger et al.	2004	Baboon ($n = 4$)	Pig	Heart, RBCs, PBMCs	–	–	–	–	[5]
Key et al.	2004	Cynomolgus monkey ($n = 52$)	Pig	Kidney	1–53 days	–	–	–	[6]
Choi et al.	2014	Rhesus monkey ($n = 5$)	Pig	Cornea	≥49 days	Rhesus monkey ($n = 5$)	Cornea	35 > 324 days	[7]
Albritton et al.	2014	Baboon ($n = 4$)	Pig	Skin	11–13 days	Baboon ($n = 4$)	Skin	10–14 days	[8]

This group also tested sera from baboons that had rejected a WT pig liver ($n = 2$) or heart ($n = 1$) graft against a panel of baboon lymphocytes ($n = 6$). No cytotoxicity of the baboon lymphocytes was documented, supporting a conclusion that sensitization to pig antigens did not result in allosensitization.

In 2004, Baertschiger et al. [5] carried out in vitro studies using baboon serum and peripheral blood mononuclear cells (PBMC) exposed to WT pig antigens. Serum and PBMCs from four groups of baboons were studied: (i) naïve baboons ($n = 4$); (ii) baboons sensitized to galactose-α1,3-galactose (Gal) antigens ($n = 2$) by prior in vivo exposure to a pig heart or pig red blood cells; (iii) baboons sensitized to Gal+nonGal antigens ($n = 2$) by prior in vivo exposure to a pig heart or pig PBMC; and (iv) baboons sensitized to alloantigens ($n = 2$). In an antibody assay, baboon serum containing anti-pig xenoantibodies was cultured with baboon or pig PBMCs. There was no cross-reactivity between xenoantibodies and alloantigens. A complement-dependent cytotoxicity assay indicated no killing of baboon PBMC, and mixed lymphocyte reaction suggested no increased T-cell proliferative response to alloantigens. Although the study was limited, it produced no evidence that a previous pig xenograft would be detrimental to a secondary allograft.

Key et al. [6] studied 52 conventionally immunosuppressed cynomolgus monkeys that had received pig kidneys from hCD55 (human decay-accelerating factor; hDAF) transgenic pigs to investigate whether anti-swine leukocyte antigen (SLA) antibodies cross-reacted with human leukocyte antigens (HLA). Graft survival was for a mean of 20 days (range 1–53 days). Pretransplant and post-transplant serum from each monkey was incubated with "pooled, purified" HLA. There was no detectable increase in anti-HLA IgG antibodies after pig kidney transplantation.

In a WT pig-to-Chinese rhesus monkey decellularized corneal xenotransplantation model, porcine corneal lamellar (anterior partial thickness) xenografts were followed by full-thickness corneal allografts ($n = 5$) [7]. All recipients received immunosuppressive therapy (topical prednisolone acetate, subconjunctival dexamethasone, and systemic methylprednisolone). Only one of five pig corneal grafts was rejected, with four grafts surviving for 7–13 months, at which time the monkey received an allograft. On in vitro assays, there was no evidence that any humoral or cellular immune response to the xenograft adversely affected the survival of the allograft. However, the fact that 4 of the 5 xenografts were not rejected suggests that no xenoreactive antibodies developed, thus reducing the likelihood of an immune response to the allograft. Furthermore, there are differences in the mechanism of rejection between an organ and a cornea.

In a pig-to-baboon skin xenotransplantation model, Albritton et al. [8] studied four groups: (i) a primary baboon skin allograft (survival for 12–13 days) followed by a secondary skin graft from an α1,3-galactosyltransferase gene-knockout (GTKO) pig (survival for 10–13 days); (ii) a primary GTKO pig skin xenograft (survival for 11–13 days) followed by a secondary baboon skin allograft (survival for 10–14 days); (iii) a primary GTKO pig xenograft (survival for 7–11 days) followed by a secondary GTKO pig xenograft (survival for 1 day); and (iv) a primary allograft (survival for 7–11 days) followed by a secondary allograft (survival for 4 days). These results suggested that primary allograft or xenograft rejection did not accelerate the rejection of, respectively, a subsequent xenograft or allograft. However, initial sensitization to xenoantigens or alloantigens accelerated rejection of, respectively, a subsequent xenograft or allograft. A primary skin allograft was associated with the production of anti-allogeneic antibody, but not anti-xenogeneic antibody, and a primary xenograft did not induce antibodies directed to an allograft. In the highly immunogenic skin transplant model, therefore, there was no cross-reactivity between xenoantibodies and alloantigens or between alloantibodies and xenoantigens.

In order to add to the experience on this topic, we have carried out a further in vitro experiment. Flow cytometry of serum IgM and IgG binding to $CD3^+$T cells (gated from PBMC) from either a GTKO/CD46 pig or a baboon was carried out. Serum was taken from (i) naïve baboons ($n = 8$), (ii) baboons exposed to pig antigens (in the form of an organ or artery patch graft) that had not become sensitized ($n = 4$), and (iii) baboons exposed to pig antigens that had become sensitized ($n = 4$). Although there was minimal antibody binding of naïve and nonsensitized sera to GTKO/CD46 pig PBMCs, there was significant binding of sensitized serum (both IgM and IgG) to these cells ($p < 0.05$) (Fig. 3.1). In contrast, there was no significant binding of any baboon serum to baboon PBMCs. This small study strengthens the conclusion that prior sensitization to a pig xenograft would not be detrimental to a subsequent allograft.

Additional Study of Relevance

Recently (2016), Kim et al. [9] transplanted pig islets into mice that had previously been sensitized to either pig or mouse antigens. Survival of islet *allografts* in naïve mice (15.5 ± 2.38 days) was no different from that of islet allografts in mice previously sensitized to pig antigens (14.4 ± 1.41 days). Furthermore, there was no difference in survival of pig islets in naïve mice (5.8 ± 2.04 days) or in allo-sensitized mice (6.4 ± 2.26 days). In vitro assays indicated that (i) xenosensitized mice did not induce anti-allogeneic antibody, and there was (ii) no cross-reactivity between xenoantibody and alloantigens, and (iii) no accelerated cellular response to a subsequent allotransplant.

Experimental Secondary Allotransplantation After Xenosensitization in Concordant Models

Although today clinical concordant xenotransplantation, that is, between closely related species where hyperacute rejection would not be anticipated, for example, an NHP-to-human [3], is not being considered, there have been a few experimental studies of secondary allotransplantation in an NHP after an initial concordant xenograft that are of interest. Given the closer evolutionary relationship between the initial donor and recipient species, with the likelihood of more conserved antigen structure, it might be anticipated that there would be greater cross-reactivity of antibodies that develop after rejection of a concordant xenograft.

However, three groups provided evidence to suggest that primary concordant xenografts in immunosuppressed NHPs did *not* induce a humoral or an accelerated cell-mediated immune response that jeopardized the survival of a secondary allograft (Table 3.2) [4, 10, 11].

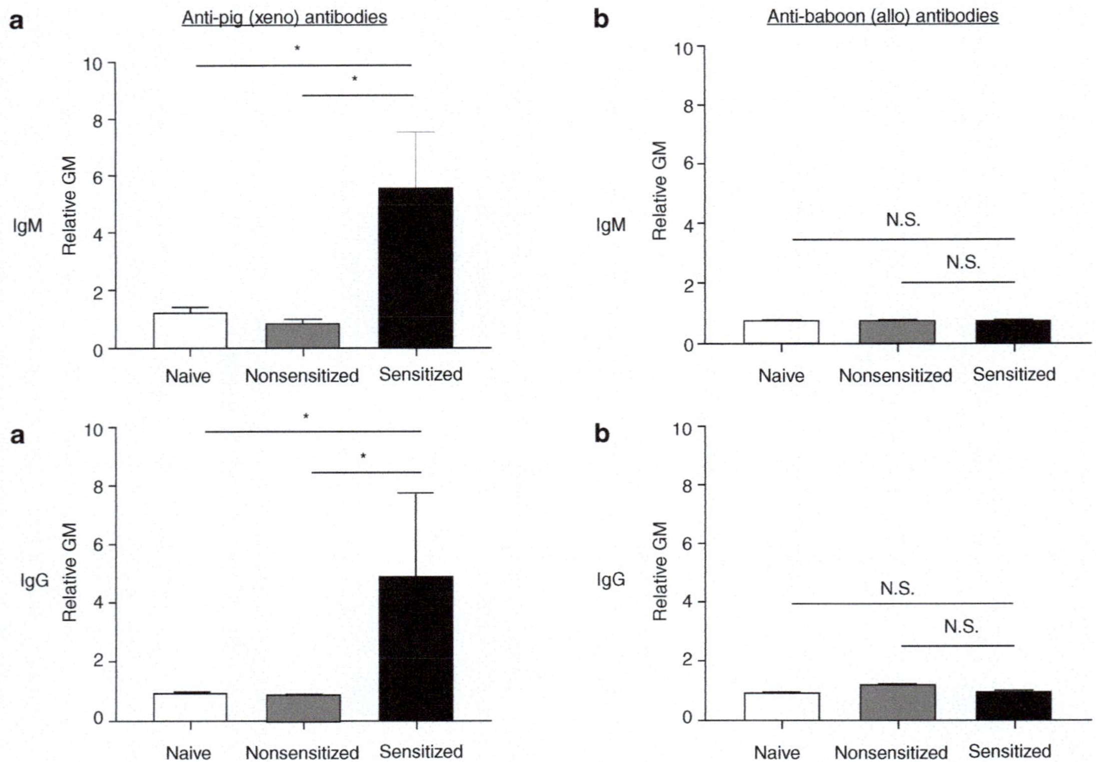

Fig. 3.1 Flow cytometry of serum IgM (top) and IgG (bottom) binding to $CD3^+$T cells (gated from PBMC) from (**a**) a GTKO/CD46 pig, and (**b**) a panel of cells from 4 baboons, using naïve baboon serum ($n = 8$), nonsensitized baboon serum ($n = 4$), and sensitized baboon serum ($n = 4$). Although the sensitized baboon sera showed high IgM and IgG antibody binding to GTKO/CD46 pig PBMC, there was no increased binding to any of the panel of 4 baboon PBMC. (NS = not significant; $*P < 0.05$)

Table 3.2 Secondary allotransplantation after xenosensitization in concordant NHP models

Author	Year	Recipient (*n*)	Primary xenograft donor	Primary organ/tissue/cell	Graft survival	Secondary allograft donor (*n*)	Secondary organ/tissue/cell	Graft survival	Reference
Alonso de Begona et al.[a]	1992	Baboon ($n = 5$)	African green monkey	Heart	5–65 days	Baboon ($n = 5$)	Heart	10 ≥ 198 days	[10]
Ye et al.[b]	1995	Baboon ($n = 6$)	African green monkey	Liver	10–120 days	Baboon ($n = 1$)	Heart	>30 days	[4]
Michler et al.[c]	1996	Baboon ($n = 4$)	Cynomolgus monkey	Heart	>14 days	Baboon ($n = 4$)	Heart	>56 days	[11]

[a]Two baboons with allografts survived for 164 and 198 days (until they were euthanized) without evidence of rejection

[b]The baboon in which the xenograft survived 120 days received a secondary baboon heart allotransplant, which was followed for >30 days without features of rejection (until elective euthanasia). At the time of allotransplantation, a lymphocytotoxicity assay carried out with serum from the recipient and cells from the baboon heart donor was negative

[c]Despite the fact that, after xenotransplantation, 50% of the baboons developed cytotoxic antibodies against the MHC class II-like antigens expressed by lymphocytes of more than half of a panel of 12 baboons, neither the presence of these antibodies nor the severity of the prior xenograft rejection impacted the histology of allograft rejection. When T cell lines were developed from T cells isolated from xenograft biopsies, none demonstrated cell-mediated proliferative or cytotoxic activity against cells from the secondary allograft donor. These data suggested that a prior concordant xenograft was not detrimental to a subsequent allograft

Additional Studies of Relevance

There have also been studies in other animal models that provide further evidence.

The groups of Gannedahl et al. [12], Chice et al. [13] and Di Stefano et al. [14], respectively using initial mouse-to-Lewis rat, hamster-to Lewis rat, and hamster-to Lewis rat models followed by allotransplantation, independently concluded that an initial xenograft was not detrimental to a subsequent allograft.

In contrast, two groups provided data to suggest that secondary allografts are at risk after transplantation into a recipient previously sensitized to a xenograft, though these studies were not in the clinically more relevant pig-to-NHP model.

Hammer et al. [15] carried out allogeneic and xenogeneic heterotopic heart transplantation. Six dog recipients accepted primary allogeneic hearts (with immunosuppressive therapy in the form of cyclosporine, azathioprine, and corticosteroids) with a mean allograft survival of 18 days. Heart xenotransplants between donor foxes and recipient dogs (under identical immunosuppressive therapy) were rejected in a mean of 10 days, and subsequent dog allograft hearts (under the same immunosuppressive regimen) were rejected in a mean of 5 days. Therefore, rejection of the concordant xenograft reduced survival of a subsequent allograft (from a mean of 18 days to 5 days).

In addition, Etheredge et al. [16] reported accelerated skin allograft rejection following xenogeneic sensitization in dog- (a distantly related species) and guinea pig- (a closely related species) to-rabbit models. Five rabbit recipients accepted a primary allogeneic skin graft with a mean survival of 10 days. Fifteen rabbits received an allograft after being sensitized to a guinea pig skin xenograft, with rejection occurring in a mean of 7 days. Eighteen rabbits received an allograft following a dog skin xenograft, with rejection in a mean of 7 days. It would therefore appear that, in these models, an initial skin xenograft (whether discordant or concordant) was detrimental to survival of a subsequent skin allograft.

Clinical Allotransplantation After (or Before) Exposure to Pig Antigens

Six clinical studies are of relevance. Three patients undergoing dialysis were exposed to pig kidneys that were inserted into the dialysis circuit, and there were three studies related to bridging a patient in hepatic failure to liver allotransplantation. In addition, there was one study in which patients with renal allografts were subsequently sensitized to pig antigens.

Kidney Allotransplantation After Exposure to Pig Antigens (Table 3.3a)

After a course of plasmapheresis to remove anti-pig antibodies, Welsh and his colleagues exposed a patient undergoing dialysis to WT pig kidneys on two

Table 3.3a Clinical allotransplantation after (or before) exposure to pig antigens. Clinical allotransplantation after exposure to pig antigens

Author	Year	Patients (*n*)	Primary perfusion	Primary tissue	Secondary allograft donor	Secondary organ/tissue/cell	Reference
Welsh et al.	1991	1	Pig	Kidney	Human	Kidney	[17, 18]
Breimer et al.	1996	2	Pig	Kidney	Human	Kidney	[19–21]
Chari et al.	1994	1	Pig	Liver	Human	Liver	[22]
Baquerizo et al.	1999	8	Pig	Liver	Human	Liver	[23]
Levy et al.	2000	2	Pig	Liver	Human	Liver	[24]

occasions approximately 1 month apart [17, 18]. The patient received conventional immunosuppressive therapy to cover the "experimental period." The two pig kidneys were perfused for 6 and 1.5 hours, respectively. There was only a weak anti-pig immune response. The patient successfully underwent renal allotransplantation.

In 1996, Breimer, Rydberg, and colleagues [19–21] reported extracorporeal blood perfusion of WT pig kidneys (after plasmapheresis but in the absence of any immunosuppressive therapy) in two patients undergoing dialysis. Patient 1 (HLA-sensitized with panel-reactive antibodies of 85%) was exposed to the pig kidney for 65 minutes (before hyperacute rejection occurred). The patient developed a high level of anti-pig antibodies [20, 21], but there was no change in anti-HLA antibodies (panel-reactive antibodies remaining at 85%) [20]. Three years later, the patient received a cadaveric renal allograft (with a negative cytotoxic crossmatch). Graft function was excellent for 2 years when, unexpectedly, function rapidly declined (for unclear reasons). Microscopy showed thrombotic microangiopathy, but chronic antibody-mediated rejection could not be excluded.

Patient 2 (HLA-nonsensitized) was exposed to a WT pig kidney for 15 minutes, when he developed anaphylaxis, necessitating discontinuation of the perfusion. He recovered quickly, and developed a weak anti-pig antibody response [20, 21]. No anti-HLA antibodies developed. Four years later, he received a cadaveric renal allograft, which had to be removed 18 days later for persistent bleeding associated with thrombocytopenia which had been problematic for years, even before exposure to the pig kidney [19]. It therefore did not appear to be a complication of sensitization to the pig kidney. Microscopy showed features of cellular and vascular rejection.

Based on this, it can be concluded that the perfusion experiments were not detrimental to the patients in obtaining subsequent renal allografts, though there may be some doubt in the second patient. It should be borne in mind that a large number of passenger leukocytes are transferred from the organ to the recipient during ex vivo organ perfusion [25], increasing the risk of immunization to pig antigens.

Liver Allotransplantation After Exposure to Pig Antigens (Table 3.3a)

In 1994, Chari et al. [22] reported a successful liver allotransplantation in a patient who had undergone ex vivo perfusion of five pig livers during the previous few days, though no details were given on the anti-pig or anti-HLA antibody responses.

Baquerizo et al. [23] provided data in 1999 from eight patients bridged by a bioartificial liver (BAL) that incorporated pig hepatocytes, who subsequently underwent successful liver allotransplantation. When BAL treatment was performed only once, there was no increase in anti-pig antibody. After two or more BAL treatments, however, there was a significant increase in anti-Gal IgG antibody, though no antibodies developed to non-Gal pig specificities. Sensitization to pig antigens appeared to have no detrimental effect on the outcome of the subsequent liver allografts.

In 2000, Levy et al. [24] reported bridging of two patients to successful liver allotransplantation by ex vivo extracorporeal blood perfusion through livers from pigs transgenic for the human complement-regulatory proteins CD55 and CD59. The periods of perfusion were only 6.5 and 10 hours, respectively, because deceased human donor livers became available. Nevertheless, despite treatment of the recipient with tacrolimus-based immunosuppressive therapy, the level of anti-Gal antibody initially markedly increased (IgM 10-fold, IgG 25-fold). At 60 days after the transplant, IgG remained 20% higher than the pretransplantation level. However, no anti-HLA antibodies developed.

Although the periods of organ perfusion were short in several of these reported cases, it should be remembered that livers and, perhaps to a less extent, kidneys contain large numbers of passenger leukocytes that enter the patient's circulation, increasing the possibility of sensitization.

Exposure to Pig Antigens after, or at the Time of, Kidney Allotransplantation (Table 3.3b)

There is one other clinical study of relevance. Groth and coworkers [26] reported on 10 patients with type 1 diabetes with long-standing renal allografts who received WT fetal pig islet-like cell clusters intraportally or under the kidney capsule (Table 3.3b). All patients developed xenoreactive antibodies against Gal antigens, which remained high for up to 6–8 years [27], but there was no increase in panel-reactive antibodies, and the kidney grafts continued to function well [28].

Table 3.3b Clinical allotransplantation after (or before) exposure to pig antigens. Exposure to pig antigens after clinical allotransplantation

Author	Year	Patients (*n*)	Primary allograft donor	Primary organ	Secondary xenograft donor	Secondary organ/tissue/cell	Reference
Groth et al.	1994	10	Human	Kidney	Pig	Islet	[26]

Conclusions

The data from discordant experimental pig-to-NHP models of xenotransplantation indicated that a primary xenograft did not induce a humoral or accelerated cell-mediated immune response that jeopardized the survival of a secondary allograft. The clinical experience, though also very limited, supports this conclusion.

Patients with a high level of anti-HLA antibodies (calculated panel-reactive antibodies) often wait many years before an organ from a deceased human donor becomes available. If there are no antibodies that cross-react between HLA and SLA, pig xenotransplantation would alleviate this problem. Several groups have investigated whether highly allosensitized human serum cross-reacts with SLA (reviewed in [29]) (Table 3.4). The majority of groups have concluded that human anti-HLA antibodies can cross-react with SLA and thus jeopardize the survival of a pig graft [30, 32, 41]. A smaller number of groups, however, have found no evidence of cross-reactivity between antibodies directed to HLA and SLA (Table 3.4).

If, indeed, HLA-specific antibodies do recognize SLA, then the question needs to be asked as to why no groups have reported a detrimental effect of initial exposure to SLA on the outcome of subsequent allotransplantation, as reviewed in the present report. HLA and SLA genes encode proteins on the cell surface (antigens), but the HLA genes are at least 100 kbp longer than SLA, and are greater than SLA in number [42]. As the HLA system is more complex than the SLA system, the greater number and complexity of anti-HLA antibodies might result in their recognition of SLA. In contrast, anti-SLA antibodies may be insufficient in variety and number to recognize HLA. This hypothesis requires investigation.

The results of our review must be interpreted cautiously as not only are the numbers of reports very few, and in some cases almost anecdotal, but exposure to pig

Table 3.4 Studies relating to cross-reactivity between the anti-HLA immune response and pig antigens to investigate whether HLA sensitization is detrimental to pig xenotransplantation

HLA sensitization *is* detrimental	
Author/year	Reference
Naziruddin et al. (1998)	[30]
Taylor et al. (1998)	[32]
Barreau et al. (2000)	[34]
Popma et al. (2000)	[36]
Mulder et al. (2000)	[37]
Oostingh et al. (2002)	[38]
Varela et al. (2003)	[39]
Mulder et al. (2010)	[40]
Martens et al. (2017)	[41]

HLA sensitization *is not* detrimental	
Author/year	Reference
Bartholomew et al. (1997)	[31]
Wong et al. (2006)	[33]
Hara et al. (2006)	[35]
Zhang et al. (2018)	[29]

antigens was at times relatively brief. There were also varying periods between developing sensitization to pig antigens and subsequent exposure to alloantigens. With initial clinical trials of pig organ transplantation drawing closer, more data from the important pig-to-NHP model are required to allow a deeper understanding of the topic.

Acknowledgments We thank renal pathologist, J. Mölne, for re-evaluation of the pig kidney histology from the two patients investigated in Sweden [19]. Work on xenotransplantation at the University of Alabama at Birmingham is supported in part by NIH NIAID U19 grant AI090959.

Conflict of Interest No author declares a conflict of interest.

References

1. Hart A, Smith J, Skeans M, et al. OPTN/SRTR 2015 annual data report: kidney. Am J Transplant. 2017;17:21–116.
2. Cooper DKC, Ekser B, Tector A. Immunobiological barriers to xenotransplantation. Int J Surg. 2015;23:211–6.
3. Calne RY. Organ transplantation between widely disparate species. Transplant Proc. 1970;2:550–6.
4. Ye Y, Luo Y, Kobayashi T, et al. Secondary organ allografting after a primary "bridging" xenotransplant. Transplantation. 1995;60:19–22.
5. Baertschiger R, Dor F, Prabharasuth D, Kuwaki K, Cooper DKC. Absence of humoral and cellular alloreactivity in baboons sensitized to pig antigens. Xenotransplantation. 2004;11:27–32.
6. Key T, Schuurman H, Taylor C. Does exposure to swine leukocyte antigens after pig-to-nonhuman primate xenotransplantation provoke antibodies that cross-react with human leukocyte antigens? Xenotransplantation. 2004;11:452–6.
7. Choi H, Lee J, Kim M, et al. Cross-reactivity between decellularized porcine corneal lamellae for corneal xenobridging and subsequent corneal allotransplants. Xenotransplantation. 2014;21:115–23.
8. Albritton A, Leonard D, Barone A, et al. Lack of cross-sensitization between α-1,3-galactosyltransferase knockout porcine and allogeneic skin grafts permits serial grafting. Transplantation. 2014;97:1209–15.
9. Kim HJ, Byun N, Kwon O, Park CG. Cross-sensitization between xeno-and allo-antigens on subsequent allogeneic and xenogeneic pancreatic islet transplantation in a murine model. Biochem Biophys Res Commun. 2016;480:474–8.
10. de Begona A, Gundry S, Kawauchi M, et al. Assessment of baboon lymphocyte subsets and activity in cardiac xenobridging to allotransplantation. Transplant Proc. 1992;24:453–4.
11. Michler RE, Shah AS, Itescu S, et al. The influence of concordant xenografts on the humoral and cell-mediated immune responses to subsequent allografts in primates. J Thorac Cardiovasc Surg. 1996;112:1002–9.
12. Gannedahl G, Tufveson G. The impact of xenograft rejection of future grafting. Transplant Proc. 1992;24:276–8.
13. Chice L, Bittard H, Adam R, Castaing D, Bismuth H. Xenogeneic liver transplantation: its influence on secondary allogeneic graft rejection. Transplant Proc. 1990;22:2333.
14. Di Stefano R, Carobbi A, Frosini F, et al. Secondary allogeneic challenge of xenotransplant recipients. Transplant Proc. 1992;24:694.
15. Hammer C, Schutz A, Pratschke J, et al. Bridging to transplant: allogeneic heart transplantation after xenografting. J Heart Lung Transplant. 1992;11:182–8.

16. Etheredge EE, Shons AR, Najarian JS. Accelerated skin allograft rejection following xenogeneic sensitization. Transplant Proc. 1971;3:1585.
17. Cairns TDH, Taube DH, Stevens N, et al. Xenografts-future prospects for clinical transplantation. Immunol Lett. 1991;29:167–70.
18. Welsh KL, Taube DH, Thick M, et al. Human antibodies to pig determinants and their association with hyperacute rejection of xenografts. In: Xenotransplantation. Heidelberg: Springer Berlin Heidelberg; 1991. p. 501–10.
19. Breimer ME, Björck S, Svalander C, et al. Extracorporeal ("ex vivo") connection of pig kidneys to humans. I. Clinical data and studies of platelet destruction. Xenotransplantation. 1996;3:328–39.
20. Rydberg L, Björck S, Hallberg E, et al. Extracorporeal ("ex vivo") connection of pig kidneys to humans. II. The anti-pig antibody response. Xenotransplantation. 1996;3:340–53.
21. Rydberg L, Breimer ME, Nilsson K, et al. An ELISA technique for quantitation of human xenoantibodies binding to pig cells: application in patients with pig kidneys extracorporeally connected to the circulation. Xenotransplantation. 1998;5:105–10.
22. Chari RS, Collins BH, Magee JC, et al. Treatment of hepatic failure with ex vivo pig-liver perfusion followed by liver transplantation. N Engl J Med. 1994;331:234–7.
23. Baquerizo A, Mhoyan A, Kearns-Jonker M, et al. Characterization of human xeno-reactive antibodies in liver failure patients exposed to pig hepatocytes after bioartificial liver treatment: an ex vivo model of pig to human xenotransplantation. Transplantation. 1999;67:5–18.
24. Levy MF, Crippin J, Sutton S, et al. Liver allotransplantation after extracorporeal hepatic support with transgenic (hCD55/hCD59) porcine livers: clinical results and lack of pig-to-human transmission of the porcine endogenous retrovirus. Transplantation. 2000;69:272–80.
25. Magnusson S, Mansson JE, Strokan V, et al. Release of pig leukocyte during pig kidney perfusion and characterization of pig lymphocyte carbohydrate xenoantigens. Xenotransplantation. 2003;10:432–45.
26. Groth CG, Korsgren O, Tibell A, et al. Transplantation of porcine fetal pancreas to diabetic patients. Lancet. 1994;344:1402–4.
27. Groth CG, Tibell A, Wennberg L, et al. Clinical aspects and perspectives in islet xenotransplantation. J Hepato-Biliary-Pancreat Surg. 2000;7:364–9.
28. Satake M, Kumagai-Braesch M, Kawagishi N, et al. Kinetics and character of xenoantibody formation in diabetic patients transplanted with fetal porcine islet cell clusters. Xenotransplantation. 1994;1:24–35.
29. Zhang Z, Hara H, Long C, et al. Immune responses of HLA-highly-sensitized and non-sensitized patients to genetically engineered pig cells. Transplantation. 2018;102:e195.
30. Naziruddin B, Durriya S, Phelan D, et al. HLA antibodies present in the sera of sensitized patients awaiting renal transplant are also reactive to swine leukocyte antigens. Transplantation. 1998;66:1074–80.
31. Bartholomew A, Latinne D, Sachs DH, et al. Utility of xenografts: lack of correlation between PRA and natural antibodies to swine. Xenotransplantation. 1997;4:34–9.
32. Taylor C, Tang K, Smith S, White D, Davies H. HLA-specific antibodies in highly sensitized patients can cause a positive crossmatch against pig lymphocytes. Transplantation. 1998;65:1634–41.
33. Wong BS, Yamada K, Okumi M, et al. Allosensitization does not increase the risk of xenoreactivity to alpha 1,3-galactosyltransferase gene-knockout miniature swine in patients on transplantation waiting lists. Transplantation. 2006;82:314–9.
34. Barreau N, Godfrin Y, Bouhours JF, et al. Interaction of anti-HLA antibodies with pig xenoantigens. Transplantation. 2000;69:148–56.
35. Hara H, Ezzelarab M, Rood PPM, et al. Allosensitized humans are at no greater risk of humoral rejection of GT-KO pig organs than other humans. Xenotransplantation. 2006;13:357–65.
36. Popma SR, Krasinskas AM, Kreisel D, et al. Allosensitization increases human anti-pig cellular xenoreactivity. Transplant Proc. 2000;32:954–5.
37. Mulder A, Kardol MJ, Niterink JGS, et al. Reactivity of human monoclonal HLA antibodies with porcine lymphocytes [abstract]. Hum Immunol. 2000;61:27.

38. Oostingh GJ, Davies HFS, Tang KCG, et al. Sensitization to swine leukocyte antigens in patients with broadly reactive HLA specific antibodies. Am J Transplant. 2002;2:267–73.
39. Varela ID, Mozo PS, Cortes AC, Blanco CA, Canedo FV. Cross-reactivity between swine leukocyte antigen and human anti-HLA-specific antibodies in sensitized patients awaiting renal transplantation. J Am Soc Nephrol. 2003;14:2677–83.
40. Mulder A, Kardol MJ, Arn JS, et al. Human monoclonal HLA antibodies reveal interspecies crossreactive swine MHC class I epitopes relevant for xenotransplantation. Mol Immunol. 2010;47:809–15.
41. Martens GR, Reyes LM, Butler JR, et al. Humoral reactivity of renal transplant-waitlisted patients to cells from GGTA1/CMAH/B4GalNT2, and SLA Class I knockout pigs. Transplantation. 2017;101:86–92.
42. Lunney JK. Current status of the swine leukocyte antigen complex. Vet Immunol Immunopathol. 1994;43:19–28.

Sensitization to Human Leukocyte Antigens and Xenotransplantation

4

Guerard Byrne

Abbreviations

CMAH	Cytidine monophosphate-N-acetylneuraminic acid hydroxylase
Gal	Terminal galactose-α1,3-galactose glycans
GTKO	GGTA-1 mutated
Neu5Gc	N-glycolylneuraminic acid
Sd^a	Glycans with terminal GalNAcβ1,4 (Neu5Ac α2,3) Gal β-R groups
SLA	Swine leukocyte antigens
β4GalNT2	Beta-1,4-N-acetyl-galactosaminyltransferase 2

Introduction

The classical major histocompatibility complex (MHC) molecules are highly polymorphic glycoproteins which play a central role in adaptive immunity by capturing and presenting peptide antigens to the T-cell receptor expressed on T lymphocytes [1]. There are two major classes of human MHC, the human leukocyte antigen (HLA) class I and class II proteins. The class I proteins are expressed widely on nucleated cells and present antigen in association with beta-2-microglobulin to T-cell receptors on $CD8^{+}$T cells. The three major class I loci (A, B, and C) account

Based on a paper published in *Xenotransplantation* May/June *2018 (*DOI: 10.1111/xen.12411*)* with the editor's permission.

G. Byrne (✉)
Institute of Cardiovascular Science, University College London, London, UK

Department of Surgery, University of Alabama at Birmingham, Birmingham, AL, USA
e-mail: guerard.byrne@ucl.ac.uk

D. K. C. Cooper, G. Byrne (eds.), *Clinical Xenotransplantation*,
https://doi.org/10.1007/978-3-030-49127-7_4

for over 12,000 alleles. The HLA class II proteins are expressed on antigen-presenting cells (B cells, dendritic cells, and macrophages) and present peptide antigen to $CD4^+$ T cells. There are fewer class II genes, with 4802 listed in the International Immunogenetics and HLA database (IMGT/HLA) [2].

Because of a high level of polymeric amino acid variation, human class I and II proteins have long been recognized as the major transplantation antigens that stimulate allograft organ rejection [3]. The majority of amino acid variations occur in the regions of the proteins which form the peptide-binding site for antigen presentation [4]. This polymorphism allows for a high diversity of peptide presentation, but also creates antigenic diversity between individuals. Sensitization to HLA gene products occurs as an induced immune response when patients are challenged through blood transfusions, pregnancies, or failed organ transplants. For patients awaiting kidney transplantation, sensitization is commonly due to the relatively high frequency of dialysis-related blood transfusion. Highly sensitized patients remain longer on the transplant waiting list and, when they are transplanted, are at higher risk of early graft injury, rejection, and reduced graft survival [5, 6].

The efficacy of preclinical xenotransplantation has recently improved with heterotopic pig-to-nonhuman primate cardiac xenotransplantation [7–10] now measured in years, encouraging early success in orthotopic cardiac transplantation [11–13], and major improvements in life-supporting renal xenotransplantation [14, 15] with recipient survival beyond 1 year. These results are spurring renewed interest in moving toward clinical xenotransplantation. In addition to increasing the overall supply of organs for transplantation, successful clinical xenotransplantation may be particularly helpful to sensitized patients if increased antibody reactivity to human HLA antigens does not also increase antibody reactivity to porcine donor organs. This review summarizes the literature which has examined the potential of anti-HLA antibody in allosensitized patients to cross-react with porcine cells.

The body of evidence from these studies suggests that, at the current level of sensitivity, most transplant patients and patients with moderate allosensitization show minimal human antibody reactivity to pig cells when these cells lack the three known xenogeneic antigens, galactose-α1,3-galactose (Gal), N-glycolylneuraminic acid (Neu5Gc)-modified glycans, and porcine β4GalNT2-dependent SDa glycans [16, 17] (Table 4.1). For highly sensitized patients, there is often, but not always, an

Table 4.1 Known carbohydrate xenoantigens expressed on pig cells

Glycan (abbreviation)	Enzyme	Gene-knockout pig
Galactose-α1,3-galactose (Gal)	GGTA-1[a]	GTKO
N-glycolylneuraminic acid (Neu5Gc)	CMAH[b]	CMAHKO
GalNAc-β1,4-(Neu5Ac α2,3)-Gal-β (SDa)	B4GALNT2[c]	β4GalNT2KO

[a]α-1,3-Galactosyltransferase
[b]Cytidine monophosphate-N-acetylneuraminic acid hydroxylase
[c]β-1,4-N-acetyl-galactosaminyltransferase 2

increase in anti-pig antibody reactivity, which could affect xenotransplant survival. Recent analysis suggests that stringent patient cross-matching, and/or elimination of a limited set of specific porcine class I swine leukocyte antigen (SLA) alleles, can further minimize anti-pig reactivity such that future clinical xenotransplantation may be appropriate even for highly sensitized patients.

Detecting HLA Sensitization and Allotransplantation

Early clinical transplantation programs screened donor and recipients for matching blood type, but did not routinely screen for evidence of sensitization to other donor antigens. In a landmark study [18], Patel and Terasaki demonstrated that a complement-dependent cytotoxicity (CDC) test of a patient's serum against a panel of unrelated donor lymphocytes could be used to detect allosensitization. By analyzing 248 renal transplants, they showed that 80% of recipients with a positive-panel reactive antibody (PRA) had immediate graft failure compared to only 2.4% of recipients without a donor-specific antibody cross-match. Adoption of this assay almost completely eliminated hyperacute antibody-mediated allograft rejection and quickly became the early gold standard for detecting donor-specific HLA antibodies.

Technical improvements to the CDC assay and development of new assays using flow cytometry [19], solid phase ELISA, or single-antigen bead assays [20] based on HLA proteins and peptides, have further increased sensitivity and specificity for detecting allosensitization [21]. This led to the current calculated panel-reactive antibody (CPRA) score used in kidney allocation, which is an estimate of the percent of deceased donors that would be cross-match-incompatible based on the identification of unacceptable HLA antigens and their frequency in a large regional pool of donors. In the Organ Procurement and Transplantation Network (OPTN) database, there are currently 95,562 kidney transplant candidates on the waiting list (Fig. 4.1). Of these, 40% have some degree of sensitization with a CPRA >1. About 31% of sensitized patients, however, have a CPRA >80%.

These advanced anti-HLA antibody detection methods have improved donor-recipient matching and provided more detailed monitoring and analysis of the immune response in transplant recipients. Most highly sensitized patients produce anti-HLA antibody which reacts with shared public epitopes present on a variety of HLA alleles. It is now clear that these antibodies are binding the defined peptide sequences and topographies shared between different HLA proteins. There is sequence homology between human and swine leukocyte antigens (SLA), and some anti-HLA monoclonal antibodies do cross-react with SLA [22].

So the question arises, do patients sensitized to HLA antigens also show increased antibody reactivity to porcine cells? If this is the case, then xenotransplantation may not be an advantageous source of organs for highly sensitized patients, but, if there is not a concomitant increase in anti-pig antibody in patients with HLA sensitization, then xenotransplantation may be an important alternative source of organs for these patients.

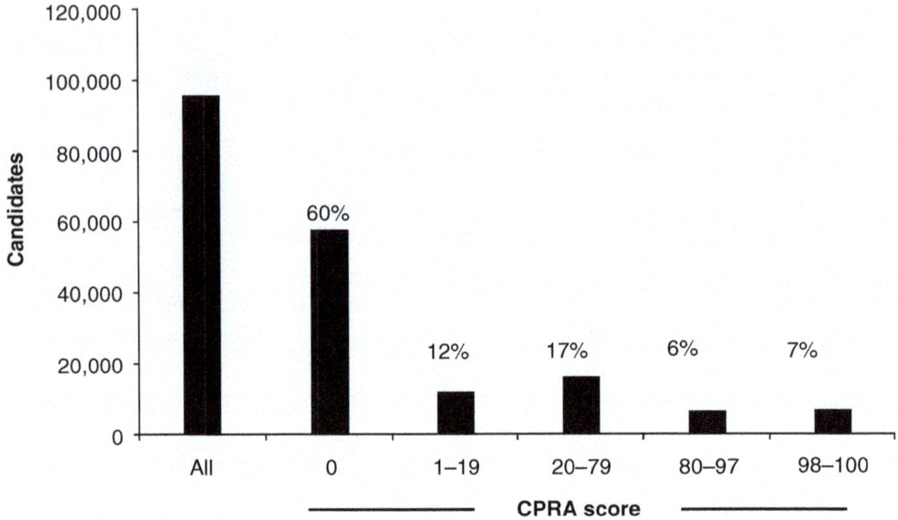

Fig. 4.1 UNOS data showing the number of kidney transplant candidates on the waiting list and the breakout of patients based on CPRA. There were 95,562 candidates for kidney transplantation listed in January 2018. The percentage values represent the total percentage for each CPRA grouping; 60% of candidates had a CPRC of zero. Within the sensitized patient group ($n = 40{,}128$), 31% ($n = 12{,}532$) have a CPRA >80%

Strategies for Detecting HLA Cross-Reactivity for Xenotransplantation

Xenograft rejection is recognized as an overwhelmingly antibody-driven process due to the very high level of anti-pig antibodies naturally present in human serum. The bulk of these antibodies are not directed to SLA, but bind to the major xenogeneic glycan, galactose-α1,3-galactose (Gal) [23]. With the advent of pigs engineered with a GGTA-1 mutation [24, 25], which eliminates Gal expression (α1,3-galactosyltransfearse gene-knockout [GTKO] pigs), the impact of other antibodies directed to non-Gal antigens, including SLA-I, has become more apparent [16, 26–28]. There have been a limited number of studies designed to determine whether sensitization to HLA results in enhanced antibody reactivity to pig cells [16, 29–37]. These studies have been performed over a 20-year period and, as such, span a range of technological developments both for defining allosensitization and in technologies to detect xenoreactive antibody.

In this review, the studies are presented as four basic research strategies based on the use of whole serum (Type I), anti-Gal-depleted serum (Type II), and anti-Gal- and anti-Class I-depleted serum (Type III). The fourth study type largely used whole human serum, but measured patient antibody reactivity to genetically modified porcine cells lacking the three-known xenogeneic glycans (Gal, Neu5Gc modified glycans, and SDa) with and without deletion of SLA-I genes.

Type I Studies

The earliest study [29] screened 105 wait-listed patient sera against Gal-expressing porcine peripheral blood lymphocytes representing the three known haplotypes for NIH miniature swine. They demonstrated that patient PRA, measured by CDC, was not correlated to the level of anti-pig antibody titer. Moreover, most human anti-pig reactivity was IgM, whereas anti-HLA was dominantly IgG.

Wong et al. [34] extended this type of analysis by comparing antibody binding of sensitized patient sera to Gal-positive wild-type and GTKO miniature swine peripheral blood mononuclear cells (PBMCs). There was a clear reduction in antibody reactivity and CDC to GTKO cells, consistent with the loss of the Gal antigen, but no correlation between antibody reactivity AND cytotoxicity to PRA level from 88 wait-listed patient sera for either wild-type or GTKO cell type. Cytotoxicity to porcine GTKO cells was mainly mediated by IgM antibody, in contrast to anti-human cytotoxicity which was predominantly IgG-dependent. Similar results were found analyzing wild-type and GTKO pig cells from animals with a commercial agricultural background (Large/White, Landrace, Duroc) [36]. A recent study using cells from GTKO and GTKO/CMAHKO pigs expressing human CD46 also failed to find enhanced antibody reactivity in 10 highly sensitized wait-listed patient serum samples [37].

Collectively, these studies concluded that patients with high PRA sera do not necessarily produce a correspondingly high titer of anti-pig antibody or a high level of anti-pig cytotoxicity. Thus, allosensitized patients would not be at greater risk of xenograft rejection.

Type II and III Studies

Type II studies used porcine red blood cells (RBCs), which do *not* express SLA-I, to deplete patient serum of anti-Gal antibody. Type III studies used a combination of porcine RBCs and porcine platelets, which *do* express SLA-I, to deplete both anti-Gal and anti-SLA antibody. When only anti-Gal antibody was depleted, Oostingh et al. [33] found that some highly sensitized patient sera showed a correlation between serum PRA and anti-pig antibody reactivity. This study of 82 patient serum samples used both CDC and more sensitive flow cross-match with class I beads to define PRA levels, identifying 12 samples with 0% PRA, 50 samples with a PRA from 11% to 84%, and 20 samples with PRA >84%.

PBMCs from 23 Gal-positive pigs transgenic for human CD55 (decay accelerating factor) with known lineage, selected to represent a broad diversity of swine SLA-I, were used as target cells. In the 1884 cross-match combinations, about 20% retained antibody reactivity to porcine PBMCs after anti-Gal antibody depletion. When the serum samples were stratified for PRA, the majority of this Gal-independent cross-reactivity was present in serum with PRAs >64%. Similar results were shown by Varela et al. using non transgenic pig cells [34]. These studies concluded that human sera with broad panel reactivity (PRA >64%) can, but do not always, exhibit increased cross-reactivity to porcine PBMCs.

When allosensitized wait-listed patient serum is progressively absorbed with RBCs and porcine platelets to deplete both anti-Gal and anti-SLA-I, it is evident that the residual Gal-independent antibody reactivity present in some highly sensitized patients reacts with SLA. Naziruddin et al. [30] demonstrated that affinity-purified anti-Gal antibody binding to pig PBMCs was effectively blocked by saturating levels of the Gal-specific lectin, GSIB-4, but that antibody reactivity to pig PBMCs in patients with medium-to-high PRA was Gal-independent. This antibody reactivity was also depleted by porcine platelets, and reacted in Western blot to porcine SLA-1 heavy chain. Likewise, platelet depletion was shown by Taylor et al. [31] to eliminate antibody reactivity to porcine cells in high PRA patient serum. Importantly, they demonstrated that the loss of antibody binding after platelet absorption was specific, as porcine platelet absorption did not affect allo-specific anti-HLA binding to human cells.

Similar results were observed in a unique study of ex vivo porcine kidney perfusion where both anti-Gal and anti-SLA-I antibodies were recovered from the perfused organ of some, but not all, plasma samples [32]. These absorption studies clearly indicate that some high PRA patient sera exhibit cross-reactivity to porcine SLA-I, suggesting that at least some broadly reactive anti-HLA antibody cross-reacts to a restricted number of conserved serologic groups, shared between human and porcine MHC class I. It is worth noting that porcine RBCs express the Gal antigen, but, unappreciated at the time of these studies, also express additional xenogeneic glycans. The depletion of additional anti-glycan antibody reactivity may have contributed to the success of detecting cross-reactive anti-SLA antibody.

Type IV Studies

The most recent studies are based on a series of genetically modified pigs which progressively eliminate expression of the known xenogeneic glycans. Tector and colleagues developed a series of pigs with mutations in GGTA-1, eliminating Gal expression (single knockouts), GGTA-1 and cytidine monophosphate-N-acetylneuraminic acid hydroxylase (CMAH) eliminating both Gal and the synthesis of Neu5Gc (double-knockouts), and GGTA-1, CMAH, and β4GalNT2, eliminating expression of Gal, Neu5Gc modified glycans, and SDa glycans (triple-knockout) [16, 38, 39]. Human serum shows progressively less antibody reactivity to PBMCs from these pigs, with approximately 60% of 820 wait-listed samples negative for IgG binding and 30% showing only background reactivity for both IgM and IgG when tested on triple-knockout cells.

Serum samples with residual antibody reactivity were further analyzed by absorption with porcine RBCs and used to stain SLA-I-positive and SLA-I-negative porcine cells. A small subset of wait-listed sera (9 of 119) showed clear SLA-I-specific reactivity. A similar SLA-I-specific analysis of RBC-absorbed serum from patients with PRA >80% identified 13 of 22 with SLA class I-specific IgG, and 4 of

17 with SLA class-I-specific IgM. When an antibody from a highly sensitized serum bound to human and porcine PBMCs was recovered, single-antigen bead analysis demonstrated that porcine-specific IgG reactivity was limited to common epitope restricted targets present on a restricted set of HLA-I antigens.

These studies confirm earlier reports that some, but not all, highly sensitized patient sera contain SLA-I-reactive antibody. Importantly, these latest studies identify for the first time the MHC cross-reactive groups present on SLA-I, making possible further genetic modification or selection to eliminate these alleles and minimize antibody reactivity even for highly sensitized patients.

Conclusions

The prospects for clinical xenotransplantation have improved significantly due to recent increases in preclinical nonhuman primate xenograft survival. While the ideal donor organ is not universally defined, and may be different for different organs, it seems likely that donor organs with minimal antigenicity (GGTA1/CMAH/β4GalNT2), which minimize human antibody reactivity, will make a prominent contribution. Additional genetic modifications to regulate complement and coagulation may also be used, but inclusion of these human transgenes should not affect antibody reactivity or tissue antigenicity. For most human sera and wait-listed patients with zero to moderate HLA sensitization, there appears to be minimal antibody reactivity to these triple-knockout pig cells, suggesting that future clinical xenotransplantation will be broadly applicable to most patients.

Highly sensitized patients (PRA >80%) can produce antibody that cross-reacts with SLA-I, but this is not an obligate condition. Whether the SLA-I cross-reactive antibody in highly sensitized patients has immediate impact on xenograft survival will depend on the antibody titer, affinity, and level of SLA-I expression. Since recent studies suggest that cross-reactive anti-HLA antibody is directed to a limited set of HLA antigens, modern genetic screening and modification methods may be used to select for pigs with minimal antibody reactivity even for highly sensitized patients. Patient cross-matching is a cornerstone of successful allotransplantation and will undoubtedly play no less of a role in future clinical xenotransplantation.

Acknowledgments This work was supported by funding from the Medical Research Council of the UK (MR/R006393/1).

References

1. Rudolph MG, Stanfield RL, Wilson IA. How TCRs bind MHCs, peptides, and coreceptors. Annu Rev Immunol. 2006;24:419–66.
2. Robinson J, Mistry K, McWilliam H, et al. The IMGT/HLA database. Nucleic Acids Res. 2011;39(Database issue):D1171–6.
3. Mehra NK, Baranwal AK. Clinical and immunological relevance of antibodies in solid organ transplantation. Int J Immunogenet. 2016;43:351–68.

4. Reche PA, Reinherz EL. Sequence variability analysis of human class I and class II MHC molecules: functional and structural correlates of amino acid polymorphisms. J Mol Biol. 2003;331:623–41.
5. Bostock IC, Alberu J, Arvizu A, et al. Probability of deceased donor kidney transplantation based on % PRA. Transpl Immunol. 2013;28:154–8.
6. Loupy A, Lefaucheur C, Vernerey D, et al. Complement-binding anti-HLA antibodies and kidney-allograft survival. N Engl J Med. 2013;369:1215–26.
7. Mohiuddin MM, Corcoran PC, Singh AK, et al. B-cell depletion extends the survival of GTKO.hCD46Tg pig heart xenografts in baboons for up to 8 months. Am J Transplant. 2012;12:763–71.
8. Mohiuddin MM, Singh AK, Corcoran PC, et al. Genetically engineered pigs and target-specific immunomodulation provide significant graft survival and hope for clinical cardiac xenotransplantation. J Thorac Cardiovasc Surg. 2014;148:1106–13; discussion 1113–1114.
9. Mohiuddin MM, Singh AK, Corcoran PC, et al. One-year heterotopic cardiac xenograft survival in a pig to baboon model. Am J Transplant. 2014;14:488–9.
10. Mohiuddin MM, Singh AK, Corcoran PC, et al. Chimeric 2C10R4 anti-CD40 antibody therapy is critical for long-term survival of GTKO.hCD46.hTBM pig-to-primate cardiac xenograft. Nat Commun. 2016;7:11138.
11. Byrne GW, Du Z, Sun Z, Asmann YW, McGregor CG. Changes in cardiac gene expression after pig-to-primate orthotopic xenotransplantation. Xenotransplantation. 2011;18:14–27.
12. McGregor CGA, Byrne GW, Vlasin M, et al. Cardiac function after preclinical orthotopic cardiac xenotransplanation. Am J Transplant. 2009;9:380.
13. McGregor CGA, Davies WR, Oi K, et al. Recovery of cardiac function after pig-to-primate orthotopic heart transplant. (Abstr 98). Am J Transplant. 2008;8:205–6.
14. Iwase H, Hara H, Ezzelarab M, et al. Immunological and physiological observations in baboons with life-supporting genetically engineered pig kidney grafts. Xenotransplantation. 2017;24.
15. Higginbotham L, Mathews D, Breeden CA, et al. Pre-transplant antibody screening and anti-CD154 costimulation blockade promote long-term xenograft survival in a pig-to-primate kidney transplant model. Xenotransplantation. 2015;22:221–30.
16. Martens, G.R., Reyes, L.M., Butler, J.R., et al. Humoral reactivity of renal transplant-waitlisted patients to cells from GGTA1/CMAH/B4GalNT2, and SLA class I knockout pigs. Transplantation 2017; 101:e86–e92.
17. Wang ZY, Li P, Butler JR, et al. Immunogenicity of renal microvascular endothelial cells from genetically modified pigs. Transplantation. 2016;100:533–7.
18. Patel R, Terasaki PI. Significance of the positive crossmatch test in kidney transplantation. N Engl J Med. 1969;280:735–9.
19. Cicciarelli J, Helstab K, Mendez R. Flow cytometry PRA, a new test that is highly correlated with graft survival. Clin Transpl. 1992;6:159–64.
20. Pei R, Wang G, Tarsitani C, et al. Simultaneous HLA class I and class II antibodies screening with flow cytometry. Hum Immunol. 1998;59:313–22.
21. Gebel HM, Bray RA. The evolution and clinical impact of human leukocyte antigen technology. Curr Opin Nephrol Hypertens. 2010;19:598–602.
22. Mulder A, Kardol MJ, Arn JS, et al. Human monoclonal HLA antibodies reveal interspecies crossreactive swine MHC class I epitopes relevant for xenotransplantation. Mol Immunol. 2010;47:809–15.
23. Cooper DKC, Good AH, Koren E, et al. Identification of alpha-galactosyl and other carbohydrate epitopes that are bound by human anti-pig antibodies: relevance to discordant xenografting in man. Transplant Immunol. 1993;1:198–205.
24. Lai L, Kolber-Simonds D, Park K-W, et al. Production of a-1,3-galactosyltransferase knockout pigs by nuclear transfer cloning. Science. 2002;295:1089–92.
25. Phelps CJ, Koike C, Vaught TD, et al. Production of alpha 1,3-galactosyltransferase-deficient pigs. Science. 2003;299:411–4.

26. Padler-Karavani V, Varki A. Potential impact of the non-human sialic acid N-glycolylneuraminic acid on transplant rejection risk. Xenotransplantation. 2011;18:1–5.
27. Byrne GW, Stalboerger PG, Du Z, Davis TR, McGregor CG. Identification of new carbohydrate and membrane protein antigens in cardiac xenotransplantation. Transplantation. 2011;91:287–92.
28. Byrne GW, Du Z, Stalboerger P, Kogelberg H, McGregor CG. Cloning and expression of porcine beta1,4 N-acetylgalactosaminyl transferase encoding a new xenoreactive antigen. Xenotransplantation. 2014;21:543–54.
29. Bartholomew A, Latinne D, Sachs D, et al. Utility of xenografts: lack of correlation between PRA and natural antibodies to swine. Xenotransplantation. 1997;4:34–9.
30. Naziruddin B, Durriya S, Phelan D, et al. HLA antibodies present in the sera of sensitized patients awaiting renal transplant are also reactive to swine leukocyte antigens. Transplantation. 1998;66:1074–80.
31. Taylor CJ, Tang KG, Smith SI, White DJ, Davies HF. HLA-specific antibodies in highly sensitized patients can cause a positive crossmatch against pig lymphocytes. Transplantation. 1998;65:1634–41.
32. Barreau N, Godfrin Y, Bouhours J-F, et al. Interaction of anti-HLA antibodies with pig xenoantigens. Transplantation. 2000;69:148–56.
33. Oostingh GJ, Davies HFS, Tang KCG, Bradley JA, Taylor CL. Sensitisation to swine leukocte antigens in patients with broadly reactive HLA specific antibodies. Am J Transplant. 2002;2:267–73.
34. Diaz VI, Sanchez MP, Centeno CA, Alonso BC, Valdes CF. Cross-reactivity between swine leukocyte antigen and human anti-HLA-specific antibodies in sensitized patients awaiting renal transplantation. J Am Soc Nephrol. 2003;14:2677–83.
35. Wong BS, Yamada K, Okumi M, et al. Allosensitization does not increase the risk of xenoreactivity to alpha1,3-galactosyltransferase gene-knockout miniature swine in patients on transplantation waiting lists. Transplantation. 2006;82:314–9.
36. Hara H, Ezzelarab M, Rood PP, et al. Allosensitized humans are at no greater risk of humoral rejection of GT-KO pig organs than other humans. Xenotransplantation. 2006;13:357–65.
37. Zhang Z, Hara H, Long C, et al. Immune responses of HLA-highly-sensitized and nonsensitized patients to genetically-engineered pig cells. Transplantation. 2017;102:e195.
38. Estrada JL, Martens G, Li P, et al. Evaluation of human and non-human primate antibody binding to pig cells lacking GGTA1/CMAH/beta4GalNT2 genes. Xenotransplantation. 2015;22:194–202.
39. Wang ZY, Martens GR, Blankenship RL, et al. Eliminating xenoantigen expression on swine RBC. Transplantation. 2017;101:517–23.

Part II

Pig Kidney and Heart Xenotransplantation in Nonhuman Primates: The Present Position

Kidney Xenotransplantation in Nonhuman Primates

5

David K. C. Cooper, Hidetaka Hara, Hayato Iwase, Takayuki Yamamoto, Abhijit Jagdale, Douglas J. Anderson, David Ayares, and Devin E. Eckhoff

Abbreviations

GTKO	α1,3-galactosyltransferase gene-knockout
Neu5Gc	N-glycolylneuraminic acid
NHP	Nonhuman primate
RBCs	Red blood cells
SLA	Swine leukocyte antigens
TKO	Triple-gene knockout

Introduction

The human immune response to a pig organ (a xenograft) is more complex than that to a human organ (an allograft). It involves binding of preformed (natural) anti-pig antibodies to the vascular endothelial cells of the graft, activation of the complement and coagulation cascades, activation of innate immune cells (e.g., polymorphonuclear leukocytes, monocytes/macrophages, natural killer [NK] cells),

D. K. C. Cooper (✉)
Xenotransplantation Program, Division of Transplantation, Department of Surgery, University of Alabama at Birmingham (UAB), Birmingham, AL, USA
e-mail: dkcooper@uabmc.edu

H. Hara · H. Iwase · T. Yamamoto · A. Jagdale · D. J. Anderson · D. E. Eckhoff
Xenotransplantation Program, Department of Surgery, University of Alabama at Birmingham (UAB), Birmingham, AL, USA

D. Ayares
Revivicor, Blacksburg, VA, USA

D. K. C. Cooper, G. Byrne (eds.), *Clinical Xenotransplantation*,
https://doi.org/10.1007/978-3-030-49127-7_5

Table 5.1 Timeline for application of evolving techniques for genetic engineering of pigs employed in xenotransplantation

Year	Technique
1992	Microinjection of randomly integrating transgenes
2000	Somatic cell nuclear transfer (SCNT)
2002	Homologous recombination
2011	Zinc finger nucleases (ZFNs)
2013	Transcription activator-like effector nucleases (TALENs)
2014	CRISPR/Cas9[a]

[a]CRISPR/Cas9 clustered randomly interspaced short palindromic repeats and the associated protein 9

inflammation, and activation of the adaptive immune response (T- and B-cell infiltration and elicited antibody formation) (reviewed in Chap. 2) [1]. Hyperacute rejection is common when a wild-type pig organ is transplanted into a nonhuman primate (NHP) [2–4].

Nevertheless, xenotransplantation provides us with our first real opportunity in transplantation *to modify the 'donor'*, rather than just treat the recipient of an organ graft.

This is being achieved by genetic engineering of the organ-source pig. The two main approaches have been (i) deletion of known pig xenoantigens on the pig vascular endothelial (and other) cells and (ii) introduction of human 'protective' transgenes into the same pig cells (Chap. 7) [5]. Over the past three decades, the techniques of genetic-engineering have steadily become simpler, cheaper, and more efficient, with CRISPR (clustered regularly interspaced short palindromic repeat) allowing multiple modifications to be made simultaneously (Table 5.1). There are now at least 25 different genetic manipulations that have been demonstrated to be of potential benefit to survival of a pig xenograft. Pigs are now available with nine genetic modifications, all aimed at protecting the cell from the human immune response [6, 7].

Early Studies in Pig-to-Nonhuman Primate (NHP) Models

After transplantation into NHPs, organs from wild-type (i.e., genetically *un*modified) pigs underwent hyperacute rejection (i.e., within 24 hours) in the majority of cases, even when intensive conventional pharmacologic immunosuppressive therapy had been administered (e.g., cyclosporine- or tacrolimus-based). Furthermore, unless the therapy was very intensive, sensitization to pig antigens developed [8]. When therapy was with costimulation pathway (signal 2) blockade using an anti-CD154 monoclonal antibody (mAb), first tested in pig hematopoietic cell xenotransplantation by Buhler in 2000 [8], pig cell survival was minimally prolonged but, more importantly, sensitization to the pig cells was prevented.

The first xenotransplants using organs from genetically engineered pigs were those by White and his colleagues in the UK [9, 10] and by Fodor and his colleagues in the USA [11]. In both cases, the pig expressed a human complement-regulatory protein.

Over a number of years, using intensive conventional immunosuppressive therapy, Cozzi and his colleagues extended CD55-transgenic pig kidney graft survival in NHPs to a maximum of 90 days, though mean survival remained at 24 days [12].

Recent Progress in Pig-to-NHP Models

With improvements in the genetic engineering of pigs, e.g., knockout of xenoantigens and introduction of additional complement- and coagulation-regulatory proteins, survival of life-supporting pig kidneys, heterotopically placed (non-life-supporting) pig hearts and, more recently, orthotopically placed (life-supporting) pig hearts (see Chap. 6) has steadily increased, particularly during the past few years. However, these encouraging results have only been obtained when costimulation blockade has been employed and not when intensive conventional pharmacologic therapy has been administered (Fig. 5.1) [13]. Furthermore, blockade of the CD40-CD154 pathway is required, whereas blockade of the CD28-B7 (CD80/86) pathway is insufficient [14–16].

The initial encouraging results of the transplantation of α1,3-galactosyltransferase gene-knockout (GTKO) pig organs were obtained in baboons selected for low anti-pig (nonGal) antibody levels [17–19]. Pig heterotopic heart transplant survival was extended to a maximum of 179 days [17, 18], and life-supporting pig kidney transplantation to 83 days [19]. The importance of selecting recipients with low anti-pig antibody levels was confirmed more recently by Higginbotham [16]. A GTKO/CD55 pig kidney survived for <1 week in a rhesus monkey selected for a high anti-pig antibody level, whereas survival was extended to >133 and > 126 days in two rhesus monkeys selected for low antibody levels. However, these good results could

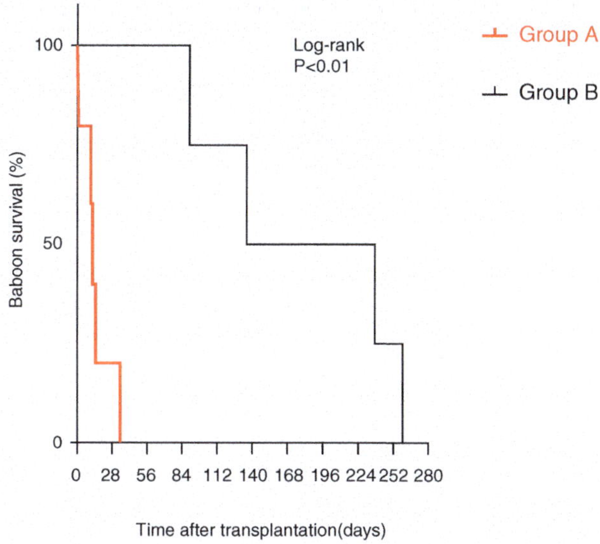

Fig. 5.1 Pig kidney graft survival in baboons receiving either conventional (tacrolimus-based; *Group A*) or anti-CD40mAb-based (*Group B*) immunosuppressive therapy. Median pig kidney graft survival in *Group B* (186 days) was significantly longer than in *Group A* (14 days) ($p < 0.01$). (Reproduced with permission from reference [13])

only be achieved when an anti-CD154mAb formed the basis of the immunosuppressive regimen, but not when CTLA4-Ig was administered to two rhesus monkeys (when graft survival was reduced to 14 and 21 days, respectively).

The importance of expression of a human coagulation-regulatory protein in kidney xenotransplantation was illustrated by a study by Iwase et al. in baboons [20]. When the GTKO/CD46 pig kidney expressed a human coagulation-regulatory protein (endothelial protein C receptor [EPCR], tissue factor pathway inhibitor [TFPI]), graft survival exceeded 6 months ($n = 2$), whereas when there was no expression of these in the kidney, the grafts failed from a consumptive coagulopathy within 2 weeks (Fig. 5.2).

In a more recent study of the transplantation of kidneys from GTKO pigs in which a second xenoantigen, Sda, had also been deleted (but with *no* expression of any human protective transgenes), graft survival in rhesus monkeys was <1 week in three cases (associated with IgM-mediated rejection), and 35, 100, and 435 days in three other cases, despite the fact that anti-CD154mAb-based immunosuppressive therapy had been administered [21]. These results suggest that expression of a human complement-regulatory protein (+/− a human coagulation-regulatory protein) contributes significantly to the success of xenografting in this model.

It therefore appears that relevant genetic engineering of the pig is required, e.g., expression of a human coagulation-regulatory protein, but also the administration of an agent that blocks the CD40-CD154 pathway. Yamamoto et al. illustrated this point when they demonstrated that, in conventionally immunosuppressed baboons (even when CTLA4-Ig was administered *in addition to* pharmacologic immunosuppressive drugs), graft failure was associated with the development of a thrombotic microangiopathy [13] (Fig. 5.3). However, the outcome may be different when

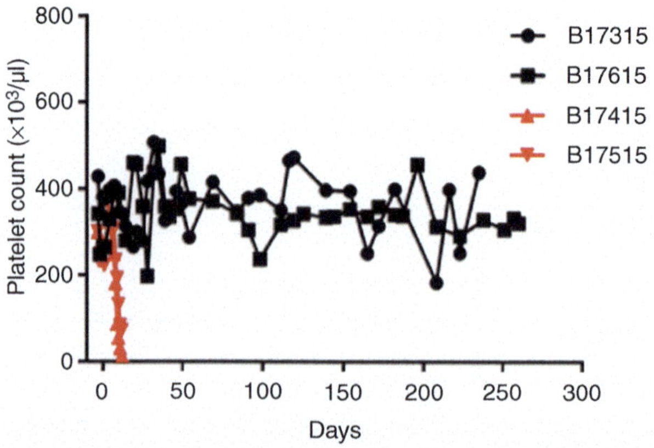

Fig. 5.2 Rapid development of thrombocytopenia (consumptive coagulopathy, a reliable indicator of graft rejection/failure) in two baboons with life-supporting GTKO/CD46 pig kidney grafts (indicated in red), and maintenance of normal platelet counts in two baboons (treated identically) with life-supporting GTKO/CD46/CD55/endothelial cell protein C receptor (EPCR)/tissue factor pathway inhibitor (TFPI)/CD47 pig kidney grafts (indicated in black). (Modified from reference [20])

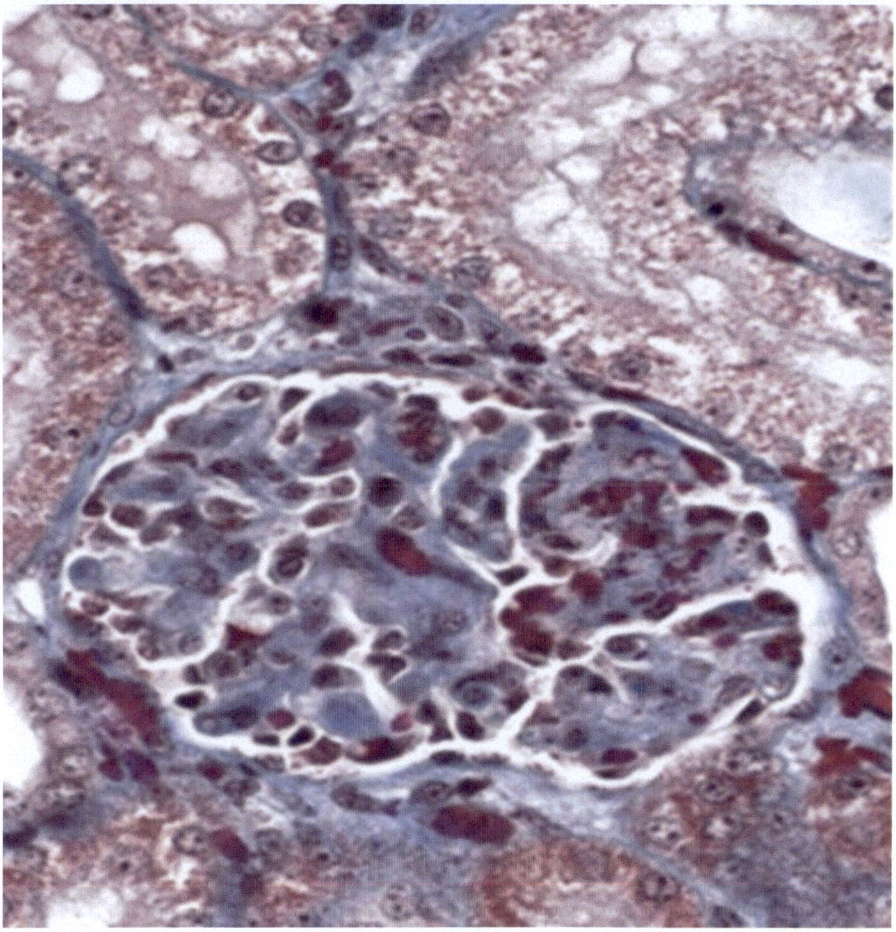

Fig. 5.3 Histopathology of thrombotic glomerulopathy in a pig kidney with conventional immunosuppressive therapy. Glomerular histopathology included thrombotic microangiopathic glomerulopathy, glomerular thrombi, mesangial thickening, and glomerular edema [expansion of Bowman's space]. (H&E stain, original magnification ×400.) (Reproduced with permission from reference [13])

organs are transplanted from pigs in which all three xenoantigens have been deleted and in which additional human transgenes are expressed [6, 7], though we are not optimistic in this respect.

Much effort has been made to induce a state of immunological tolerance in the recipient NHP to the pig graft. Most efforts have been directed to establishing a state of hematopoietic cell chimerism in the recipient [22, 23] or by the concomitant transplantation of donor-specific thymus tissue [19]. Tolerance is clearly the ultimate goal in both allotransplantation and xenotransplantation but, to date, no approach has proved successful in xenotransplantation in large animals. Although the additional pretransplant immunosuppression required, e.g., total body irradiation, may have contributed to suppression of the recipient's immune system, conclusive evidence of tolerance has not been achieved.

In summary the factors that have been demonstrated to contribute to successful pig organ graft survival in NHPs are (i) no or low anti-pig antibody levels in the recipient, (ii) an immunosuppressive regimen based on blockade of the CD40-CD154 costimulation pathway, (iii) deletion of known carbohydrate xenoantigens in the pig, and (iv) expression of at least one human complement-regulatory protein and at least one coagulation-regulatory protein in the pig.

Current and Future Developments

Almost all of the genetic manipulations carried out in the pigs to date have been aimed at reducing, or protecting from, the innate immune response. Fortunately, some of these also reduce the adaptive immune response [24, 25]. Future efforts will be directed to manipulations aimed at reducing the adaptive immune response. In this respect, there have already been pigs produced in which (i) expression of swine leukocyte antigens (SLA) has been inhibited, e.g., SLA class I-knockout [26, 27], SLA class II-knockdown [28], (ii) expression of PD-L1 has been downregulated [29–31], (iii) an immunosuppressive agent, e.g., CTLA4-Ig, has been expressed [32, 33], and (iv) HLA-E/G (that inhibits NK cell activity) has been expressed [34].

Ultimately, little or no exogenous immunosuppressive therapy will be required to prevent T-cell-mediated rejection. At that stage of development, the possibility of obtaining immunological tolerance to the pig graft, particularly in neonates and infants [35, 36], will increase considerably.

Nonimmunological Barriers

In the pig-to-NHP model, there have been several observations made that appear to be unrelated to the immune response.

Rapid Growth of the Pig Graft

The first has been a rapid growth of the pig organ in the first few weeks after transplantation (Fig. 5.4a). It was first reported by Soin as early as 2000 [37]. Iwase et al. observed rapid growth of a kidney [20, 38, 39], and this observation has subsequently been reported independently by others [40]. It appears that it has only been seen when kidneys from very young pigs (i.e., <2–3 months of age) have been transplanted, as it has not been reported when pigs >30 kg have been the sources of the organs (A. Adams, personal communication), though there may be differences in immunosuppressive therapy and management that may contribute to this discrepancy.

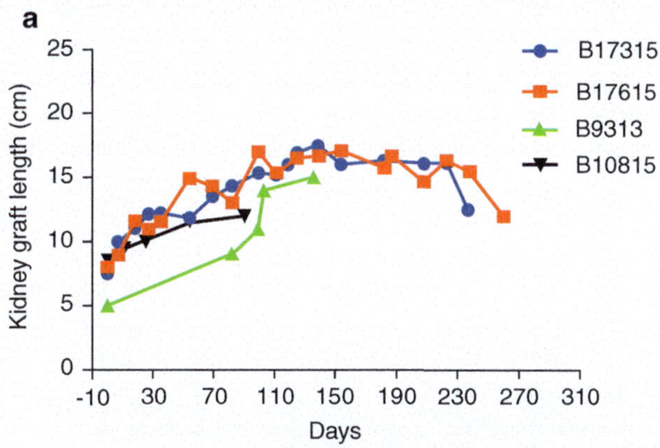

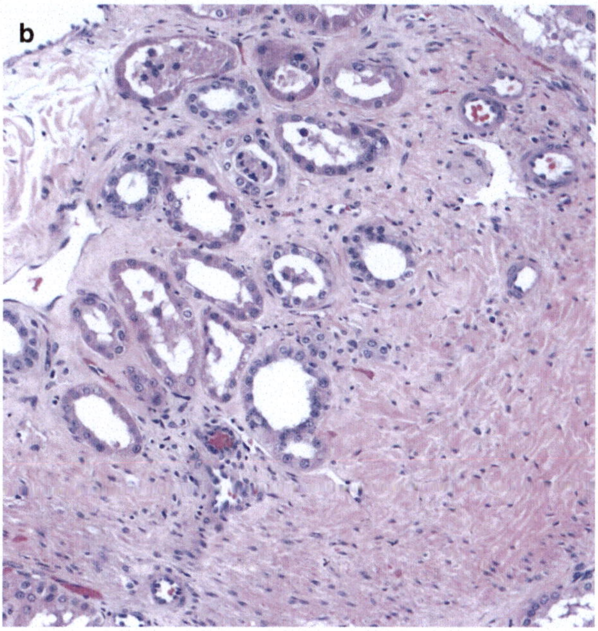

Fig. 5.4 (**a**) Increases in the lengths of the kidneys in four baboons with genetically engineered pig kidney grafts that functioned for 90, >136, >237, and >260 days, respectively. (**b**) Microscopic appearance of a pig kidney 3 months after transplantation into a baboon. The expansion of interstitial tissue can be clearly seen. (Reproduced with permission from reference [20])

Rapid growth of the transplanted organ has also been reported in the Large White-to-miniature pig allotransplantation model [40], suggesting that it is not solely a phenomenon of xenotransplantation. As a young pig grows at a much faster rate than a young baboon, it has been suggested that this phenomenon relates to the presence of an intrinsic factor, such as the pig growth hormone. It may be that, when a 'growth' factor is no longer present in the graft, the organ now falls under the control of the recipient baboon growth hormone, and thus grows at the same rate as the recipient. The expansion of tissue in the kidney appears to involve only the interstitial tissues and does not lead to an increase in the number of glomeruli (Fig. 5.4b).

Rapid growth of a transplanted pig heart was not reported when the heart was transplanted *heterotopically* into the abdomen (and therefore was not supporting the circulation) [14, 41], but it has been reported after *orthotopic* heart transplantation, where the heart supports the circulation [42]. In these cases, however, the growth has been mainly a hypertrophy of the ventricles, whereas the kidneys have doubled in length and mass. The mechanisms involved, therefore, may be different. Treatment of the baboon recipient of a pig heart graft with antihypertensive agents has reduced the degree of hypertrophy, suggesting that it may be a response to the greater work required of the heart to support the circulation in an NHP where the peripheral vascular resistance is higher than in the pig [42]. There was some evidence that the administration of an mTOR inhibitor reduced the 'growth' of the heart.

Rapid growth of a transplanted pig kidney should not be problematic within the more 'elastic' confines of the abdomen, but rapid growth of the heart could well be detrimental within the more restricted confines of the chest. If the rapid rate of growth of the pig organ becomes a major barrier to successful xenotransplantation (which seems unlikely), then the pig could be genetically engineered to reduce the rate of growth [43], or a genetically engineered miniature pig (which grows more slowly) could be selected as the organ source.

The Hypovolemia/Dehydration Syndrome

In several baboons with functioning pig kidneys in the absence of features of an immune response, we have observed a sudden increase in serum creatinine (Fig. 5.5) [44]. On occasion, to exclude rejection, we have taken the baboon to the operating room to take a needle biopsy of the kidney. Histological examination of the biopsy did not indicate any rejection or other pathology. While under

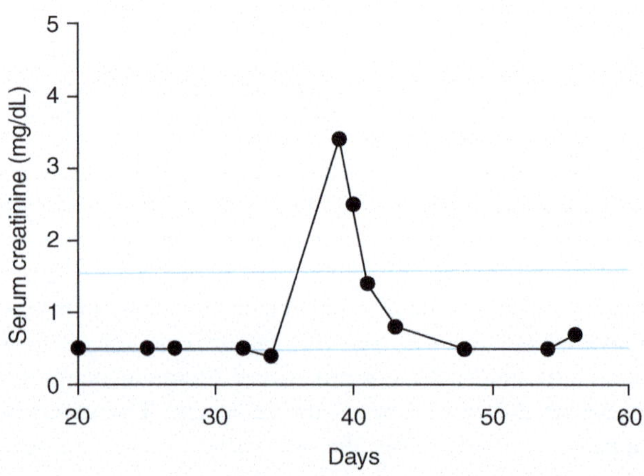

Fig. 5.5 Sudden increase in serum creatinine (in the absence of histopathological features of rejection on biopsy) in a baboon with a life-supporting, hitherto well-functioning pig kidney graft. After i.v. saline infusion, the creatinine immediately normalized. (Modified from reference [44])

anesthesia, however, we noted very low central venous (<2 mm Hg) and arterial (systolic <50–60 mm Hg) blood pressures, indicating that the baboon was hypovolemic. We also observed that the tissues appeared to be dehydrated. After the intravenous (or subcutaneous) infusion of normal saline (in quite large quantities, e.g., 30–50 mL/kg), the serum creatinine immediately returned to the normal range.

Although the baboon appeared to have been maintaining a good fluid intake, and passing adequate amounts of urine, it was clearly unaware that it was becoming hypovolemic/dehydrated. In recent experiments, we have infused normal saline subcutaneously at weekly or twice weekly intervals, and we have observed a reduced incidence of this phenomenon.

The cause of the hypovolemia remains uncertain. However, pig renin is said to be unable to cleave human angiotensinogen [45], (and so possibly also baboon angiotensinogen). This may result in an inability for vasoconstriction, and an inability to retain fluid, resulting in hypovolemia and/or dehydration. Although we have no definitive evidence, we suggest, that the syndrome relates to an abnormality of, or absence of, renin function. We are currently investigating this hypothesis.

In clinical kidney transplantation, the recipient's native kidneys are usually left in situ, and therefore the complication may not develop. Even if the native kidneys are removed, this syndrome could be avoided by ensuring that the patient maintains a high oral fluid intake. If problematic, the organ-source pig could be genetically engineered to produce human renin as well as pig renin.

The Current 'Optimal' Genetically Engineered Pig

We currently have available to us (from Revivicor, Blacksburg, VA) pigs with nine genetic manipulations. All three of the known xenoantigens (i.e., galactose-α1,3-galactose [Gal], N-glycolylneuraminic acid [Neu5Gc], and Sda) have been deleted (triple-knockout [TKO] pigs) (Table 5.2). In addition, the pigs express six human transgenic proteins – two complement-regulatory proteins (CD46, CD55), two coagulation-regulatory proteins (thrombomodulin, endothelial protein C receptor)

Table 5.2 Known carbohydrate xenoantigens expressed on pig cells

Carbohydrate pig	Responsible enzyme	Gene-knockout
1. Galactose-α1,3-galactose (Gal)	α1,3-Galactosyltransferase	GTKO
2. N-glycolylneuraminic acid (Neu5Gc)	CMAH	CMAHKO
3. Sd[a]	β4GalNT2	β4GalNT2KO

β4GalNT2 β-1,4N-acetylgalactosaminyltransferase [53, 54], *CMAH* cytidine monophosphate-N-acetylneuraminic acid hydroxylase [54–57], *GTKO* α1,3-galactosyltransferase gene-knockout [50–52] the anti-inflammatory gene, hemeoxygenase-1 (HO-1), and CD47 (that has some effect on the innate cellular response) [6, 7].

Many healthy human subjects demonstrate no serum IgM or IgG antibody binding to these 'nine-gene' pig cells (Fig. 5.6a) [7, 46, 47]. Indeed, antibody binding to TKO pig red blood cells (RBCs express only carbohydrate antigens) is no greater than to human blood type O-negative RBCs. Serum cytotoxicity is also not

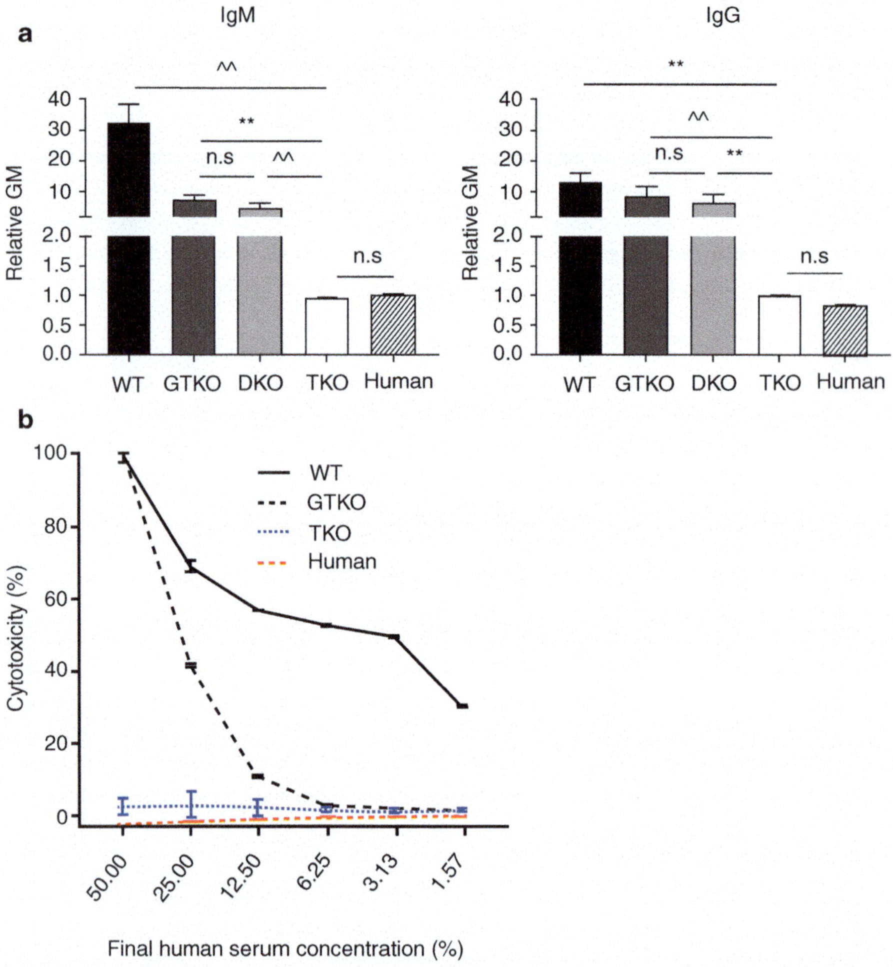

Figs. 5.6 (**a**) Human IgM (left) and IgG (right) antibody binding to wild-type (WT), GTKO, double-knockout (DKO, i.e., deletion of expression of Gal and Sda), and TKO (GTKO/β4GalKO/CMAHKO) pig red blood cells (RBCs). Binding to TKO pig RBCs was not significantly different from human IgM and IgG binding to human RBCs of blood type O. (Modified and reproduced with permission from reference [7]). (**b**) Pooled human serum complement-dependent cytotoxicity (hemolysis) to WT, GTKO, and TKO pig RBCs was performed. Cytotoxicity of the same serum to autologous human O RBCs was tested as a control. (Reproduced with permission from reference [7])

statistically greater to these pig RBCs than to human O-negative RBCs (Fig. 5.6b). These pigs would therefore appear to be ideal as the sources of organs for the initial clinical trials.

However, there is one remaining experimental barrier. Although many humans have *no* antibody to TKO pig RBCs, all baboons *do* (Fig. 5.7a). As baboons and all Old World monkeys, e.g., rhesus monkeys, express Neu5Gc (like pigs), knockout of Neu5Gc in the pig appears to expose a new xenoantigen (sometimes referred to as the 'fourth xenoantigen') against which Old World NHPs have antibodies (Fig. 5.7b). The cytotoxicity associated with these antibodies is high (Fig. 5.8a). Although the nine-gene pigs might provide ideal organs for transplantation into humans,

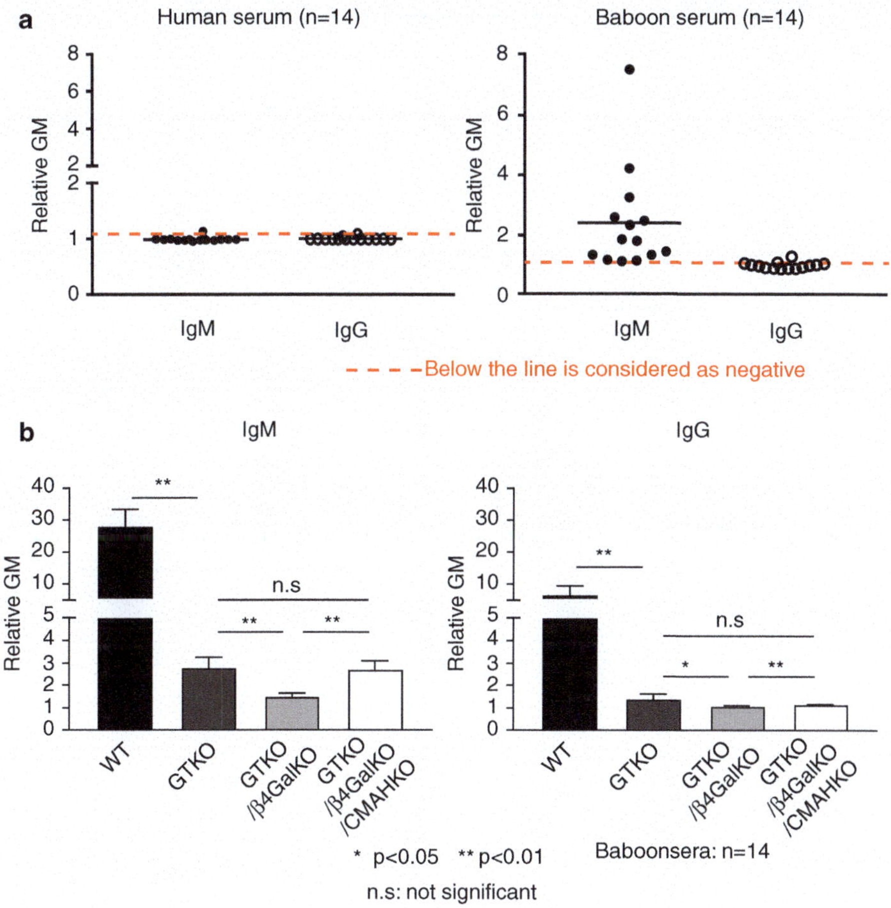

Figs. 5.7 (**a**) Comparison of human ($n = 14$, left) and baboon ($n = 14$, right) serum IgM and IgG antibody binding to TKO pig RBCs. Some baboons had significant IgM binding to TKO pig RBCs, but virtually no humans showed any binding. (**b**) Baboon serum IgM (left) and IgG (right) binding to WT, GTKO, DKO (GTKO + βGal2NT4-KO), and TKO pig RBCs. Some baboons had significant IgM binding to TKO pig RBCs. (Modified from reference [48])

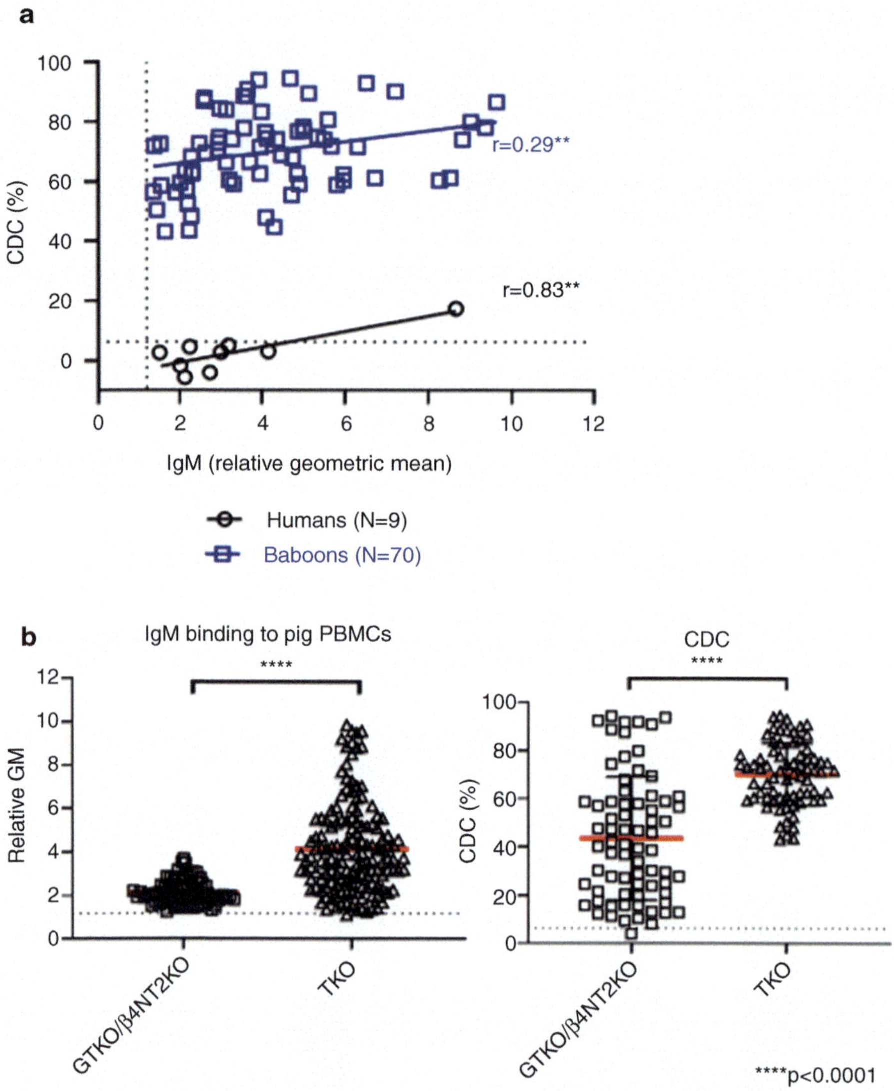

Figs. 5.8 (**a**) Correlation of human ($n = 9$) and baboon ($n = 70$) serum IgM antibody binding with serum complement-dependent cytotoxicity (CDC, at 50% serum concentration) to TKO pig PBMCs. In both humans and baboons, there was a significant increase in cytotoxicity as IgM antibody binding to TKO pig PBMCs increased. In baboons, however, cytotoxicity was high whether IgM binding was high (e.g., 75% cytotoxicity at an rGM of 8) or relatively lower (e.g., 65% at an rGM of 2). (**b**) Baboon serum IgM binding (left) and cytotoxicity (right) to GTKO/β4GalNT2KO and TKO pig PBMCs. IgM binding and cytotoxicity are much less to GTKO/β4GalNT2-KO cells Baboons with low IgM binding and low cytotoxicity to GTKO/β4GalNT2KO pig cells could be selected for experimental studies, whereas all baboons with low antibody binding to TKO cells had high cytotoxicity to these cells. (Reproduced with permission from reference [46])

eight-gene pigs (in which Neu5Gc is not deleted) would be preferable for baboons (Fig. 5.8b).

The transplantation of a TKO pig organ into an Old World NHP therefore does *not* mimic the situation in humans. It will be impossible to provide the national regulatory authorities, e.g., the Food and Drug Administration in the USA, with in vivo evidence in the pig-to-Old World NHP model that will be relevant to the pig-to-human model [46, 47].

Although New World monkeys more closely mimic humans in their response to TKO pig cells [48], they are generally too small to be recipients of pig organ grafts, although they will prove valuable in studies involving TKO pig cell (e.g., pancreatic islets) and tissue transplantation (e.g., skin, corneas).

One important observation is that if the recipient becomes sensitized to pig xeno-antigens, this may not be detrimental to the success of a subsequent allotransplant (Chap. 3) [49].

Conclusions

The currently available genetically engineered pigs would appear to be suitable as sources of organs for the first clinical trial in a small number of carefully selected patients with end-stage renal failure. However, because of species differences, it may be difficult to convince the national regulatory authorities of the genetically engineered pig's suitability with evidence from experimental studies in pig-to-NHP models. Pig renal function in a primate host is largely satisfactory, but attention will need to be paid to maintaining a normal state of hydration.

Acknowledgments The authors thank their many colleagues, past and present, who have contributed to the studies reviewed in this chapter. They gratefully recognize grant funding over many years from the NIH NIAID, and from Revivicor, Blacksburg, VA, and United Therapeutics, Silver Spring, MD.

Conflict of Interest The authors have no conflicts of interest to report.

References

1. Cooper DKC, Ezzelarab MB, Hara H, et al. The pathobiology of pig-to-primate xenotransplantation: a historical review. Xenotransplantation. 2016;23:83–105.
2. Lexer G, Cooper DKC, Rose AG, et al. Hyperacute rejection in a discordant (pig to baboon) cardiac xenograft model. J Heart Transplant. 1986;5:411–8.
3. Cooper DKC, Human PA, Lexer G, et al. Effects of cyclosporine and antibody. Adsorption on pig cardiac xenograft survival in the baboon. J Heart Transplant. 1988;7:238–46.
4. Lambrigts D, Sachs DH, Cooper DKC. Discordant organ xenotransplantation in primates – world experience and current status. Transplantation. 1998;66:547–61.
5. Cooper DKC, Ekser B, Ramsoondar J, Phelps C, Ayares D. The role of genetically-engineered pigs in xenotransplantation research. J Pathol. 2016;238:288–99.

6. Cooper DKC, Ezzelarab MB, Iwase H, Hara H. Perspectives on the optimal genetically-engineered pig in 2018 for initial clinical trials of kidney or heart xenotransplantation. Transplantation. 2018;102:1974–82.
7. Cooper DKC, Hara H, Iwase H, et al. Justification of specific genetic modifications in pigs for clinical organ xenotransplantation. Xenotransplantation. 2019;26:e12516. https://doi.org/10.1111/xen.12516. [Epub ahead of print].
8. Bühler L, Awwad M, Basker M, et al. High-dose porcine hematopoietic cell transplantation combined with CD40 ligand blockade in baboons prevents an induced anti-pig humoral response. Transplantation. 2000;69:2296–304.
9. White DJG, Langford GA, Cozzi E, et al. Production of pigs trangenic for human DAF: a strategy for xenotransplantation. Xenotransplantation. 1995;2:213–7.
10. Cozzi E, White DJ. The generation of transgenic pigs as potential organ donors for humans. Nat Med. 1995;1:964–6. [Epub ahead of print].
11. Fodor WL, Williams BL, Matis LA, et al. Expression of a functional human complement inhibitor in a transgenic pig as a model for the prevention of xenogeneic hyperacute organ rejection. Proc Natl Acad Sci USA. 1994;91:11153–7.
12. Baldan N, Rigotti P, Calabrese F, et al. Ureteral stenosis in HDAF pig-to-primate renal xenotransplantation: a phenomenon related to immunological events? Am J Transplant. 2004;4:475–81.
13. Yamamoto T, Hara H, Foote J, et al. Life-supporting kidney xenotransplantation from genetically-engineered pigs in baboons: a comparison of two immunosuppressive regimens. Transplantation. 2019;103:2090–104.
14. Iwase H, Ekser B, Satyananda V, et al. Pig-to-baboon heterotopic heart transplantation – exploratory preliminary experience with pigs transgenic for human thrombomodulin and comparison of three costimulation blockade-based regimens. Xenotransplantation. 2015;22:211–20.
15. Iwase H, Ekser B, Satyananda V, et al. Initial in vivo experience of pig artery patch transplantation in baboons using mutant MHC (CIITA-DN) pigs. Transpl Immunol. 2015;32:99–108.
16. Higginbotham L, Mathews D, Breeden CA, et al. Pre-transplant antibody screening and anti-CD154 costimulation blockade promote long-term xenograft survival in a pig-to-primate kidney transplant model. Xenotransplantation. 2015;22:221–30.
17. Kuwaki K, Tseng YL, Dor FJ, et al. Heart transplantation in baboons using α1,3-galactosyltransferase gene-knockout pigs as donors: initial experience. Nat Med. 2005;11:29–31.
18. Tseng YL, Kuwaki K, Dor FJ, et al. α 1,3-galactosyltransferase gene-knockout pig heart transplantation in baboons with survival approaching six months. Am J Transplant. 2004;4(8):450–1. Transplantation. 2005; 80:1493-1500.
19. Yamada K, Yazawa K, Shimizu A, et al. Marked prolongation of porcine renal xenograft survival in baboons through the use of α1,3- galactosyltransferase gene-knockout donors and the cotransplantation of vascularized thymic tissue. Nat Med. 2005;11:32–4.
20. Iwase H, Hara H, Ezzelarab M, et al. Immunological and physiological observations in baboons with life-supporting genetically engineered pig kidney grafts. Xenotransplantation. 2017;24:1–13.
21. Adams AB, Kim SC, Martens GR, et al. Xenoantigen deletion and chemical immunosuppression can prolong renal xenograft survival. Ann Surg. 2018;268:564–73.
22. Tseng YL, Sachs DH, Cooper DKC. Porcine hematopoietic progenitor cell transplantation in nonhuman primates: a review of progress. Transplantation. 2005;79:1–9.
23. Tena AA, Sachs DH, Mallard C, et al. Prolonged survival of pig skin on baboons after administration of pig cells expressing human CD47. Transplantation. 2017;101:316–21.
24. Wilhite T, Ezzelarab C, Hara H, et al. The effect of Gal expression on pig cells on the human T-cell xenoresponse. Xenotransplantation. 2012;19:56–63.
25. Ezzelarab MB, Ayares D, Cooper DKC. Transgenic expression of human CD46: does it reduce the primate T-cell response to pig endothelial cells? Xenotransplantation. 2015;22:487–9.
26. Reyes LM, Estrada JL, Wang ZY, et al. Creating class I MHC-null pigs using guide RNA and the Cas9 endonuclease. J Immunol. 2014;193:5751–7.

27. Martens GR, Reyes LM, Butler JR, et al. Humoral reactivity of renal transplant-waitlisted patients to cells from GGTA1/CMAH/B4GalNT2, and SLA class I knockout pigs. Transplantation. 2017;101:e86–92.
28. Hara H, Witt W, Crossley T, et al. Human dominant-negative class II transactivator transgenic pigs – effect on the human anti-pig T cell immune response and immune status. Immunology. 2013;140:39–46.
29. Jeon DH, Oh K, Oh BC, et al. Porcine PD-L1: cloning, characterization, and implications during xenotransplantation. Xenotransplantation. 2007;14:236–42.
30. Plege A, Borns K, Beer L, Baars W, Klempnauer J, Schwinzer R. Downregulation of cytolytic activity of human effector cells by transgenic expression of human PD-ligand-1 on porcine target cells. Transpl Int. 2010;23:1293–300.
31. Buermann A, Petkov S, Petersen B, et al. Pigs expressing the human inhibitory ligand PD-L1 (CD 274) provide a new source of xenogeneic cells and tissues with low immunogenic properties. Xenotransplantation. 2018;25:e12387. https://doi.org/10.1111/xen.12387.
32. Phelps CJ, Ball SF, Vaught TD, et al. Production and characterization of transgenic pigs expressing porcine CTLA4-Ig. Xenotransplantation. 2009;16:477–85.
33. Klymiuk N, van Buerck L, Bähr A, et al. Xenografted islet cell clusters from INSLEA29Y transgenic pigs rescue diabetes and prevent immune rejection in humanized mice. Diabetes. 2012;61:1527–32.
34. Burdorf L, Azimzadeh AM, Pierson RN. Progress and challenges in lung xenotransplantation: an update. Curr Opin Organ Transplant. 2018;23:621–7.
35. Cooper DKC, Hara H, Iwase H, Banks CA, Cleveland D. An approach to induction of tolerance to pig cardiac xenografts in neonates. Xenotransplantation. 2018;25:e12454. https://doi.org/10.1111/xen.12454.
36. Cleveland D, Banks CA, Hara H, Carlos W, Mauchley D, Cooper DKC. The case for cardiac xenotransplantation in neonates: is now the time to reconsider xenotransplantation for hypoplastic left heart syndrome? Pediatr Cardiol. 2018; https://doi.org/10.1007/s00246-018-1998-1. (Pediatr Cardiol. 2019; 40:437–444. doi: 10.1007/s00246-018-1998-1.)
37. Soin B, Ostlie D, Cozzi E, et al. Growth of porcine kidneys in their native and xenograft environment. Xenotransplantation. 2000;7:96–100.
38. Iwase H, Liu H, Wijkstrom M, et al. Pig kidney graft survival in a baboon for 136 days: longest life-supporting organ graft survival to date. Xenotransplantation. 2015;22:302–9.
39. Iwase H, Klein E, Cooper DKC. Physiological aspects of pig kidney transplantation in primates. Comp Med. 2018;68:332–40.
40. Tanabe T, Watanabe H, Shah JA, et al. Role of intrinsic (graft) versus extrinsic (host) factors in the growth of transplanted organs following allogeneic and xenogeneic transplantation. Am J Transplant. 2017;17:1778–90.
41. Mohiuddin MM, Singh AK, Corcoran PC, et al. Chimeric 2C10R4 anti-CD40 antibody therapy is critical for long-term survival of GTKO.hCD46.hTBM pig-to-primate cardiac xenograft. Nat Commun. 2016;7:11138.
42. Langin M, Mayr T, Reichart B, et al. Consistent success in life-supporting porcine cardiac xenotransplantation. Nature. 2018;564:430–3.
43. Hinrichs A, Klymiuk N, Reichart B, et al. Inactivation of the GHR gene – a strategy to overcome excess growth of orthotopic pig-to-baboon cardiac xenografts? Xenotransplantation. 2017;24:39–40. Abstract (#O6.2).
44. Iwase H, Yamamoto T, Cooper DKC. Episodes of hypovolemia/dehydration in baboons with pig kidney transplants: a new syndrome of clinical importance? Xenotransplantation. 2018:e12472. https://doi.org/10.1111/xen.12472. [Epub ahead of print]. (2019; 26:e12472. doi: 10.1111/xen.12472).
45. Soin B, Smith KG, Zaidi A, et al. Physiological aspects of pig-to-primate renal xenotransplantation. Kidney Int. 2001;60:1592–7.
46. Yamamoto T, Hara H, Iwase H, et al. The final obstacle to a successful pre-clinical xenotransplantation. 2019. In press.

47. Yamamoto T, Iwase H, Patel D, et al. Old World monkeys are less than ideal transplantation models for testing pig organs lacking three carbohydrate antigens (triple-knockout). Sci Rep. 2020. In press.
48. Li Q, Shaikh S, Iwase H, et al. Carbohydrate antigen expression and anti-pig antibodies in New World capuchin monkeys: relevance to studies of xenotransplantation. Xenotransplantation. 2019;26:e12498. https://doi.org/10.1111/xen.12498.
49. Li Q, Hara H, Breimer ME, Wang Y, Cooper DKC. Is sensitization to pig antigens detrimental to subsequent allotransplantation? Xenotransplantation. 2018;25:e12393. https://doi.org/10.1111/xen.12393.
50. Cooper DKC, Koren E, Oriol R. Genetically engineered pigs. Lancet. 1993;342:682–3.
51. Phelps CJ, Koike C, Vaught TD, et al. Production of alpha 1,3-galactosyltransferase-deficient pigs. Science. 2003;299:411–4.
52. Kolber-Simonds D, Lai L, Watt SR, et al. Production of alpha-1,3-galactosyltransferase null pigs by means of nuclear transfer with fibroblasts bearing loss of heterozygosity mutations. Proc Natl Acad Sci U S A. 2004;101:7335–40.
53. Byrne GW, Du Z, Stalboerger P, Kogelberg H, McGregor CG. Cloning and expression of porcine beta1,4 N-acetylgalactosaminyl transferase encoding a new xenoreactive antigen. Xenotransplantation. 2014;21:543–54.
54. Estrada JL, Martens G, Li P, et al. Evaluation of human and non-human primate antibody binding to pig cells lacking GGTA1/CMAH/β4GalNT2 genes. Xenotransplantation. 2015;22:194–202.
55. Bouhours D, Pourcel C, Bouhours JE. Simultaneous expression by porcine aorta endothelial cells of glycosphingolipids bearing the major epitope for human xenoreactive antibodies (Gal alpha1-3Gal), blood group H determinant and N-glycolylneuraminic acid. Glycoconj J. 1996;13:947–53.
56. Zhu A, Hurst R. Anti-N-glycolylneuraminic acid antibodies identified in healthy human serum. Xenotransplantation. 2002;9:376–81.
57. Lutz AJ, Li P, Estrada JL, et al. Double knockout pigs deficient in N-glycolylneuraminic acid and galactose α-1,3-galactose reduce the humoral barrier to xenotransplantation. Xenotransplantation. 2013;20:27–35.

Cardiac Xenotransplantation in Nonhuman Primates

6

David K. C. Cooper

Abbreviations

CRP	Complement-regulatory protein
Gal	Galactose -α1,3-galactose
GTKO	α1,3-galactosyltransferase gene-knockout
Neu5Gc	N-glycolylneuraminic acid
NHP	Nonhuman primate
TBM	Thrombomodulin

Introduction

Approximately five million people in the United States suffer from heart failure, and approximately 400,000 new cases are added to this pool every year [1, 2]. Approximately half of these patients will die within 5 years [1]. Thus, many thousands of patients might benefit from heart transplantation. However, at present only approximately 3000 cardiac transplants are performed annually in North America and less than 3000 in the remainder of the world (Fig. 6.1) [3]. The shortage in the availability of suitable deceased human donor hearts will almost certainly increase in the future. The limitation of organs for clinical transplantation has renewed interest in the potential of xenotransplantation, particularly with the pig being the organ-source [4].

D. K. C. Cooper (✉)
Xenotransplantation Program, Division of Transplantation, Department of Surgery, University of Alabama at Birmingham (UAB), Birmingham, AL, USA
e-mail: dkcooper@uabmc.edu

D. K. C. Cooper, G. Byrne (eds.), *Clinical Xenotransplantation*,
https://doi.org/10.1007/978-3-030-49127-7_6

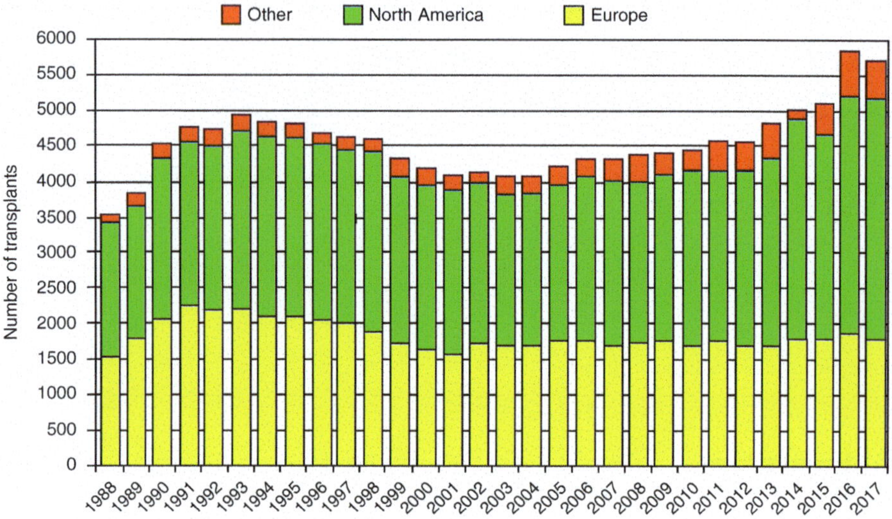

Fig. 6.1 Number of heart transplants (adult and pediatric) by year (transplants 1988–2017) and geographic region (ISHLT database). (Source: Khush et al. [3])

Historical Background

Interest in the field of xenotransplantation continued sporadically throughout the late twentieth century. In the majority of clinical xenotransplants, hearts from non-human primates (NHPs) were selected, with a minority using hearts from other mammals (Chap. 1). In particular, well-known surgeons James Hardy, Donald Ross, Denton Cooley, Christiaan Barnard, and Leonard Bailey made unsuccessful attempts to provide animal hearts for dying patients (Chap. 1). There were two reports of pig hearts being transplanted [one in Poland [5] and one in India [6]], but details of the latter cases were scarce and are mainly available through the lay press.

From these early experiences, it was clear that the immune response to a nonprimate mammalian heart, such as from a pig, was much stronger than to an NHP heart. Nevertheless, for a number of reasons, the pig was selected as the most likely source of organs for clinical transplantation (Chap. 1).

The early experience in pig-to-NHP heart transplantation models was extensively reviewed by Lambrigts et al. [7]. Subsequent experimental experience has also been fully reviewed [8].

Significant milestones included (i) the first description of hyperacute rejection [9, 10], (ii) the identification that immunoadsorption of anti-pig antibody could delay rejection [11], but (iii) conventional immunosuppressive therapy alone had little or no effect [11], (iv) the demonstration that complement inhibition could delay rejection [12], (v) the first transplantation of a genetically modified pig heart (expressing the human complement-regulatory protein, CD55) with extended survival (reviewed in [13]), (vi) the first relative success after orthotopic heart transplantation [14], and (vii) the first transplant of hearts from α1,3-galactosyltransferase gene-knockout (GTKO) pigs [15, 16]. In addition, the first use of anti-CD154mAb [17] and anti-CD40mAb [18, 19] as immunosuppressive agents in pig heart xenotransplantation was noteworthy and advanced the field.

Surgical Techniques of Experimental Heart Transplantation

The early development of these techniques in the laboratory has been reviewed previously [20]. Heterotopic transplantation indicates placing the heart in an ectopic position without removing the native heart. Heterotopic techniques can be categorized as "working" or "nonworking" models. In orthotopic transplantation, the recipient's own heart is removed and replaced by the donor heart.

"Non-working" Heterotopic Heart Transplantation (HHT)

In non-working models, the donor heart is perfused and beats, but does not contribute support of the recipient's circulation. These models have been applied extensively for the study of the immunopathology of graft rejection and the efficacy of immunosuppressive therapies. The techniques have proved valuable in xenotransplantation (when graft survival initially proved to be very short, i.e., minutes, rather than hours or days). The site of the graft can be in the abdomen or neck. Compared with orthotopic heart transplantation, the procedure is technically simpler, less expensive, allows better access for myocardial biopsies, and allows survival of the recipient even in the event of graft rejection. Excision of the graft allows continuing monitoring of the immune response to the graft.

When carried out in the abdomen, the donor aorta is anastomosed end-to-side to the host's abdominal aorta (thus establishing a coronary circulation) and the donor pulmonary artery is anastomosed to the recipient inferior vena cava (IVC) (thus permitting emptying of the venous return of the coronary sinus) (Fig. 6.2a). This technique has formed the basis for many studies. An alternative site, if the recipient is large, is in the neck. The donor aorta is anastomosed to the common carotid artery and the pulmonary artery to the external jugular vein.

Intrathoracic "Working" Heterotopic Heart Transplantation

The transplanted heart contributes to the cardiac output of the recipient. Several different techniques have been demonstrated, but the one in which there is most experience is that introduced by Losman and Barnard, which the Cape Town group utilized in the clinic for allotransplantation for several years (Fig. 6.2b) [21, 22]. Reichart and his colleagues in Germany attempted this technique in the pig-to-baboon model, but with poor results [23–25], largely because the pig heart enlarged rapidly through

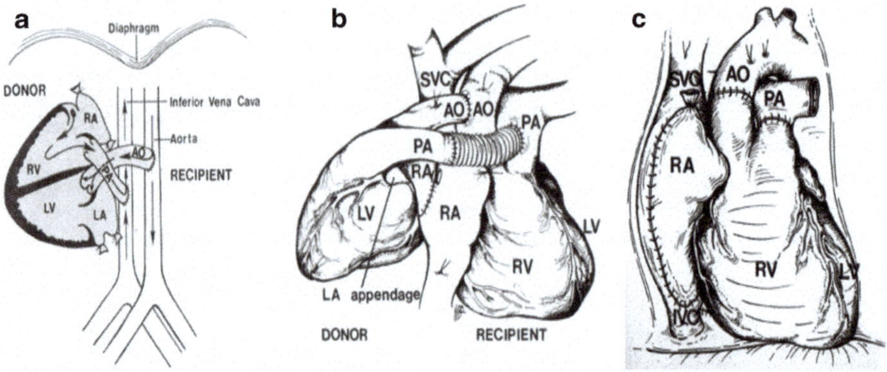

Fig. 6.2 Surgical techniques of heart transplantation used in the experimental laboratory. (**a**) Nonworking heterotopic heart transplantation in the abdomen. (**b**) Intrathoracic working heterotopic heart transplantation (technique described by Barnard and Losman [21] and [22]). (**c**) Orthotopic heart transplantation. (Reproduced with permission from reference [53])

either the effects of rejection, or ventricular hypertrophy, or abnormal growth (see below). This enlargement distorted the anatomy of the two hearts and disrupted the flow of blood through the heterotopic pig heart.

Life-Supporting Orthotopic Heart Transplantation (OHT)

A more rigorous test of the function of a donor heart is OHTx [26] (Fig. 6.2c). There are no significant differences in the anatomy of the pig heart that would prove problematic using the current surgical techniques for OHTx. It is almost certain that the regulatory authorities will require evidence of the ability of a pig heart to support an NHP recipient for several months as a preliminary to a clinical trial. Three months' survival was recommended by the International Society for Heart and Lung Transplantation (ISHLT) in 2000 [27], but it is likely the regulatory authorities may require a longer period of follow-up.

The Immunobiological Barriers of Heart Xenotransplantation and Methods of Overcoming Them

The patterns of rejection of a pig heart graft are similar to those of a pig kidney graft (Chap. 2) and will not be described again here. Rejection can be largely antibody-mediated, e.g., hyperacute or acute (humoral, vascular, delayed) rejection. Acute cellular rejection, with intense cellular infiltration of the graft, is a common occurrence early after heart allotransplantation but, perhaps surprisingly, has relatively rarely been described after cardiac xenotransplantation. This is possibly because the humoral response occurs more rapidly and, if treated successfully, prevents an intense cellular response. Nevertheless, some T cells are often seen in the graft and, if the adaptive (T cell) response is not adequately controlled by immunosuppressive therapy, an elicited anti-pig antibody response can develop that almost always results in graft failure [28].

Pharmacologic immunosuppressive therapy alone (in the absence of a genetically engineered pig graft) has never proved sufficient to protect a pig graft from immune destruction [11]. Nevertheless, it is essential to prevent the T-cell response. Conventional immunosuppressive regimens, e.g., based on cyclosporine or tacrolimus, have been associated with relatively successful prolongation of graft function, but only if administered in higher dosages than are required for allotransplantation [29]. Not surprisingly, this has been associated with a high incidence of infectious and other complications.

More encouraging results were achieved when novel "costimulation-blockade" agents were introduced, first utilized in the pig-to NHP model by Buhler et al. in 2000 [30]. The initial agent, an anti-CD154 monoclonal antibody (mAb), was highly effective at preventing a T-cell response, but was found to be thrombogenic and is currently not available for clinical use. However, an anti-CD40mAb, which also blocks the CD40/CD154 pathway, appears almost equally effective [19, 31]. Blockade of the CD28/B7 co-stimulation pathway with agents such as belatacept is insufficient [32].

In addition to induction therapy with an antithymocyte globulin to deplete T cells, additional induction therapy with an anti-CD20mAb to deplete B cells has been reported to be beneficial [18, 19, 29, 31]. There is no definitive evidence yet that agents that deplete antibody-producing plasma cells, e.g., bortezomib, are effective in xenotransplantation models.

The cellular response can also be inhibited by genetic manipulation of the pigs, e.g., by transgenic endogenous expression of an immunosuppressive agent [33] or by a mutant MHC class II gene [34]. Even the absence of expression of the major pig glycan (galactose α1,3 galactose (Gal)), reduces the T-cell response to pig cells [35], as does the expression of a human complement-regulatory protein [36].

Coagulation dysfunction and a systemic inflammatory response have proved problematic, as they have after pig kidney xenotransplantation (Chap. 2), but have largely been resolved by genetic engineering of the pig (Chap. 7) and judicious drug therapy.

Pig Heart Graft Survival in NHPs

Based on the above developments, and particularly on the genetically engineered pigs that became available, Murthy and his colleagues loosely divided progress in nonworking pig HHTx in NHPs into three eras (Table 6.1, Fig. 6.3). Graft survival in *Era 1* reached a maximum of 99 days, whereas in *Era 2* it was extended to 179 days and in *Era 3 to* >2 years [reviewed in 7, 8, 37]. The maximum survival after heterotopic (working) thoracic heart transplantation has been reported to be 50 days (see above) [24].

Table 6.1 Progress in the results of nonworking pig heterotopic heart transplantation (HHTx) in nonhuman primates, based on availability of genetically engineered pigs

Era 1 (1986–2003): Pre-α1,3-galactosyltrasferase gene-knockout (GTKO)
Era 2 (2004–2011): GTKO+/−human complement-regulatory protein expression (CRP), e.g., CD46, CD55
Era 3 (2012-present): GTKO/CRP + expression of one or more human coagulation-regulatory (antithrombotic) proteins, e.g., thrombomodulin (TBM), endothelial protein C receptor (EPCR)

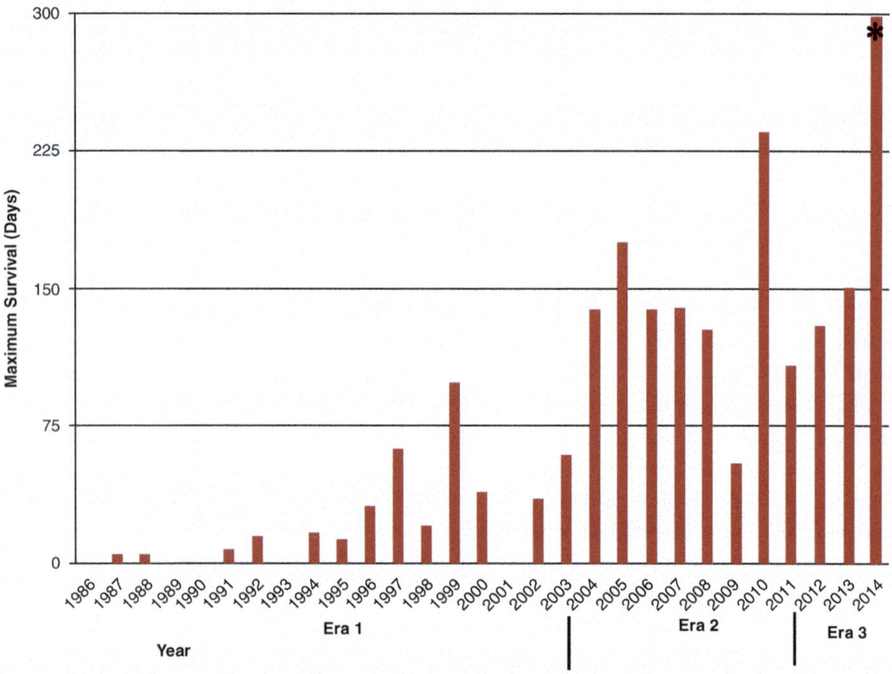

Fig. 6.3 Maximum pig nonworking heterotopic heart graft survival by year in *Era 1* – pre-GTKO (1986–2003), *Era 2* – GTKO+/-hCRP (2004–2011), and *Era 3* – GTKO/hCRP/human coagulation-regulatory protein (2012–2014). In 2014, graft survival extended to >2 years. Since then, most attention has been directed to orthotopic heart transplantation. (Reproduced with permission from reference [53])

The number of studies of pig OHTx in NHPs has been relatively small, but again graft and recipient survival have increased from a maximum of 39 days in *Era 1* to 6 months in *Era 3* [7, 8, 37]. After OHTx, a high mortality within the first post-transplant 48 hours was reported by Byrne and his colleagues and was termed "peri-operative cardiac xenograft dysfunction" (PCXD) [38–40]. This occurred in a reported 40%–60% of cases by Byrne and others [37]. The exact cause of PCXD remains uncertain, but the histological features in the graft indicate it is not from hyperacute rejection or, indeed, from any form of immune response. It appears to be related more to a sensitivity of the pig myocardium to ischemia or to other insult during the transfer of the heart from donor to recipient. Mitochondrial dysfunction related to low triiodothyronine levels after cardiopulmonary bypass [41–43], and also seen after pig kidney xenotransplantation in NHPs [44, 45], is one possibility that might play a role. Limited data indicate that, if a mild form of PCXD occurs early after the heart transplant, it can recover within days [40]. Even though the exact causative factors remain poorly understood, PCXD has recently been overcome by employing more effective methods of protecting the myocardium from ischemic insult.

The most important study in recent years was that by Langin et al. [46], who demonstrated life-supporting pig hearts that functioned well for up to 6 months, at which time the recipient baboons were electively euthanized. This group overcame the early graft dysfunction reported by themselves and others by utilizing a rather complex perfusion system to preserve the heart during transplantation [46, 47]. Although this proved a major breakthrough, whether this system is essential has been questioned by encouraging results by Cleveland and his colleagues at UAB using a much simpler system of cardiac graft protection (Cleveland D, et al., unpublished data).

Langin and his colleagues used hearts from α1,3-galactosyltransferase gene-knockout (GTKO) pigs that expressed the human complement-regulatory protein, CD46, and the human coagulation-regulatory protein, thrombomodulin (GTKO.CD46.TBM pigs). The fact that, if PCXD was avoided, these hearts supported the baboon's circulation for elective periods of 3 or 6 months encourages us that clinical pig OHTx will be successful if the adaptive immune response is controlled by adequate immunosuppressive therapy. The immunosuppressive regimen used by Langin and his colleagues was that introduced by Mohiuddin et al. [19]. However, there is some evidence that GTKO.CD46.TBM pigs are not the optimal sources of organs for transplantation into humans, as opposed to transplantation into NHPs [48].

The function of a life-supporting pig graft in a primate recipient has been discussed elsewhere [49, 50], but the above data indicate fairly conclusively that, when the immunological problems are successfully controlled, a pig heart will support life in a primate host. Nevertheless, there is one problem that requires further investigation and resolution, and that is the problem of rapid growth of the heart within the first few weeks after transplantation.

Rapid Growth of Pig Heart Grafts After Transplantation into NHPs

The topic of rapid growth of a pig organ after transplantation into an NHP has been discussed in Chap. 5, and so details will not be repeated here. However, in contrast to the kidney, the heart is placed within the restricted confines of the chest. Rapid growth of the heart could therefore compress it significantly and thus impair its function (or possibly that of the lungs). As stated above, this was observed after intrathoracic HHT, in which case, of course, there are two hearts (donor pig and recipient baboon) within the confines of the chest.

Rapid growth was not reported after pig HHTx in the abdomen [15, 16, 18, 19, 31, 51]. Whether this was related to the fact that the heart was not carrying a workload, as it was not supporting the circulation of the recipient NHP, remains uncertain but seems likely. After intrathoracic HHTx and after OHTx, hypertrophy of the ventricular myocardium has been the major observation, and the mechanism by which this occurs may be different from that of growth after pig kidney transplantation. A higher peripheral vascular resistance in baboons than in pigs could result in the development of myocardial hypertrophy, as the pig heart struggles to work against this resistance.

Indeed, because of the higher systolic blood pressure observed in baboons (approximately 120 mm Hg) than in pigs (approximately 80 mm Hg), Langin et al. attempted to reduce the development of ventricular hypertrophy by administering antihypertensive therapy to the recipient baboons, with some – though not complete – success [46]. The inclusion of rapamycin in the immunosuppressive regimen may be a factor in reducing growth of the organ [50, 52].

Clinical Prospects

Before a clinical trial can be undertaken with a realistic chance of success (either as a bridge to allotransplantation or as destination therapy), Murthy et al. [53] suggested that the following milestones need to be achieved.

(i) Relatively consistent survival of the NHP recipient of an orthotopically transplanted pig heart for at least 3 months with some evidence of survival for 6 months in the absence of major complications related to the intensity of the immunosuppressive regimen, e.g., infection [27].
(ii) The availability of a facility in which genetically engineered pigs can be bred and housed under isolation conditions that will protect them from being exposed to infectious agents, including viruses, that could be transferred with the graft to the recipient (designated pathogen-free pigs) (Chap. 8).
(iii) A continuing absence of evidence that the inevitable transfer of porcine endogenous retroviruses (that are present in the genome of every pig cell) will not prove to be a health risk to the recipient and, more importantly, to his/her close contacts, e.g., family, medical staff, and thus to the community (Chap. 17).

A final point to be considered is whether, in the event that a pig heart is transplanted as a bridge to allotransplantation, an immune response to the xenograft will sensitize the recipient to a subsequent allograft. The present very limited data suggest that this will *not* occur [54, 55].

Although complications of ventricular assist devices are not uncommon, the results are steadily improving, and a question that needs to be addressed is which patients, if any, could be considered for an initial clinical trial of pig heart transplantation, and what form the trial should take, i.e., should it be a bridging trial or destination trial (Chaps. 15 and 16). A bridging trial would be more realistic and ethically justified.

Even if a pig graft does not survive as long as an allograft, there will be no limitation on retransplantation, as there will be no ethical dilemma as to whether a patient should undergo retransplantation at the expense of another patient who awaits his/her first transplant. Furthermore, patients will not languish in an intensive care unit for weeks waiting for a suitable allograft; the transplant will be carried out as soon as the decision to transplant has been made.

Acknowledgments The author gratefully recognizes grant funding over many years from the NIH NIAID, and from Revivicor, Blacksburg, VA, and United Therapeutics, Silver Spring, MD.

Conflict of Interest The author reports no conflict of interest.

References

1. Writing Group Members, Mozaffarian D, Benjamin EJ, et al. Heart disease and stroke statistics-2016 update: a report from the American Heart Association. Circulation. 2016;133(4):e38–360. https://doi.org/10.1161/CIR.0000000000000350.
2. DeSalvatore S, Segreto A, Chiusaroli A, Congiu S, Bizzarri F. The role of xenotransplantation in cardiac transplantation. J Card Surg. 2015;30:111–6.
3. Khush KK, Cherikh WS, Chambers DC, et al. The International Thoracic Organ Transplant Registry of the International Society for Heart and Lung Transplantation: Thirty-sixth adult heart transplantation report – 2019; focus theme: donor and recipient size match. J Heart Lung Transplant. 2019;38:1056–66.
4. Cooper DKC. The case for xenotransplantation. Clin Transpl. 2015;29:288–93.
5. Czaplicki J, Blonska B, Religa Z. The lack of hyperacute xenogeneic heart transplant rejection in a human. J Heart Lung Transplant. 1992;11(2 Pt 1):393–39. New India Express. 02/1999
6. New India Express. 1999.
7. Lambrigts D, Sachs DH, Cooper DKC. Discordant organ xenotransplantation in primates: world experience and current status. Transplantation. 1998;66:547–61.
8. Cooper DKC, Satyananda V, Ekser B, et al. Progress in pig-to-nonhuman primate transplantation models (1998–2013): a comprehensive review of the literature. Xenotransplantation. 2014;21:397–419.
9. Lexer G, Cooper DKC, Rose AG, et al. Hyperacute rejection in a discordant (pig to baboon) cardiac xenograft model. J Heart Transplant. 1986;5:411–8.
10. Rose AG, Cooper DKC, Human PA, Reichenspurner H, Reichart B. Histopathology of hyperacute rejection of the heart: experimental and clinical observations in allografts and xenografts. J Heart Lung Transplant. 1991;10:223–34.
11. Cooper DKC, Human PA, Lexer G, et al. Effects of cyclosporine and antibody adsorption on pig cardiac xenograft survival in the baboon. J Heart Transplant. 1988;7:238–46.
12. Leventhal JR, Dalmasso AP, Cromwell JW, et al. Prolongation of cardiac xenograft survival by depletion of complement. Transplantation. 1993;55:857–65; discussion 865–856.
13. Cozzi E, Yannoutsos N, Langford GA, Pino-Chavez G, Wallwork J, White DJG. Effects of transgenic expression of human decay accelerating factor on the inhibition of hyperacute rejection of pig organs. In: Cooper DKC, Kemp E, Platt JL, White DJG, editors. Xenotransplantation – the transplantation of organs and tissues between species. Berlin: Springer; 1997. p. 665–82.

14. Vial CM, Ostlie DJ, Bhatti FN, et al. Life supporting function for over one month of a transgenic porcine heart in a baboon. J Heart Lung Transplant. 2000;19:224–9.
15. Kuwaki K, Tseng YL, Dor FJ, et al. Heart transplantation in baboons using α1,3-galactosyltransferase gene-knockout pigs as donors: initial experience. Nat Med. 2005;11:29–31.
16. Tseng YL, Kuwaki K, Dor FJ, et al. α 1,3-galactosyltransferase gene-knockout pig heart transplantation in baboons with survival approaching six months. Transplantation. 2005;80:1493–500.
17. Bühler L, Basker M, Alwayn IPJ, et al. Coagulation and thrombotic disorders associated with pig organ and hematopoietic cell transplantation in nonhuman primates. Transplantation. 2000;70:1323–31.
18. Mohiuddin MM, Singh AK, Corcoran PC, et al. Genetically engineered pigs and target-specific immunomodulation provide significant graft survival and hope for clinical cardiac xenotransplantation. J Thorac Cardiovasc Surg. 2014;148:1106–13; discussion 1113–1104.
19. Mohiuddin MM, Singh AK, Corcoran PC, et al. Role of anti-CD40 antibody-mediated costimulation blockade on non-Gal antibody production and heterotopic cardiac xenograft survival in a GTKO.hCD46Tg pig-to-baboon model. Xenotransplantation. 2014;21:35–45.
20. Cooper DKC. Experimental development of cardiac transplantation. Br Med J. 1968;4:174–81.
21. Barnard CN, Losman JG. Left ventricular bypass. S Afr Med J. 1975;49:303–12.
22. Novitzky D, Cooper DKC, Barnard CN. The surgical technique of heterotopic heart transplantation. Ann Thorac Surg. 1983;36:476–82.
23. Bauer A, Postrach J, Thormann M, et al. First experience with heterotopic thoracic pig-to-baboon cardiac xenotransplantation. Xenotransplantation. 2010;17:243–9.
24. Abicht JM, Mayr T, Reichart B, et al. Pre-clinical heterotopic intrathoracic heart xenotransplantation: a possibly useful clinical technique. Xenotransplantation. 2015;22:427–42.
25. Mayr T, Bauer A, Reichart B, et al. Hemodynamic and perioperative management in two different preclinical pig-to-baboon cardiac xenotransplantation models. Xenotransplantation. 2017;24(3). https://doi.org/10.1111/xen.12295.
26. Barnard CN. The operation. A human cardiac transplant: an interim report of a successful operation performed at Groote Schuur hospital, Cape town. S Afr Med J. 1967;41:1271–4.
27. Cooper DKC, Keogh AM, Brink J, et al. Report of the xenotransplantation advisory committee of the international society for heart and lung transplantation: the present status of xenotransplantation and its potential role in the treatment of end-stage cardiac and pulmonary diseases. J Heart Lung Transplant. 2000;19:1125–65.
28. Chen G, Qian H, Starzl T, et al. Acute rejection is associated with antibodies to non-gal antigens in baboons using gal-knockout pig kidneys. Nat Med. 2005;11:1295–8.
29. McGregor CGA, Davies WR, Oi K, et al. Cardiac xenotransplantation: recent preclinical progress with 3-month median survival. J Thorac Cardiovasc Surg. 2005;130:844–51.
30. Buhler L, Awwad M, Basker M, et al. High-dose porcine hematopoietic cell transplantation combined with CD40 ligand blockade in baboons prevents an induced anti-pig humoral response. Transplantation. 2000;69:2296–304.
31. Mohiuddin MM, Singh AK, Corcoran PC, et al. Chimeric 2C10R4 anti-CD40 antibody therapy is critical for long-term survival of GTKO.hCD46.hTBM pig-to-primate cardiac xenograft. Nat Commun. 2016;7:11138. https://doi.org/10.1038/ncomms11138.
32. Ezzelarab MB, Ekser B, Echeverri G, et al. Costimulation blockade in pig artery patch xenotransplantation – a simple model to monitor the adaptive immune response in nonhuman primates. Xenotransplantation. 2012;19:221–32.
33. Phelps CJ, Ball SF, Vaught TD, et al. Production and characterization of transgenic pigs expressing porcine ctla4-ig. Xenotransplantation. 2009;16:477–85.
34. Hara H, Witt W, Crossley T, et al. Human dominant-negative class ii transactivator transgenic pigs – effect on the human anti-pig T-cell immune response and immune status. Immunology. 2013;140:39–46.
35. Wilhite T, Ezzelarab C, Hara H, et al. The effect of gal expression on pig cells on the human T-cell xenoresponse. Xenotransplantation. 2012;19:56–63.

36. Ezzelarab MB, Ayares D, Cooper DKC. Transgenic expression of human CD46: does it reduce the primate T-cell response to pig endothelial cells? Xenotransplantation. 2015;22:487–9.
37. Mohiuddin MM, Reichart B, Byrne GW, McGregor CGA. Current status of pig heart xenotransplantation. Int J Surg. 2015;23(Pt B):234–9.
38. Byrne GW, Du Z, Sun Z, Asmann YW, McGregor CGA. Changes in cardiac gene expression after pig-to-primate orthotopic xenotransplantation. Xenotransplantation. 2011;18:14–27.
39. Byrne GW, McGregor CGA. Cardiac xenotransplantation: progress and challenges. Curr Opin Organ Transplant. 2012;17:148–54.
40. McGregor CGA, Byrne GW. Porcine to human heart transplantation: is clinical application now appropriate? J Immunol Res. 2017;2017:2534653.
41. Novitzky D, Human PA, Cooper DKC. Inotropic effect of triiodothyronine following myocardial ischaemia and cardiopulmonary bypass: an experimental study in pigs. Ann Thorac Surg. 1988;45:50–5.
42. Novitzky D, Human PA, Cooper DKC. Effect of triiodothyronine (T3) on myocardial high energy phosphates and lactate after ischemia and cardiopulmonary bypass. An experimental study in baboons. J Thorac Cardiovasc Surg. 1988;96:600–7.
43. Novitzky D, Cooper DKC, Swanepoel A. Inotropic effect of triiodothyronine (T3) in low cardiac output following cardioplegic arrest and cardiopulmonary bypass: an initial experience in patients undergoing open heart surgery. Eur J Cardiothorac Surg. 1989;3:140–5.
44. Iwase H, Ekser B, Satyananda V, Ezzelarab M, Cooper DKC. Plasma free triiodothyronine (fT3) levels in baboons undergoing pig organ transplantation: relevance to early recovery of organ function. Xenotransplantation. 2014;21:582–3.
45. Iwase H, Ekser B, Hara H, et al. Thyroid hormone: relevance to xenotransplantation. Xenotransplantation. 2016;23:293–9.
46. Längin M, Mayr T, Reichart B, et al. Consistent success in life-supporting porcine cardiac xenotransplantation. Nature. 2018;564(7736):430–3.
47. Steen S, Paskevicius A, Liao Q, Sjöberg T. Safe orthotopic transplantation of hearts harvested 24 hours after brain death and preserved for 24 hours. Scand Cardiovasc J. 2016;50:193–200.
48. Yamamoto T, Hara H, Iwase H, et al. The final obstacle to a successful pre-clinical xenotransplantation. Xenotransplantation. 2020. In press.
49. Ibrahim Z, Busch J, Awwad M, Wagner R, Wells K, Cooper DKC. Selected physiologic compatibilities and incompatibilities between human and porcine organ systems. Xenotransplantation. 2006;13:488–99.
50. Iwase H, Klein E, Cooper DKC. Physiologic aspects of pig kidney transplantation in nonhuman primates. Comp Med. 2018;68:332–40.
51. Iwase H, Ekser B, Satyananda V, et al. Pig-to-baboon heterotopic heart transplantation – exploratory preliminary experience with pigs transgenic for human thrombomodulin and comparison of three costimulation blockade-based regimens. Xenotransplantation. 2015;22:211–20.
52. Iwase H, Hara H, Ezzelarab M, et al. Immunological and physiologic observations in baboons with life-supporting genetically-engineered pig kidney grafts. Xenotransplantation. 2017;24. https://doi.org/10.1111/xen.12293.
53. Murthy R, Bajona P, Bhama JK, Cooper DKC. Heart xenotransplantation: historical background, experimental progress, and clinical prospects. Ann Thorac Surg. 2016;101:1605–13.
54. Cooper DKC, Tseng YL, Saidman SL. Alloantibody and xenoantibody cross-reactivity in transplantation. Transplantation. 2004;77:1–5.
55. Li Q, Hara H, Zhang Z, Breimer ME, Wang Y, Cooper DKC. Is sensitization to pig antigens detrimental to subsequent allotransplantation? Xenotransplantation. 2018;25:e12393. https://doi.org/10.1111/xen.12393.

Part III

Organ-Source Pig Genetic Engineering and Regulation

Gene-edited Pigs for Xenotransplantation

7

Willard Eyestone, Kent Adams, Suyapa Ball, John Bianchi, Steve Butler, Amy Dandro, Kasinath Kuravi, Maria Kokkinaki, Ana Laura Fazio, Jeff Monahan, Ben Morrill, Carol Phelps, Farzana Rahman, Jagdeece Ramsoondar, Lori Sorrells, Todd Vaught, Anneke Walters, and David Ayares

Abbreviations

CMAH	Cytidine monophosphate-N-acetylneuraminic acid hydroxylase-2
DAF	Decay-accelerating factor (CD55)
EPCR	Endothelial protein C receptor
Gal	Galactose-α1,3-galactose
GTKO	α1,3-Galactosyltransferase-knockout
HAR	Hyperacute rejection
HO1	Hemeoxygenase-1
MCV	Multicistronic vector
Neu5Gc	N-glycolylneuraminic acid
NHP	Nonhuman primate
pAECs	Porcine aortic endothelial cells
SCNT	Somatic cell nuclear transfer
TBM	Thrombomodulin
β4KO	β4GalNT2-knockout

W. Eyestone (✉) · K. Adams · S. Ball · J. Bianchi · S. Butler · A. Dandro · K. Kuravi
M. Kokkinaki · A. L. Fazio · J. Monahan · B. Morrill · C. Phelps · F. Rahman · J. Ramsoondar
L. Sorrells · T. Vaught · A. Walters · D. Ayares
Revivicor, Inc., Blacksburg, VA, USA
e-mail: weyestone@revivicor.com

D. K. C. Cooper, G. Byrne (eds.), *Clinical Xenotransplantation*,
https://doi.org/10.1007/978-3-030-49127-7_7

Introduction

The critical shortage of human organs for allotransplantation has led to intensive research into xenotransplantation, with the pig being the species of choice as an organ source. Early efforts to transplant wild-type pig organs into nonhuman primate (NHP) models were unsuccessful as the organs underwent nearly immediate hyperacute rejection (HAR) due to the presence of preformed anti-pig antibodies in the organ recipient's blood [1]. A seminal discovery in the development of xenotransplantation was the identification of galactose-α1,3-galactose (Gal) as the principal xenoantigen involved in HAR [2].

Gal is a carbohydrate moiety that decorates many glycoproteins, glycolipids, and proteoglycans in most mammalian species, with the exception of Old World NHPs and humans [3]. These species lack a functional gene for the enzyme α1,3 galactosyltransferase (GGTA1) that catalyzes the synthesis of Gal. Chronic exposure to Gal expressed by gut microflora stimulates production of anti-Gal antibodies in these species [4] to the extent that anti-Gal antibodies comprise a substantial proportion of the antibody population (~1% of the total) and are the most abundant single species of antibody in the blood [5].

Hyperacute rejection of pig organs is initiated upon reperfusion of the transplanted organ with the recipient's blood, and the subsequent binding of anti-pig antibodies to Gal and non-Gal xenoantigens on the endothelial lumen [2, 6]. This initial binding event is rapidly followed by complement activation and dissolution of the endothelium, leading to interstitial hemorrhage, coagulation, and necrosis. Early attempts to prevent HAR included transgenic expression of human complement-regulatory proteins (decay-accelerating factor, DAF) in porcine organs which prevented HAR and increased organ survival [7].

However, overcoming HAR by complement-inhibitor expression, and eventually by Gal-knockout, revealed other impediments to xenotransplantation, for example, coagulation dysregulation and inflammation. Much of the progress made to date toward successful xenotransplantation has stemmed from the generation of genetically modified pigs tailored to overcome these problems.

A major step forward in the generation of pigs as organ donors was the advent of somatic cell nuclear transfer (SCNT) [8, 9] (Fig. 7.1). In the pig, cultured fibroblasts were used as nuclear donors to replace the endogenous nuclei of porcine oocytes. Upon fusion with an enucleated oocyte, fibroblast nuclei were reprogrammed to totipotency by factors in oocyte cytoplasm. The newly reconstructed oocyte then developed into a new individual with the genetic constitution of the donor nucleus. SCNT technology opened the door for genetic modification of cultured somatic cells, which could be used to generate pigs bearing those modifications. Using this technology, Dai et al. [10] knocked out porcine GGTA1 by conventional homologous recombination and insertional mutagenesis. SCNT pigs made from these cells were GGTA1-knockouts by genotype and displayed a corresponding Gal-knockout (GTKO) phenotype. Importantly, organs from GTKO pigs did not undergo HAR when transplanted into NHPs (baboons), and posttransplant survival of GTKO organs increased from hours to weeks [6, 11].

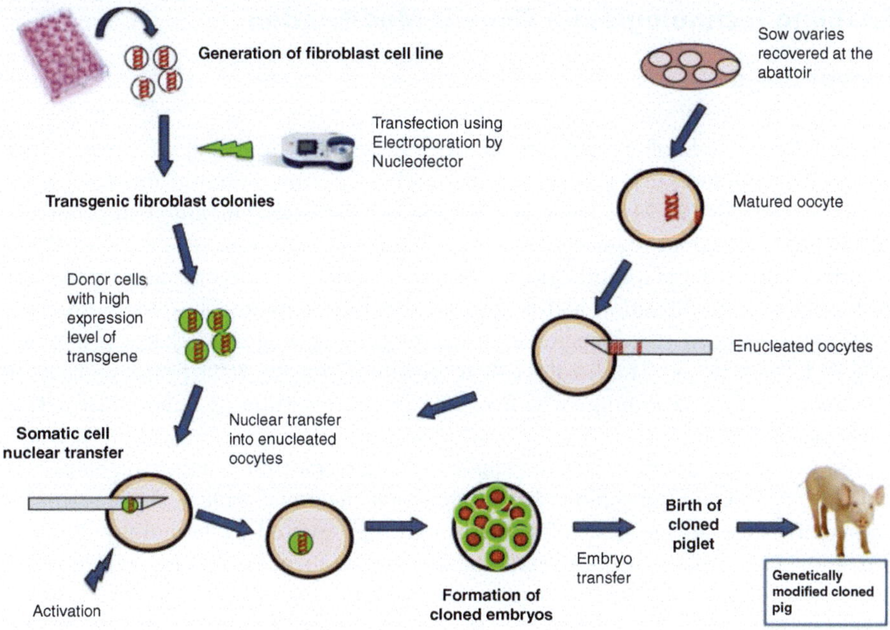

Fig. 7.1 Somatic cell nuclear transfer

However, these transplants still suffered to some degree from binding of preformed antibodies to non-Gal antigens, as well as inflammation resulting from ischemia–reperfusion injury, and thrombosis and coagulopathies [11, 12] due to molecular incompatibilities between the pig and human proteins that regulate the complement and coagulation cascades. These challenges are being addressed by expressing transgenes that encode the human counterparts of these proteins in organ-source pigs using various genetic modification technologies enabled by SCNT.

The GTKO genotype and phenotype just described have remained stable over 11 generations of natural breeding, and a New Animal Drug Application for our GTKO pigs ("GalSafe" pigs) was granted by the US Food and Drug Administration (FDA) in 2018. These GTKO pigs are the foundational genotype in our laboratory on which all additional modifications are made. Recent advances in genetic modification technologies have facilitated the rapid and efficient generation of pigs bearing multiple knockouts of genes encoding xenoantigens and knock-ins of multiple transgenes encoding key human proteins. These modifications are designed to deal primarily with rejection issues due to innate immunity. Our aim here is to review our work on genotypes that have performed well in models of xenotransplantation, supported by corresponding functional evaluations of each modification in vitro.

Enabling Technologies for Genetic Modification

CRISPR/Cas9

CRISPR/Cas9 technology permits highly efficient modification of precise loci within the genome [13]. Based on a naturally occurring antiviral mechanism in bacteria, CRISPR/Cas9 has been cleverly adapted for editing mammalian genomes. A CRISPR (clustered regularly interspaced short palindromic repeat) is an RNA sequence consisting of a short "guide" with homology to a genomic target at the site of the desired modification and a nuclease-binding domain. The guide directs a CRISPR-associated (Cas) nuclease to the target site where it creates a double-stranded DNA break. The break is then "repaired" by the error-prone process of nonhomologous end joining, which results in the insertion or deletion of a few nucleotides (indels).

Indels that occur within an exon can result in a frameshift, leading to generation of a stop codon and translation of truncated, nonfunctional protein, thus creating a knockout. Alternatively, the DNA break may be repaired by homology-directed repair in the presence of a DNA vector flanked with sequences homologous ("homology arms") to those in the target region. During homology-directed repair, homologous recombination occurs between the vector's homology arms and the host's genomic sequences, enabling the incorporation of an intervening sequence, for example, a transgene vector, into the genome. CRISPR/Cas9 technology allows for very efficient gene targeting for creating knockouts, and, when coupled with homology-directed repair, allows for transgene insertion into predetermined "landing pad" sites in the genome.

Multicistronic Vectors (MCVs)

The ability to incorporate multiple transgenes into a single vector offers multiple advantages for making organ-source pigs. First, using homology-directed repair, the vector can be targeted to specific "landing pads" known to be permissive for transgene expression. This ensures that each transgene within the vector can be expressed without interference from repressive chromatin that frequently affects randomly integrated transgenes. Second, by using self-cleaving "2A" sequences between transgene coding sequences, multiple transgenes can be expressed from a single promoter, which simplifies vector construction and keeps vector size within practical limits [14] (Fig. 7.2). Finally, all transgenes within the vector will be inherited as a single Mendelian locus to facilitate and simplify breeding and scale-up of organ-source herds.

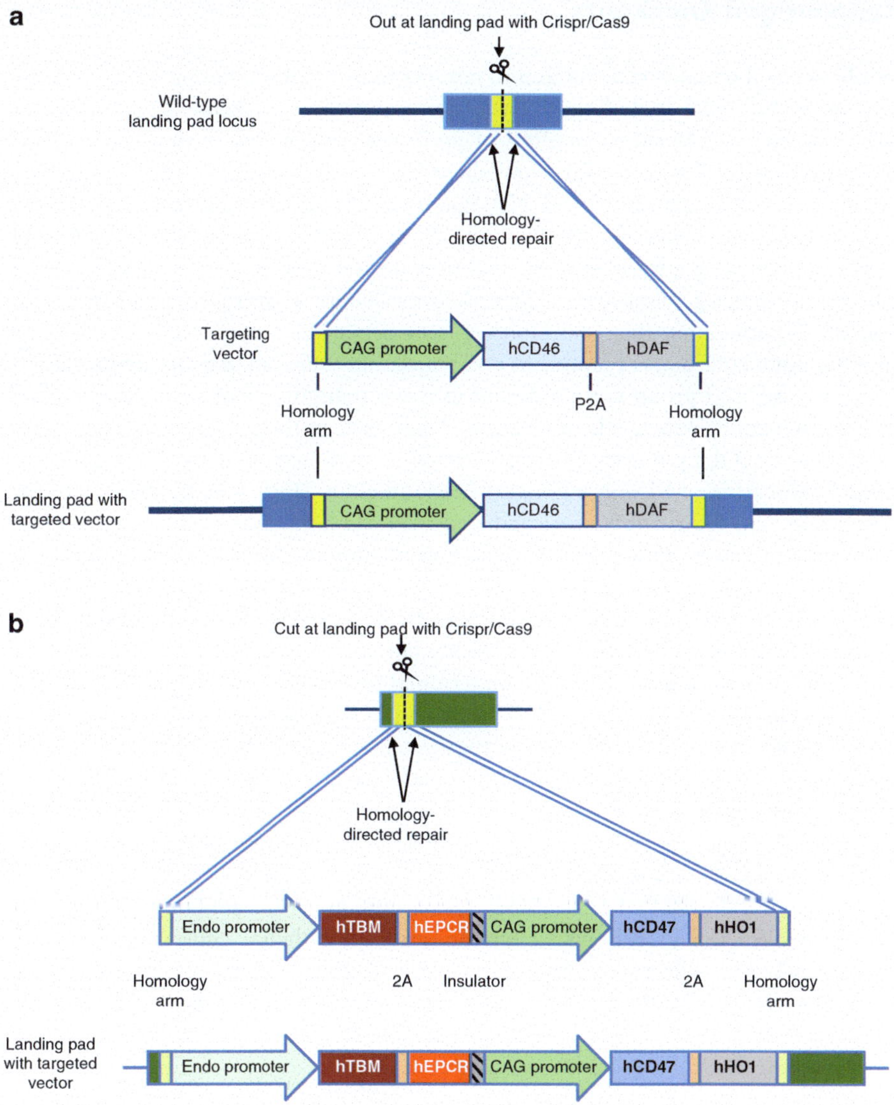

Fig. 7.2 Design and targeting strategy of multicistronic vectors (MCVs). CRISPR/Cas9 is designed to cut within an expression-permissive landing pad. Homology arms direct vector insertion to the landing pad by homology-directed repair. The CAG promoter was used to drive ubiquitous transgene expression (**a** and **b**) while one of several "Endo promoters" was used to obtain endothelial-specific expression (**b**). See text for additional details

Xenoantigen Knockouts

While Gal is the major xenoantigen in porcine tissues, other non-Gal antigens have also been identified, including N-glycolylneuraminic acid (Neu5Gc) [15] and the SDa antigen [16]. Residual binding of preformed antibodies to these antigens is likely responsible for organ rejection, albeit somewhat delayed, observed in GTKO organs after transplantation [2, 6]. Neu5Gc is not expressed in humans, but chronic dietary exposure stimulates production of anti-Neu5Gc antibodies [17]. Unlike humans, baboons express Neu5Gc, and thus do not have preformed anti-Neu5Gc antibodies, making them a less than ideal model for xenotransplantation in this respect. Complicating this issue further is evidence for a neoantigen exposed by Neu5Gc-knockout that is reactive to preformed antibodies in baboon serum [18].

SDa is a blood group antigen found in many mammals, but some humans and NHPs also express low levels of preformed anti-SDa antibody [16]. Binding studies have confirmed the presence of these antibodies in human sera and have also confirmed expression of SDa in pig vascular endothelium (Fig. 7.3). In addition to the preformed anti-SDa antibodies, induced anti-SD antibodies were detected in baboons 3–4 weeks after porcine organ transplantation. Anti-SDa antibodies have also been induced in humans by antitumor vaccines known to contain SDa. It is thus likely that anti-SDa antibodies contribute to both HAR and delayed antibody-mediated rejection in baboons, and possibly humans as well [16].

Like Gal, both of these antigens are terminal residues on sialylated glycans. The synthesis of Neu5Gc is catalyzed by cytidine monophosphate-N-acetylneuraminic acid hydroxylase (CMAH), encoded by the CMAH gene. SDa synthesis is catalyzed by β-1,4-N-acetyl-galactosaminyltransferase 2 encoded by β4GalNT2. We sought to eliminate both antigens in pigs by knockout of the CMAH (CMAHKO) and β4GalNT2 genes (β4KO) using CRISPR/Cas9 in cultured GTKO fibroblasts, followed by SCNT to make pigs.

To assess the effect of each knockout on human serum antibody binding and complement-mediated cell lysis, porcine aortic endothelial cells (pAECs) were obtained from wild-type, GTKO, GTKO+β4KO, GTKO+CMAHKO, and GTKO+β4KO+CMAHKO (TKO) pigs. Human serum IgG and IgM antibody

	% bound immunoglobulin	
Genotype	IgG	IgM
Wild type	96.7	96.2
GTKO	29.0	51.6
GTKO + β4KO	10.6	38.9
GTKO + CMAHKO	7.6	28.2
GTKO + β4KO + CMAHKO	4.6	21.6

Fig. 7.3 Human serum antibody binding to pAECs bearing various xenoantigen knockouts. Cells were incubated with sera from human donors ($n = 3$), probed with anti-IgG or IgM secondary antibody and counted by flow cytometry. Results are expressed as percent immunoglobulin bound relative to wild type

binding was assessed by flow cytometry (Fig. 7.3). GTKO alone reduced IgG binding by 68%, followed by an additional 21% with CMAHKO or 18% with β4KO. Together, all three knockouts reduced total human serum IgG binding by 92%. Thus, knockout of these three xenoantigens eliminated the majority of preformed human serum antibody binding to porcine endothelial cells.

The functional cytoprotective effect of these knockouts was assessed in complement-dependent cytotoxicity (CDC) assays (an in vitro proxy for HAR). Each successive knockout produced a significant reduction in CDC relative to wild-type pig cell. GTKO reduced the rate and level of cytotoxicity by 47%, the addition of CMAHKO by 61%, and with the addition of β4KO, by 74% (Fig. 7.4). While xenoantigen knockouts were clearly effective at reducing cytotoxicity in vitro, they alone did not offer complete protection against cytotoxicity, likely due to the existence of additional xenoantigens. Nevertheless, ex vivo lung perfusion experiments showed a survival benefit to genotypes that included β4KO [19]. Our own in vitro data showing vastly reduced binding of human serum antibodies to cells of

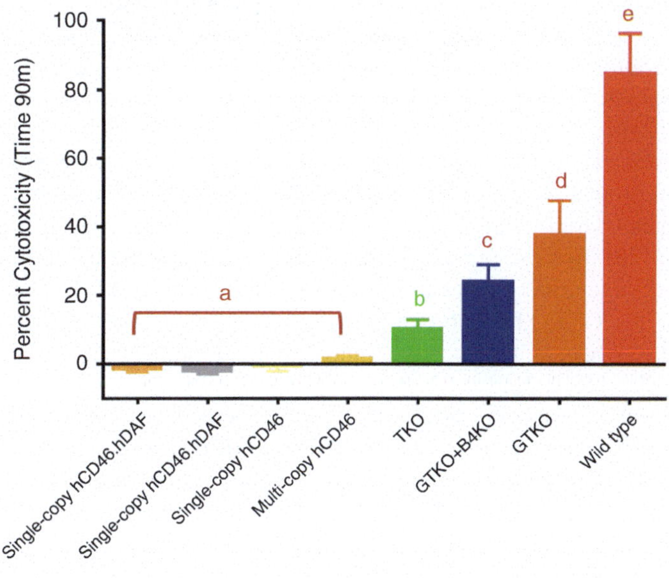

Fig. 7.4 Effect of xenoantigen knockout and expression of complement inhibitors on cytotoxicity as measured by image-based complement-dependent cytotoxicity assay. Assays were carried out using pAECs incubated with 30% pooled human serum ($n = 3$) for 30 minutes followed by exposure to 5% rabbit complement for 120 minutes. Cytotox Red Reagent (IncuCyte) was used to stain dead cells, and total cell counts were determined by high-contrast brightfield imaging using a Cytation cell imager (BioTek). Data from three replicates were expressed as the percent cytotoxicity after 90 minutes of incubation, and compared by ANOVA. Cytotoxicity decreased significantly from wild type with each additional knockout (columns with different superscripts, $P < 0.05$). Cytotoxicity was nearly eliminated altogether when complement inhibitors were expressed as either hCD46 alone from single-copy hCD46 transgene or multicopy hCD46 minigene, or from a single-copy bicistron composed of hCD46 and hDAF. All genotypes (except wild type) include a GTKO background

CMAHKO and β4KO genotypes (Fig. 7.3) argues in favor of including them in pig organ transplant donors. Recently, kidneys transplanted from GTKO+β4KO pigs to rhesus monkeys survived longer (in one case up to 435 days) than previous transplants with GTKO kidneys (2–3 weeks) [20]. In addition, the TKO genotype should eliminate delayed humoral rejection generated by an adaptive response to these antigens, as demonstrated for β4KO in baboons [16].

Complement Inhibitors

Early on, it was observed that HAR could be reduced in ex vivo perfusions of porcine organs with human blood by de-complementation [21] or by administering a complement inhibitor like cobra venom factor to the transplant recipients [22]. These observations suggested that porcine complement-regulatory proteins were unable to inhibit primate complement. This led to the development of transgenic pigs engineered to express a human complement inhibitor, hCD55 (hDAF) [7]. Organs from hDAF-expressing pigs were protected from HAR after transplantation into baboons and survived longer than those from wild-type pigs. Similarly, pigs expressing another complement inhibitor (hCD46; membrane cofactor protein) protected transplanted kidneys from HAR in baboons with high titers of serum anti-Gal antibodies [23].

We introduced a hCD46 expression vector into a line of GTKO pigs to provide a dual level of protection against HAR. The GTKO.hCD46 combination led to longer survival times of transplanted hearts and kidneys compared to GTKO alone [24], reviewed in [25]. In these pigs, hCD46 was expressed from a randomly integrated, multicopy minigene driven by the human CD46 promoter [23]. The protection afforded by the GTKO.hCD46 combination in vivo was reflected in cytotoxicity assays in vitro where hCD46 nearly eliminated cytotoxicity in pAECs from GTKO.hCD46 pigs [26] (Fig. 7.4).

To achieve consistent, predictable, and protective expression levels of complement-regulatory proteins, we built a suite of bicistronic vectors to express both hCD46 and hDAF from the CAG promoter using viral 2A technology (Fig. 7.2a). These vectors were flanked with homology arms to target integration of single copies into specific landing pads using homology-directed repair in porcine fibroblasts [18]. Single-cell clones bearing a single, targeted copy of the vector were then used in SCNT to generate pigs bearing the hCD46.hDAF transgenes on GTKO, GTKO+CMAHKO, and GTKO+CMAHKO+β4KO backgrounds. Expression of both hCD46 and hDAF was confirmed by Western blotting and immunohistochemistry (Fig. 7.5). Expression of hCD46 and hDAF was robust and comparable among individual pigs with vectors inserted at the same landing pads.

Within tissues, single-copy vectors expressed threefold to fourfold less hCD46 vs. the randomly integrated, multicopy hCD46 minigene described above (data not shown). However, in complement-dependent cytotoxicity assays, the cytoprotective effect of the single-copy hCD46.hDAF transgene was equivalent to that of the multicopy CD46 minigene (Fig. 7.4) [26]. Similarly, in transplants, the hCD46.hDAF

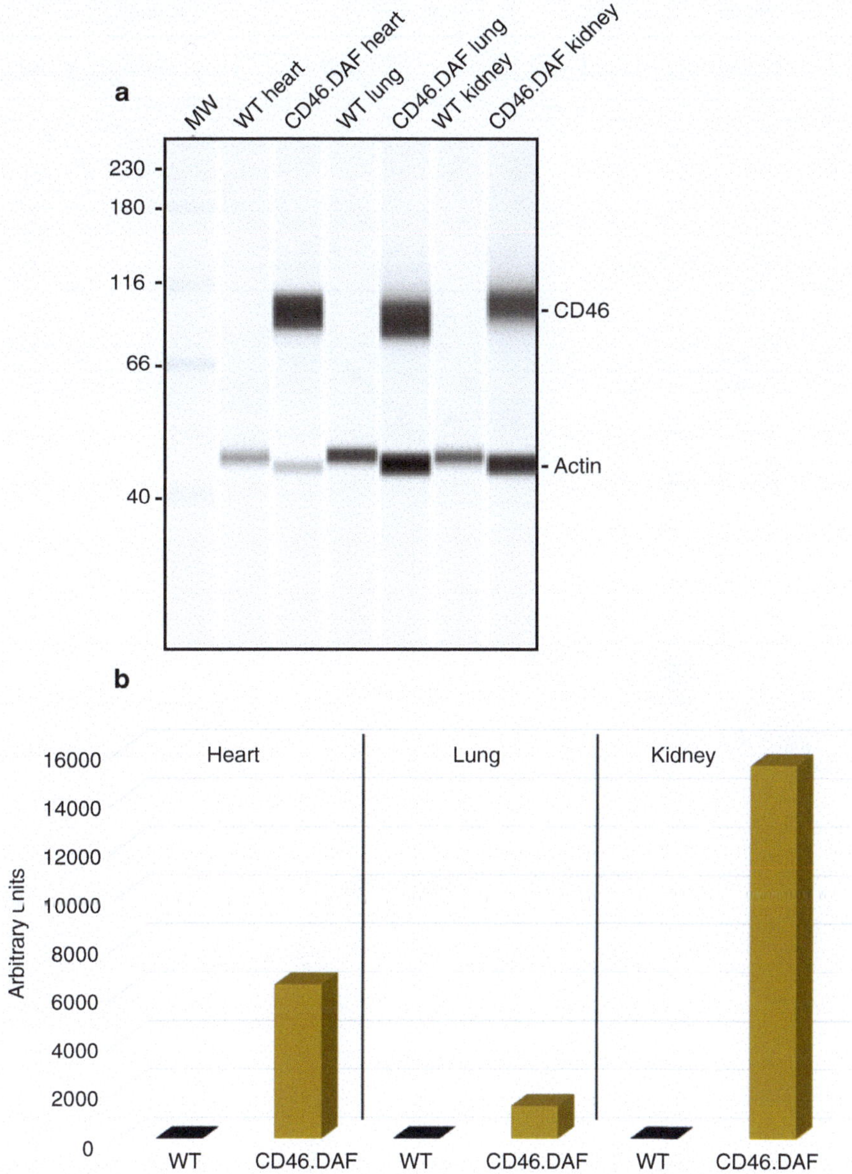

Fig. 7.5 Expression of hCD46 and hDAF from the bicistronic vector (hCD46.hDAF) described in Fig. 7.2 compared to wild-type (WT) controls. (**a**) Western blot of hCD46 expression in porcine heart, lung, and kidney; (**b**) relative quantitation of hCD46 in the blot shown in (**a**); (**c**) immunohistochemistry (IHC) showing robust, ubiquitous expression of hCD46 (insets: WT controls; 200×); (**d**) Western blot of hDAF expression; (**e**) relative quantitation of hDAF in the blot shown in (**d**); (**e**) IHC of hDAF expression in porcine tissues (insets: WT controls; 200×)

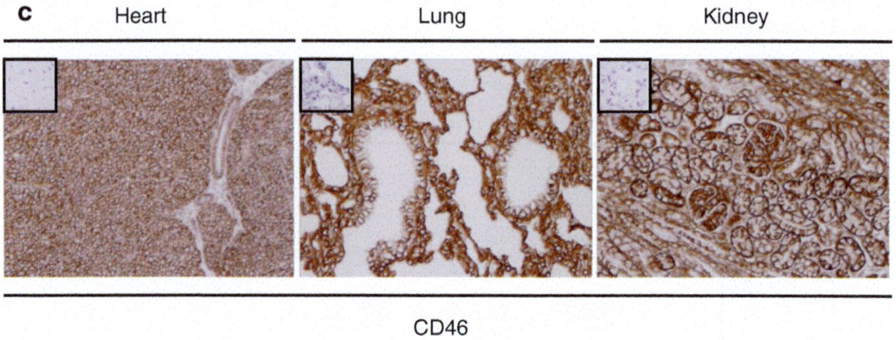

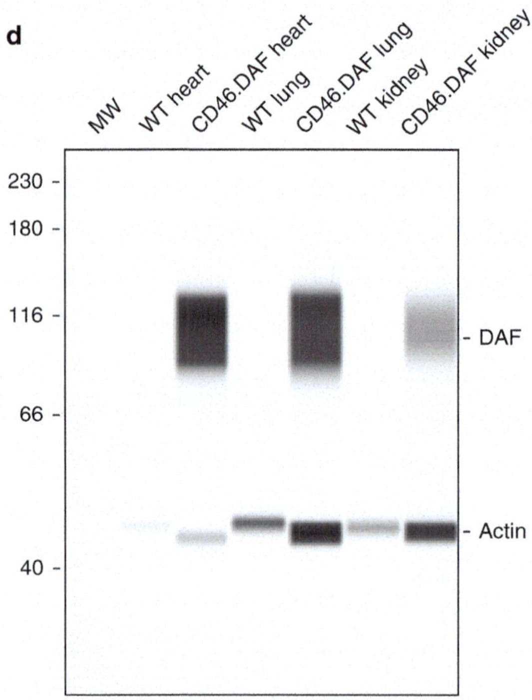

Fig. 7.5 (continued)

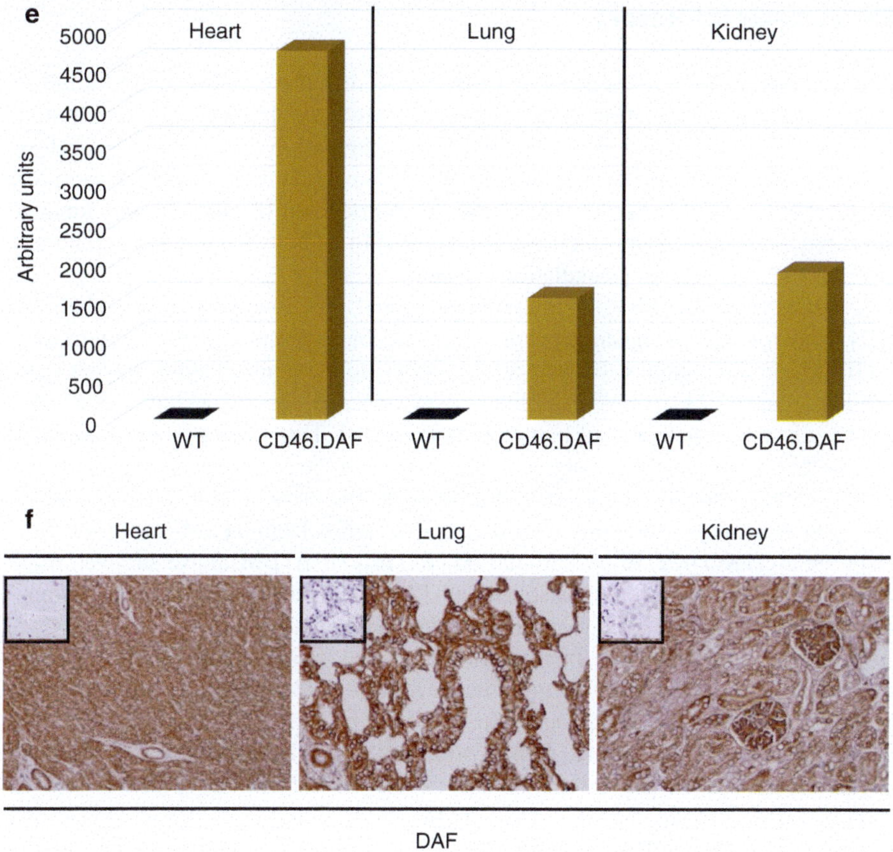

Fig. 7.5 (continued)

bicistron prevented HAR when expressed on GTKO, DKO, and TKO backgrounds (Yamamoto T, et al. unpublished observations). In other experiments, a single copy of CAG-driven hCD46 alone (without hDAF) was expressed as the only complement inhibitor, and afforded the same level of protection against cytotoxicity in vitro as the single-copy hCD46.hDAF and multicopy hCD46 minigene (Fig. 7.4) [26]. The single-copy hCD46 genotype was also protective against HAR in transplanted organs. These results indicate that even a single copy of CAG-driven hCD46 could prevent HAR in vivo and blocked cytotoxicity in vitro at levels equivalent to vectors expressing multiple copies of hCD46 as well as a single copy of a vector expressing both hCD46 and hDAF.

Thrombosis Inhibitors

While knockout of xenoantigens and expression of human complement inhibitors overcome HAR, coagulation dysregulation is frequently observed in both functioning and failing transplants (reviewed in [27]). Coagulation dysregulation manifests as thrombotic microangiopathy within the transplanted organ and consumptive coagulopathy in the recipient. These can be independent of humoral and adaptive immunity and result from functional incompatibilities between porcine and human proteins that regulate the coagulation cascade.

Thrombomodulin (TBM) is a membrane-bound protein expressed in endothelial cells. Under normal hemostatic conditions, TBM binds circulating thrombin to form a TBM:thrombin complex that activates protein C to maintain an anticoagulant state. While porcine TBM (pTBM) can bind human thrombin, the pTBM:human thrombin complex is a poor activator of human protein C [28]. This generates a procoagulant state within xenotransplanted organs leading to thrombotic microangiopathy and consumptive coagulopathy. To overcome this interspecific incompatibility, we and others have generated pigs expressing hTBM [29, 30]. Hearts from hTBM transgenic pigs, made on a GTKO.hCD46 background (GTKO.hCD46.hTBM), have survived more than 2 years after heterotopic transplantation [29] and 6 months after orthotopic transplantation [31] without indications of coagulation dysregulation.

In these pigs, hTBM was expressed from a multicopy, randomly integrated vector composed of the hTBM coding sequence and a ~ 7.6 kb fragment of genomic pig sequence cloned from the region upstream of the pTBM gene. This fragment contained the endothelial-specific pTBM promoter and other regulatory elements to ensure appropriate, physiological expression of hTBM in porcine tissues. Immunohistochemical evaluation of organs from transgenic pigs revealed that hTBM was endothelial-specific and closely resembled the pattern of endogenous TBM expression in human tissue (Fig. 7.6).

The bioactivity of hTBM expressed in porcine tissues was evaluated by testing its ability to complex with human thrombin and activate human protein C in vitro. In this assay, activated protein C cleaves a colorimetric substrate that is quantified by absorbance. As shown in Fig. 7.7, pAECs from the multicopy hTBM expressed relatively high levels of hTBM that was capable of activating protein C.

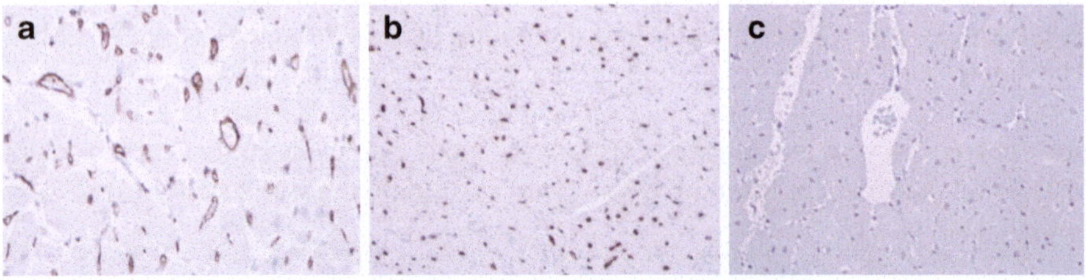

Fig. 7.6 Immunohistochemical detection of hTBM expression with human-specific TBM antibody. (**a**) Human heart expressing endogenous hTBM. (**b**) GTKO.hCD46.hTBM porcine heart expressing a randomly integrated, multicopy hTBM vector with a 7.6 kb porcine TBM promoter. (**c**) Wild-type porcine heart. hTBM expressed in an endothelial-specific manner in both human and transgenic porcine heart tissue. (Magnification, 200×)

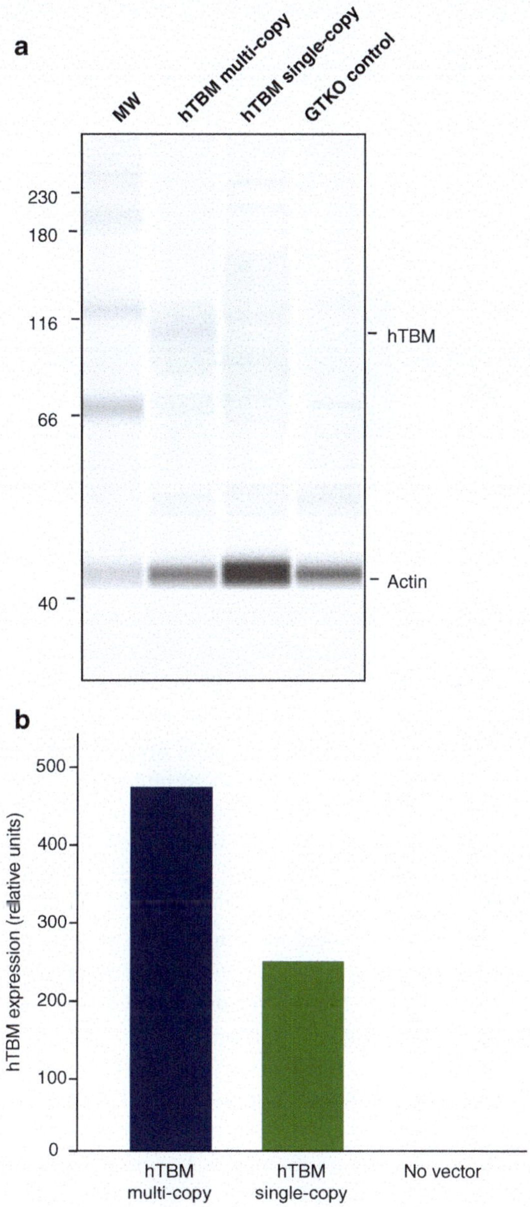

Fig. 7.7 Generation of activated protein C (APC) by cultured pAECs. (**a**) Western blot showing hTBM expression in pAECs with a multicopy hTBM minigene vector, single-copy MCV targeted to a landing pad, and control with no hTBM vector. (**b**) Relative quantitation of hTBM detected on Western blot in (**a**). (**c**) Activated protein C (APC) generated by the pAECs in (**a**) and (**b**). See text for assay details. APC levels tended to be higher in pAECs with the MCV, even though hTBM expression level was lower vs. the multicopy hTBM vector, likely due to augmentation of APC by hEPCR which is also expressed by the MCV

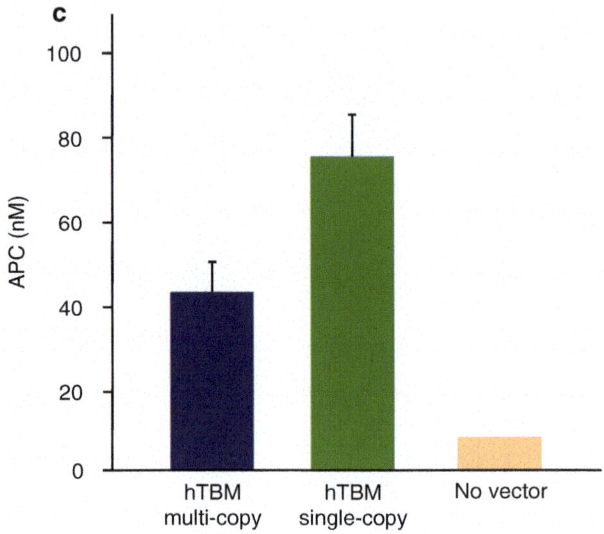

Fig. 7.7 (continued)

In keeping with our aim of generating pigs with single-copy MCVs targeted to specific landing pads, we built another suite of vectors (Fig. 7.2b) designed to express both hTBM and the human endothelial protein C receptor (hEPCR) [18, 32]. EPCR is a membrane protein that augments the activation of protein C by the TBM:thrombin complex to maintain an anticoagulant state [33]. This vector is composed of two bicistrons, the first of which includes both hTBM and hEPCR separated by a 2A sequence. Expression of both is driven by an endothelial-specific promoter. The second bicistron is separated by an insulator from the first and is composed of CAG-driven hCD47-2A-hHO1, the activities of which will be described below. This MCV is flanked with homology arms to guide its insertion as a single copy into a landing pad via homology-directed repair in a GTKO.CMAHKO cell line harboring the hCD46.hDAF vector described above and in Fig. 7.2a.

When transplanted into baboons, porcine kidneys of this genotype (GTKO. CMAHKO + hCD46.hDAF + hTBM.hEPCR.hCD47.human hemeoxygenase-1 [hHO1]) avoided HAR and thrombotic microangiopathy, and no consumptive coagulopathy was observed in the recipients. As of this writing, kidneys in one baboon have functioned in a life-supporting manner for over 6 months, demonstrating that this particular genotype is consistent with long-term xenograft survival and function. Western blot and immunohistochemistry on tissues from organ-source pigs and from clonal littermates confirmed that all six transgenes were expressed in hearts, lungs, and kidneys (Figs. 7.5 and 7.8) [32]. In addition, immunohistochemistry revealed that expression of hTBM and hEPCR was endothelial-specific, while the expression of both CAG-driven transgenes was ubiquitous (Figs. 7.5c, f and 7.8).

Overall, these analyses confirmed the expression levels and patterns expected for the vectors used to make these pigs. Moreover, the patterns and levels of expression observed should be indicative of those in the long-term, life-supporting transplanted kidneys, and thus can serve as an in vitro reference for expression consistent with successful performance in vivo.

Bioassays were conducted to confirm that functional proteins were expressed from each of the six transgenes in the GTKO.CMAHKO + hTBM.hEPCR.hCD47.hHO1 genotype. As measured by cytotoxicity assay, complement activity was essentially inhibited altogether by hCD46 and hDAF (Fig. 7.4). The function of hTBM and hEPCR was evaluated in the activated protein C assay described above. In this case, hTBM, presumably with participation from hEPCR, increased the level of activated protein C almost tenfold above that of pAECs from controls (Fig. 7.7c). Interestingly, cells with the single-copy MCV generated more activated protein C than those with the multicopy hTBM vector (Fig. 7.7c) despite lower levels of hTBM expression from the MCV (Fig. 7.7b). Higher activated protein C production in the cells with the single-copy MCV was likely greater than that from the multicopy hTBM vector due to the presence of hEPCR in the MCV. These observations are consistent with the absence of coagulation dysregulation in the transplanted kidneys and baboon recipients.

Taken together, these observations indicate that transgenes expressing human complement and coagulation inhibitors, packaged within MCVs, and inserted as single copies at expression-permissive landing pads, can be used to generate functional organ donors for successful xenotransplantation. These pigs should not only serve as reliable sources of organs, but also as good breeding stock, since all transgenes (six in this case) are confined to only two loci to facilitate efficient propagation of the genotype through conventional breeding.

Inflammation and Apoptosis Inhibitors

Inflammation is an inevitable consequence of ischemia–reperfusion injury in both allo- and xenotransplantation and, in both cases, it likely exacerbates HAR and coagulation dysregulation [12]. In addition, the presence of a transplanted organ can lead to sustained systemic inflammation in the recipient. Inflammation can thus endanger the transplanted organ and recipient. While inflammation can be controlled by systemic administration of anti-inflammatory agents, the transgenic expression of anti-inflammatory proteins in the transplanted organ has also been considered.

A number of genes are upregulated in organs after transplantation, including heme oxygenase-1 (HO1). HO1 is a ubiquitously expressed, stress-induced gene upregulated by the presence of heme that results from hemolysis associated with ischemia–reperfusion injury. While the primary function of HO1 is to catabolize heme, it has also been shown to limit the inflammatory response and prevent apoptosis and cytotoxicity after ischemia–reperfusion injury [34]. Human HO1 (hHO1) has been constitutively expressed in transgenic pigs [35]. When perfused with xenogeneic human blood for up to 6 hours, kidneys from hHO1-expressing pigs survived longer, were protected from increases in vascular resistance, expressed fewer molecular markers of vascular damage, and did not develop thrombotic microangiopathy, compared to kidneys from wild-type control pigs.

Results such as these compelled us to express hHO1 in our organ-source pigs. Accordingly, hHO1 was added to the MCV described above (Fig. 7.2b), driven by the constitutive CAG promoter to ensure ubiquitous expression to emulate its natural expression pattern. hHO1 expression was verified by Western blotting (data not shown) and immunohistochemistry (Fig. 7.8). Upon transplantation into baboons, kidneys expressing hHO1 did not undergo HAR, coagulation dysregulation, or an overt inflammatory response, and as of this writing have provided life-supporting function for more than 6 months. However, the individual contribution of hHO1 could not be assessed in vivo since it is not possible to isolate its effects from those of other transgenes in the MCV. However, we did compare the antiapoptotic

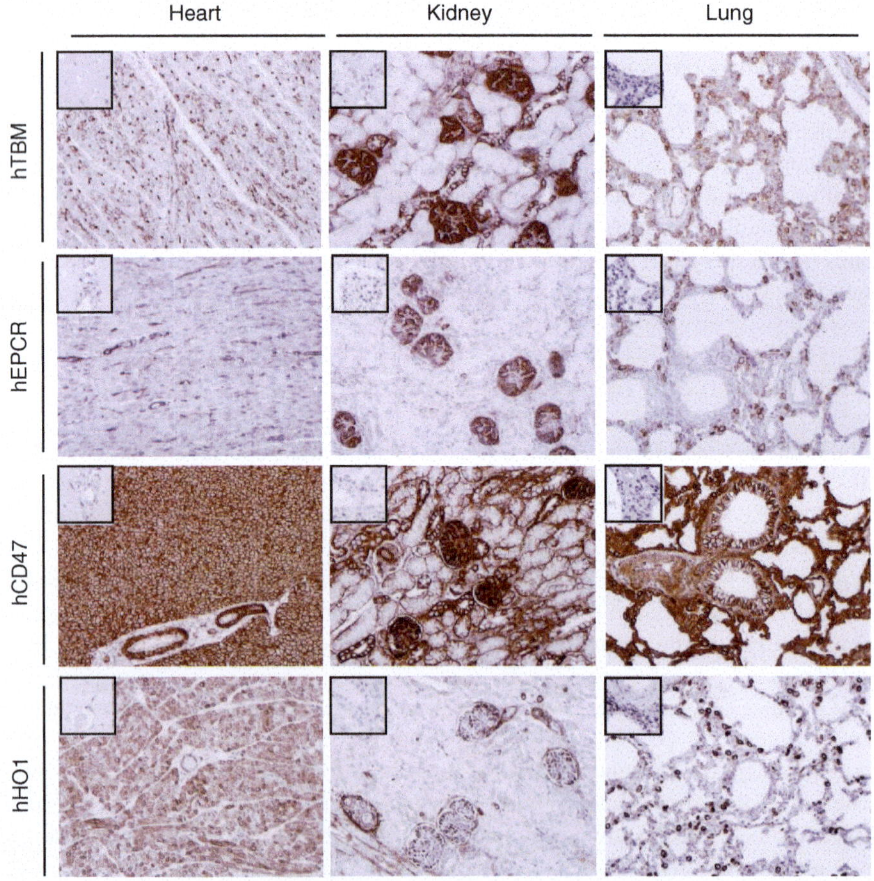

Fig. 7.8 Immunohistochemical detection of transgenes expressed in pig tissues from a single-copy MCV with targeted insertion at an expression-permissive landing pad. Human-specific antibodies were used in each case. Insets: wild-type controls. (200× magnification)

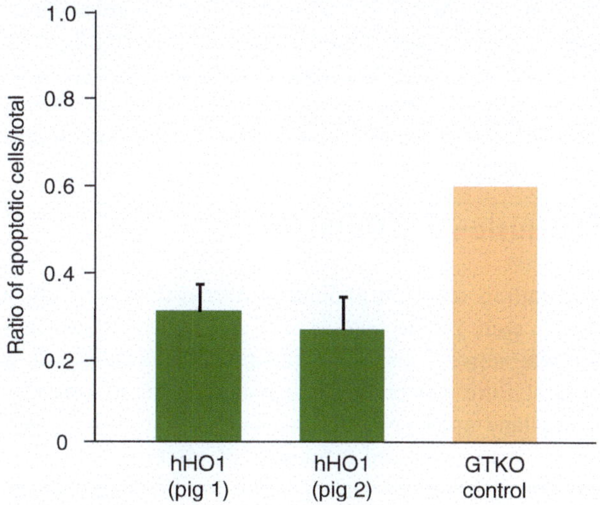

Fig. 7.9 Reduction of apoptosis in cultured pAECs expressing a hHO1 vector (pig 1 and pig 2) vs. control with no hHO1 vector. Apoptosis was induced by incubating cells in 1 μM staurosporine for 10 h. (See text for assay details.) Apoptosis was reduced in hHO1-expressing pAECs ($P < 0.05$)

function pAECs expressing hHO1 vs. controls using a real-time Caspase 3 assay [36]. Briefly, pAECs were treated with staurosporine to induce apoptosis. Cells expressing hHO1 displayed a significant reduction in apoptotic cells, indicating a potential benefit to hHO1 in vitro (Fig. 7.9).

Macrophage Inhibition

As part of the innate immune system, macrophages play an important role in the identification and elimination of foreign cells in the body. Autologous cells are protected from endogenous macrophages by CD47, a membrane protein expressed in all cells that binds and activates signal-regulatory protein alpha (SIRP-α) on macrophages to block phagocytosis [37]. Xenogeneic hematopoietic stem cells and pancreatic islets are particularly susceptible to macrophage phagocytosis due to the inability of porcine CD47 to bind primate SIRP-α [38]. This interspecies incompatibility can be overcome by transgenic expression of human CD47 (hCD47) in porcine cells to protect them from macrophage attack [38, 39]. However, it is less clear whether macrophage phagocytosis plays a role in rejection of solid organs. Nevertheless, transgenic expression of hCD47 in porcine organs should confer some degree of protection against host macrophages.

For this reason, we have included hCD47 in the four-gene MCV described above (Fig. 7.2b). In this vector, hCD47 expression is driven by the CAG promoter as part of a 2A bicistron that includes hHO1 in the second position. While we have not specifically evaluated the function of hCD47 in recipient baboons after pig organ transplantation, we have tested its ability to reduce macrophage phagocytosis in an in vitro assay [18]. Briefly, pAECs were isolated from control and hCD47-expressing pigs and transfected with a constitutive green fluorescent protein marker, and then co-cultured with human macrophages tagged with red and blue fluorescent antibodies to Major Histocompatibility Complex class II and CD14, respectively. Cells were harvested after 4 hours, and cells displaying all three fluorescent markers were counted as having undergone phagocytosis. Assay results showed a modest, but significant, reduction of phagocytosis in the hCD47-expressing cells to justify the inclusion of the hCD47 transgene in porcine organ sources.

Conclusion and Outlook

The path to xenotransplantation started with the disheartening observation that pig organs were rapidly and overtly rejected upon transplantation. However, the need for alternative sources of transplantable organs for patients suffering end-stage organ failure was (and still remains) great, so a prodigious, decades-long effort to understand and overcome the factors underlying organ rejection followed. These efforts have produced a number of effective strategies to meet the challenges of xenotransplantation. Among these are the genetic modifications to the donor organ described here, which deal primarily with the innate immune system. Others, including improved immunosuppressive protocols [40] and progress in the induction of immune tolerance to xeno organs [41], are focused on adaptive immunity.

Promising advances in all these areas are bringing clinical xenotransplantation closer to reality. In 2019, the FDA approved initiation of the first clinical xenotransplantation trial, in this case porcine skin grafts from GTKO pigs [42]. Preliminary results showed good graft acceptance at the end of the 30-day trial period, with no evidence for transmission of zoonotic disease. Ongoing, long-term survivals of solid-organ xenotransplants of genetically modified kidneys and hearts in NHPs suggest that clinical trials for these organs could begin in the near future.

Acknowledgments A portion of this work was supported by NIH NIAID U19 grant AI090959.

Conflict of Interest All authors are employees of Revivicor, Inc., Blacksburg, VA.

References

1. Rose AG, Cooper DKC, Human PA, Reichenspurner H, Reichart B. Histopathology of hyperacute rejection of the heart: experimental and clinical observations in allografts and xenografts. J Heart Lung Transplant. 1991;10:223–34.
2. Cooper DKC, Good AH, Koren E, et al. Identification of alpha-galactosyl and other carbohydrate epitopes that are bound by human anti-pig antibodies: relevance to discordant xenografting in man. Transpl Immunol. 1993;1:198–205.
3. Galili U, Shohet SB, Kobrin E, Stults CL, Macher BA. Man, apes, and Old World monkeys differ from other mammals in the expression of alpha-galactosyl epitopes on nucleated cells. J Biol Chem. 1988;263:17755–62.
4. Galili U, Mandrell RE, Hamadeh RM, Shohet SB, Griffiss JM. Interaction between human natural anti-alpha-galactosyl immunoglobulin G and bacteria of the human flora. Infect Immun. 1988;56:1730–7.

5. Galili U, Rachmilewitz EA, Peleg A, Flechner I. A unique natural human IgG antibody with anti-alpha-galactosyl specificity. J Exp Med. 1984;160:1519–31.
6. Tseng YL, Kuwaki KL, Dor FJ, et al. alpha1,3-galactosyltransferase gene-knockout pig heart transplantation in baboons with survival approaching 6 months. Transplantation. 2005;80:1493–500.
7. Schuurman HJ, Pino-Chavez G, Phillips MJ, Thomas L, White DJ, Cozzi E. Incidence of hyperacute rejection in pig-to-primate transplantation using organs from hDAF-transgenic donors. Transplantation. 2002;73:1146–51.
8. Wilmut I, Schnieke AE, McWhir J, Kind AJ, Campbell KH. Viable offspring derived from fetal and adult mammalian cells. Nature. 1997;385:810–3.
9. Polejaeva IA, Chen SH, Vaught TD, et al. Cloned pigs produced by nuclear transfer from adult somatic cells. Nature. 2000;407:86–90.
10. Dai Y, Vaught TD, Boone J, et al. Targeted disruption of the alpha1,3-galactosyltransferase gene in cloned pigs. Nat Biotechnol. 2002;20:251–5.
11. Shimizu A, Hisashi Y, Kuwaki K, et al. Thrombotic microangiopathy associated with humoral rejection of cardiac xenografts from alpha1,3-galactosyltransferase gene-knockout pigs in baboons. Am J Pathol. 2008;172:1471–81.
12. Li J, Hara H, Esmon C, Cooper DKC, Iwase H. Evidence for the important role of inflammation in xenotransplantation. J Inflamm (Lond). 2019;16:10.
13. Ran FA, Hsu PD, Wright J, Agarwala V, Scott DA, Zhang F. Genome engineering using the CRISPR-Cas9 system. Nat Protoc. 2013;8:2281–308.
14. Szymczak-Workman AL, Vignali KM, Vignali DA. Design and construction of 2A peptide-linked multicistronic vectors. Cold Spring Harb Protoc. 2012;2012(2):199–204.
15. Zhu A, Hurst R. Anti-N-glycolylneuraminic acid antibodies identified in healthy human serum. Xenotransplantation. 2002;9:376–81.
16. Byrne G, Ahmad-Villiers S, Du Z, McGregor C. B4GALNT2 and xenotransplantation: a newly appreciated xenogeneic antigen. Xenotransplantation. 2018;25:e12394.
17. Tangvoranuntakul P, Gagneux P, Diaz S, et al. Human uptake and incorporation of an immunogenic nonhuman dietary sialic acid. Proc Natl Acad Sci U S A. 2003;100:12045–50.
18. Cooper DKC, Hara H, Iwase H, et al. Justification of specific genetic modifications in pigs for clinical organ xenotransplantation. Xenotransplantation. 2019;26:e12516.
19. Burdorf L, Abady Z, Cerel B, et al. Beta4Gal is an antigen relevant to injury of pig lung xenografts in baboons. Xenotransplantation. 2019;26:Abstr. 231.3.
20. Adams AB, Kim SC, Martens GR, et al. Xenoantigen deletion and chemical immunosuppression can prolong renal xenograft survival. Ann Surg. 2018;268:564–73.
21. Moberg AW, Shons AR, Gewurz H, Mozes M, Najarian JS. Prolongation of renal xenografts by the simultaneous sequestration of preformed antibody, inhibition of complement, coagulation and antibody synthesis. Transplant Proc. 1971;3:538–41.
22. Kobayashi T, Taniguchi S, Neethling FA, et al. Delayed xenograft rejection of pig-to-baboon cardiac transplants after cobra venom factor therapy. Transplantation. 1997;64:1255–61.
23. Loveland BE, Milland J, Kyriakou P, et al. Characterization of a CD46 transgenic pig and protection of transgenic kidneys against hyperacute rejection in non-immunosuppressed baboons. Xenotransplantation. 2004;11:171–83.
24. Azimzadeh AM, Kelishadi SS, Ezzelarab MB, et al. Early graft failure of GalTKO pig organs in baboons is reduced by expression of a human complement pathway-regulatory protein. Xenotransplantation. 2015;22:310–6.
25. Cooper DKC, Ezzelarab MB, Hara H, et al. The pathobiology of pig-to-primate xenotransplantation: a historical review. Xenotransplantation. 2016;23:83–105.
26. Dandro A, Kuravi K, Butler S, et al. Real time in vitro CDC assays comparing the effect of various complement inhibitor transgene designs on complement-dependent cytotoxicity. Xenotransplantation. 2019;26:Abstr.165.
27. Iwase H, Jagdale A, Yamamoto T, et al. Indicators of impending pig kidney and heart xenograft failure: relevance to clinical organ xenotransplantation. Int J Surg. 2019;70:84–91.

28. Roussel JC, Moran CJ, Salvaris EJ, Nandurkar HH, d'Apice AJ, Cowan PJ. Pig thrombomodulin binds human thrombin but is a poor cofactor for activation of human protein C and TAFI. Am J Transplant. 2008;8:1101–12.
29. Mohiuddin MM, Singh AK, Corcoran PC, et al. Chimeric 2C10R4 anti-CD40 antibody therapy is critical for long-term survival GTKO.hCD46.hTBM pig-to-primate cardiac xenograft. Nat Commun. 2016;7:11138.
30. Wuensch A, Baehr A, Bongoni AK, et al. Regulatory sequences of the porcine THBD gene facilitate endothelial-specific expression of bioactive human thrombomodulin in single- and multitransgenic pigs. Transplantation. 2014;97:138–47.
31. Längin M, Mayr T, Reichart B, et al. Consistent success in life-supporting porcine cardiac xenotransplantation. Nature. 2018;56:430–3.
32. Kuravi K, Sorrells L, Butler S, et al. Comparison of transgene expression from different promoters in multi-gene transgenic pigs. Xenotransplantation. 2019;26:Abstr. 171.
33. Mohan RLV, Esmon CT, Pendurthi UR. Endothelial cell protein C receptor: a multiliganded and multifunctional receptor. Blood. 2014;124:1553–62.
34. Camara NO, Soares MP. Heme oxygenase-1 (HO-1), a protective gene that prevents chronic graft dysfunction. Free Radic Biol Med. 2005;38:426–35.
35. Petersen B, Ramackers W. Lucas-Hahn, et al. transgenic expression of human hemeoxygenase-1 in pigs confers resistance against xenograft rejection during ex vivo perfusion of porcine kidneys. Xenotransplantation. 2011;18:355–68.
36. Dandro A, Phelps C, Vaught T, et al. Anti-apoptotic effects of HO-1 and A20 expression in endothelial cells of multi-transgenic pigs for xenotransplantation. Xenotransplantation. 2017;24:Abstr. 7.
37. Wang H, VerHalen J, Madariaga ML, et al. Attenuation of phagocytosis of xenogeneic cells by manipulating CD47. Blood. 2007;109:836–42.
38. Ide K, Wang H, Tahara H, et al. Role for CD47-SIRPα signaling in xenograft rejection by macrophages. Proc Natl Acad Sci U S A. 2007;104:5062–6.
39. Navarro-Alvarez N, Yang Y-G. CD47: a new player in phagocytosis and xenograft rejection. Cell Mol Immunol. 2011;8:285–8.
40. Yamamoto T, Hara H, Foote J, et al. Life-supporting kidney xenotransplantation from genetically engineered pigs in baboons: a comparison of two immunosuppressive regimens. Transplantation. 2019;103:2090–104.
41. Tena AA, Sachs DH, Mallard C, et al. Prolonged survival of pig skin on baboons after administration of pig cells expressing human CD47. Transplantation. 2017;101:316–21.
42. Holzer P, Moulton K, Monroy R, et al. Preclinical characterization of safety, immunogenicity, and lack of zoonotic infectivity of vital, porcine xenotransplants in cynomolgus macaque recipients. Xenotransplantation. 2019;26:Abstr. 258.

8 Addressing Regulatory Requirements for the Organ-Source Pig – A Pragmatic Approach to Facility Design and Pathogen Prevention

Karl Kraebber and Edward Gray

Abbreviations

CBER	Center for Biologics Evaluation and Research
CFR	Code of Federal Regulations
CVM	Center for Veterinary Medicine
FDA	Food and Drug Administration
HEPA	High-efficiency particulate air
INAD	Investigational New Animal Drug
IND	Investigational New Drug
IXA	International Xenotransplantation Association
NADA	New Animal Drug Application
OCTGT	Office of Cellular, Tissue and Gene Therapies
OR	Operating Room
PHS	Public Health Service
SOP	Standard Operating Procedure

Regulatory Overview

Any xenotransplantation trial in the United States and the facility that houses the pigs and supplies the organs are regulated by two centers within the US Food and Drug Administration (FDA), namely the Center for Veterinary Medicine (CVM)

K. Kraebber (✉)
Xenotransplantation Program, Department of Surgery, University of Alabama at Birmingham, Birmingham, AL, USA
e-mail: kkraebber@uabmc.edu

E. Gray
Independent Consultant, Satellite Beach, FL, USA

D. K. C. Cooper, G. Byrne (eds.), *Clinical Xenotransplantation*,
https://doi.org/10.1007/978-3-030-49127-7_8

and the Center for Biologics Evaluation and Research (CBER). Standard operating procedures (SOPs), facility design specifications, and the pathogen detection and response system must be reviewed and approved by the FDA prior to final approval to begin the clinical trial, along with reinspections or audits during the trial.

The primary FDA regulatory guidelines for clinical trials of xenotransplantation that guide the design, construction, and operation of the facilities are described below.

Source Animal, Product, Preclinical, and Clinical Issues Concerning the Use of Xenotransplantation Products in Humans (CBER16, also known as the Guideline) [1]
This document contains guidance for the content of an Investigational New Drug (IND) application for a xenotransplantation trial. It addresses the characterization of source animals, source-animal husbandry, characterization of xenotransplantation products, xenotransplantation product manufacturing facility, appropriate preclinical models for xenotransplantation protocols, and monitoring of recipients of xenotransplantation products.

Guidance for Industry: Regulation of Genetically Engineered Animals Containing Heritable Recombinant DNA Constructs (CVM09/CVM17) [2]
This document describes the Investigational New Animal Drug (INAD) process for genetically engineered animals. The INAD must be filed with or before an IND for a clinical trial. The requirements for an INAD concern mainly shipping, labeling, and disposition of animals. All of these can be included in a default Material Transfer Agreement (MTA) for the pigs. Also described is a New Animal Drug Application (NADA), which will be needed in the future after the pilot trial and before application for final approval from the FDA.

US Public Health Service (PHS) Guideline on Infectious Disease Issues in Xenotransplantation (PHS01) [3]
Although newer methods are suggested and allowed in CBER16, the PHS01 Guideline contains recognized methods for pathogen detection. In general, FDA regulations covering the overall process of drug development, testing, and production are contained in the Code of Federal Regulations (CFR) Title 21 Food and Drugs (21CFR) [4], although information from other sections of the CFR are referenced throughout, including Title 9 Animals and Animal Products (9CFR) [5] and Title 42 Public Health (42CFR) [6].

The FDA has assigned CBER as the primary center with regulatory responsibility for the xenotransplantation product. CBER and CVM will likely work together regarding the regulation of the source-animal herd, animals for xenotransplantation, and xenotransplant product (organ). As a novel therapeutic, the IND application will be subjected to review by a review committee assembled by The Office of Tissues and Advanced Therapies (OTAT, formerly known as the Office of Cellular, Tissue and Gene Therapies, or OCTGT) within CBER. This team will include members of the xenotransplantation research community as indicated by the FDA in 2010 [7] (Tables 8.1 and 8.2).

Table 8.1 FDA CBER review organizational structure

OCTGT	CBER	FDA	Potential external consultant
Project manager Chemistry, manufacturing, and controls (CMC) Pharmacology/toxicology Clinical Virology	Compliance Product quality Clinical monitoring Veterinary Statistics Epidemiology	Additional expert Product specialist Clinical specialist Methodology expert Unique policy expert Expert on genetically engineered animals	Scientific expert (SGE) Patient Advocate
Basic review team	*Extended review team*	*Potential consults or collaborators*	*Potential consults*

Table 8.2 FDA xenotransplantation product oversight scope

Source herd	Source animal	Processing	Physician/patient
Animal origin Establishment and management of closed herd Animal housing Adventitious agent testing Animal health and husbandry Records sample retention	Quarantine Transport (if applicable) Organ harvest Adventitious agent testing Records/sample retention Disposal and use of by-products	Process validation Adventitious agent testing Product characterization Records/sample retention	Patient selection, consent, and education Protocol review Clinical site Follow-up and screening Records/sample retention

In September 2017, at the 14th Congress of the International Xenotransplantation Association (IXA), a joint symposium between the IXA and FDA was conducted. Scientists, clinicians, and regulators shared perspectives on recent advances, infection challenges, and regulatory considerations. During his presentation entitled "FDA Expectations for Source Animal Herds and Characterizations," the Director, Division of Veterinary Services, John Dennis, DVM, MS, DACLAM, shared CBERs clarification of source-animal expectations, as follows:

(i) Conventional – least clean
(ii) Specific pathogen-free
(iii) Designated pathogen-free
(iv) *Xenograft production (xenotransplantation product)*
(v) Gnotobiotic
(vi) Germ-free – most clean

Locally, the University of Alabama at Birmingham (UAB) off-campus xenotransplantation pig facility is part of the Animal Resource Program accredited by the Association for Assessment and Accreditation of Laboratory Animal Care (AAALAC) [8] and governed by the University's Animal Use and Care Committee (IACUC). Animals are accommodated according to the Guide for the Care and Use of Agricultural Animals in Research and Teaching [9]. Animal production processes and supporting facilities have been designed to meet the modest volume needed for a proposed pilot clinical trial with clinical-quality porcine kidneys that are free of disease. With success in clinical trials, it is expected that the production process used now will be translated and adapted in many ways to meet the larger number of pigs required for future clinical xenotransplantation.

Facility Design

According to Schuurman [10], barrier facility biosecurity can be divided into two categories: (i) prevention of contamination from the outside, and (ii) maintenance of "clean" animals within the barrier (Table 8.3). As shown during his presentation called "Regulatory Aspects" at the WHO Global Consultation on Regulatory

Table 8.3 A biosecure barrier facility

A. Preventing contamination from outside the barrier
Physical
Metal or stick-built outer building surrounding a concrete bunker
Entry of materials/equipment
Air filtration (high efficiency particulate air or HEPA) filter
Reverse osmosis (RO) and/or ultra-violet (UV)-treated water
Disinfection of materials (autoclave or vaporized hydrogen peroxide)
Exit of waste
Fluid waste – prevent reverse flow of any component to the inside
Prevention of local ground water contamination
People
Intensive health screening prior to gaining barrier access
Shower-in/shower-out; wearing of personal protective equipment (hats, masks, gloves, Tyvek suits over scrubs)
Occupational Health Surveillance Program for the staff, involving a vaccination program, disease monitoring, and serum banking
B. Keeping animals "clean"
Animals
Population by Caesarean section and colostrum deprivation
Closed herd production (potential for mixed genetics is possible)
Husbandry following the "Guide for lab animals" (or Guide for agriculture animals?)
Control by Institutional Animal Care and Use Committee
Accreditation by AAALAC (Association for Assessment and Accreditation of Laboratory Animal Care)
Feed
Certified free of mammalian proteins, no herbicides or pesticides, irradiated
Such feed for at least two generations
Infectious control
Disease monitoring and sentinel animal testing
Barrier rooms swabbed regularly and tested for target microorganisms
Alternatives?
Is such a barrier required if "DPF-status" can be guaranteed in another way for a specified porcine xenotransplantation product, e.g., *pancreatic islets?*

Requirements for Xenotransplantation Clinical Trials in Changsha, China, in 2019 [10], Schuurman offered strategies for prevention of outside contamination and maintenance of "clean" animals. To a large extent, we have followed these strategies in planning the clean pig facility at the UAB.

The UAB off-campus xenotransplantation research facility was previously a large animal research facility and has been upgraded and renovated to serve as the primary facility to produce clinical-quality pig kidneys for a planned pilot clinical trial. Between April 2016 and June 2017, upgrades were designed and completed in phases. Occupancy was granted in May 2016 for laboratory activities and in July 2017 for animal housing. A final list of minor tasks to be completed or corrected at the end of a project (punch list) and facility-wide decontamination were conducted by October 2017. Recipient gilts were populated into the facility in November of 2017, and the first genetically modified, cloned swine were delivered in the facility in March of 2018.

The facility is remote from large population centers and agricultural activities, and access is physically restricted by layers of access control (including a final perimeter fence). Door entry key code access is granted only to the individuals directly responsible for producing the organ-source pigs. A separate office trailer is available for visits by others, as required. The main section of the research facility is a concrete block structure serviced by central utilities (central heating, ventilation, and air conditioning [HVAC]), reverse osmosis water purification system [with local lab water deionizer], a local septic tank for sewage, and electrical service with a diesel back-up generator). The site has been prepared for future extension if needed, which could expand the existing ~6000 sq. ft. area to approximately 20,000 sq. ft.

The pig research facility is currently separated into two primary areas: (i) the cloning laboratory and (ii) the animal area (Fig. 8.1). Workrooms are maintained with positive pressure to the adjacent corridor or to the exterior and provided with high-efficiency particulate air (HEPA) filter fan units, as required. Personnel enter the facility through a single outer door controlled by key card access. Once in the building, workers enter/exit the animal area through the shower and exit through the shower (Table 8.4). Within the barrier, personal protective clothing includes hats, masks, gloves, and Tyvek® suits. Dedicated protective clothing/footwear is available in each room.

The standards for animal accommodation and animal care are provided in accordance with the Guide for the Care and Use of Laboratory Animals [9]. Animal housing is provided within custom-designed, raised-floor swine pens from Alternative Designs (Siloam Springs, AR). In-pen care (feeding, cleaning, and observation) is provided to the animals using established standard operating procedures, based on the type and number of animals within the facility. Animals are provided sterile, mammalian-byproduct-free feed. Gestating gilts receive LabDiet 5082 (LabDiet, St. Louis, MO), nursing piglets receive Esbilac (concentrated puppy milk replacer) (Pet-Ag, Inc., Hampshire, IL), and weaned piglets receive irradiated LabDiet 5080 (LabDiet, St. Louis, MO).

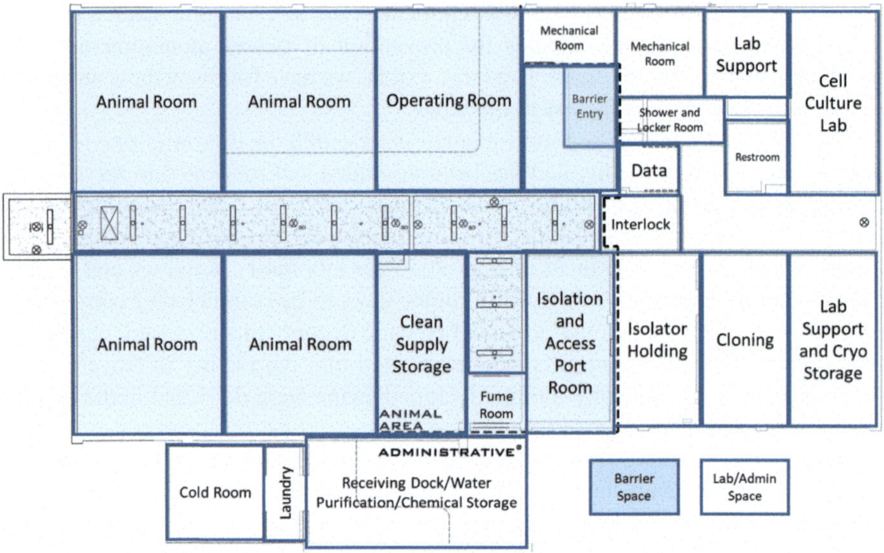

Fig. 8.1 Floorplan of the pig production barrier facility

Table 8.4 Individual room and content descriptions within the DPF pig facility

Room/function	Room description	Equipment	Remarks
Cloning laboratory support	Cold storage, hazardous materials handling, and satellite workstation	Fume hood, liquid nitrogen storage tanks, −80 ° freezer, −20 ° freezer, 4 ° refrigerator (additional items listed in room 101 section)	Cell bank and supplies for cloning lab also stored here
Cloning – cell culture	Cell culture	Biosafety cabinet (BSC), incubators, assistant work station, deionized water polisher	
Cloning – micromanipulation station	Cellular manufacturing	Micromanipulator, electroporator, microscopes	
Restroom	Unisex restroom		
Interlock	Negative space between barrier and laboratory		
Shower	Unisex shower		In/out shower for all personnel entering the barrier facilities; In-use light system indicates the status
Data	Electrical, networking and telecommunications closet		

Table 8.4 (continued)

Room/function	Room description	Equipment	Remarks
Isolator holding	Easy access to piglet husbandry isolators during first 30 days	Medical air and oxygen manifolds and tank storage	Climate-controlled space able to heat ambient temperature into the upper 80s
Barrier entry	The clean locker room, scrub sink, access to barrier mechanical room	HEPA fan-filtered space; medical air and oxygen alarm and emergency shut-off	
Surgery preparation and instrument processing	A multipurpose room used for pre-procedure anesthesia induction, necropsies, instrument washing and sterilization, sterile pass-through to laboratory via interlocked cabinet or acrylic porthole	HEPA fan-filtered space; autoclave; lift table; anesthesia machine	
Operating room	Used for embryo transfer, piglet derivation, and organ harvest	HEPA fan-filtered space; anesthesia machine, powered operating table; operating room lights on swing arm; supply storage cabinets; electrocautery and suction	
Fumigation room	Chlorine dioxide sterilization of all supplies, equipment, feed, etc., entering the facility		Gas-tight dampers and gasketed doors; direct fan exhaust to outside via stainless steel duct
Supply room	Storage and personnel workstation	HEPA fan-filtered space; ClorDiSys Minidox-M Chlorine Dioxide (ClO_2) generator	
Animal room	Donor pig finishing/isolation, gilt gestation	HEPA fan-filtered space; alternative designs (AD) custom swine accommodation; vinyl curtain for internal room division	Adjacent to the operating room, personnel flow to isolated animal via a door from the operating room

(continued)

Table 8.4 (continued)

Room/function	Room description	Equipment	Remarks
Animal room	Gilt isolation	HEPA fan-filtered space; AD custom swine accommodation	7-day quarantine, accessible via door directly from receiving area
Animal room	Gilt synchronization and gestation	HEPA fan-filtered space; AD custom swine accommodation	
Animal room	Gilt synchronization and gestation	HEPA fan-filtered space; AD custom swine accommodation	
Dock	Scissor lift added to facilitate transfer to/from various level vehicles		
Receiving	Reverse osmosis water filtration system, wash bay, backup freezer storage, chemical storage		Climate-controlled space
Cold room	4 °C cold storage for carcasses, on-site laundry equipment		
Mechanical (outside)	Water heater, electrical service, transfer switch, building automation, roof access		
Mechanical (barrier)	Vacuum pump, fire water plumbing, building automation controls, electrical panel		
Corridor	Access to animal rooms and procedural spaces, controls for mechanical dampers and variable speed HEPA filter fans, hand sink, light timer controls		

Piglets are derived through either surgical derivation or natural farrowing. For surgical derivation, piglets are maintained in self-contained isolators [designed and manufactured by ParkBioservices (Groveland, MA)] during weaning in the isolator-holding room. Access to the piglets is via gloved arm cuffs in the isolator. Piglets are raised in a gnotobiotic fashion prior to the introduction into the barrier [11].

Piglets are also delivered via natural farrowing behind the barrier. Utilizing farrowing crates from Vittetoe (Keota, IA), gestating gilts are introduced 3 days prior to induction. Piglets are farrowed and then nursed directly by the surrogate sow. Milk replacer and probiotic are provided to the piglets when the sow stops producing milk. A sterile solid feed is introduced 7–14 days after delivery to help facilitate weaning around day 28.

After weaning at 1 month of age and maturation to 4–6 months of age, pigs are relocated to a specially divided animal room immediately adjacent to the operating room. Isolated donor pig care is provided from the rear of the room, accessed through the operating room (OR) as an anteroom. Isolated donors are cared for first, followed by other pigs in the finishing room. Care continues for the pre-isolation cohort of animals that are housed on the other side of a curtain within the same room as the isolated donor (most clean to least clean within barrier facility). These animals are accessed via the main corridor, where the animal care staff don new personal protective equipment (PPE). After providing care for the animals, the staff exits to the hall and facility.

Surgical staff enter through the shower, complete preparations in the scrub room section of the barrier entry room, and directly enter the OR. Donor pigs are pre-anesthetized and prepped in the pre-harvest donor isolation area and transported via a lift table through the direct access door to the OR. Organ harvest is done within an International Standardization Organization (ISO) 5 or Class 100 Cleanroom condition curtained area. ISO 5/Class 100 conditions are defined by 240–360 air changes per hour (unidirectional airflow) and less than 100 particles per cubic foot of air [1]. After the surgery, the staff exit via the shower.

Materials, including feed, enter the barrier through the fume room or through the port between the laboratory and barrier sides of the facility. Chlorine dioxide gas is generated using a Minidox-M provided by ClorDiSys (Somerville, NJ). Items that are introduced via the interlocked pass-through are sprayed in with ethyl alcohol. An autoclave is present behind the barrier to reprocess any items that need to be sterilized but remain within the barrier. Materials exit through the fume-room, port, or via the interlock. All animal rooms have the ability to be isolated and decontaminated using the same chlorine dioxide generator as the fume-room. Doorplates facilitate the decontamination of individual rooms while the generator is housed in the common corridor. This allows for each room to be run as an independent environment and populated using an "all-in/all-out" strategy.

In addition to the facility itself, staff must (i) complete a thorough health screening before gaining access to the barrier, (ii) verify that they are fever-free within the last 24 hours, (iii) have not been in contact with pigs within the last 72 hours, and (iv) comply with a rigorous Occupational Health and Safety Surveillance Program [10].

Below is a description of the function and relevant equipment for each room within the facility.

Pathogen Screening

In the decade since the publication of the Guideline, a large body of pathogen-related work has been published. Some potential infectious risks, such as

Table 8.5 Pathogenic "agents of concern" in pigs proposed for sources of organs and cells for clinical xenotransplantation: potential sources of infection

Title (year)	Oversight/sponsorship
Second WHO global consultation on regulatory requirements for xenotransplantation clinical trials (2011) [12]	World Health Organization (WHO)
International Xenotransplantation Association consensus statement on conditions for undertaking clinical trials of porcine islet products in type 1 diabetes – Chapter 2: Source pigs. (2009) [13]	International Xenotransplantation Association (IXA)
Microbiological safety of the first clinical pig islet xenotransplantation trial in New Zealand (2014) [14]	New Zealand Ministry of Health
Trends in the emergence of swine pathogens (2012) [15]	Food and Agriculture Organization of the United Nations (UN FAO)
Evaluation of the human host range of bovine and porcine viruses that may contaminate bovine serum and porcine trypsin used in the manufacture of biological products (2011) [16]	National Institute of Health (NIH)

Transmissible spongiform encephalopathies (TSE) and porcine endogenous retrovirus (PERV), have been better quantified, and new risks, such as Ebola generally and emerging porcine pathogens specifically, have arisen.

To construct a comprehensive list of "agents of concern," three primary sources that represent (i) a global regulatory viewpoint, (ii) the technical viewpoint of the industry association, and (iii) the practical viewpoint of an actual clinical trial, were reviewed. A fourth source to address emerging diseases associated with swine was added. Finally, these sources were complemented by a fifth source specific to viral pathogens (a National Institutes of Health [NIH] intramural report on the host range of viral pathogens). These sources are listed in full in Table 8.5 [12–16].

Further expanding on the previously recommended pathogens of interest, Hartline et al., published a list of xenotransplantation pathogens of interest, along with PCR-based screening assays developed to quantitatively detect each pathogen (Table 8.6) [17].

Each pig involved (gestating gilts and organ-source pigs) will be tested for agents of concern as part of the pathogen-screening program. Pigs will be necropsied as a result of one of the following scenarios – (i) surrogate sow surgery after terminal delivery of piglets, (ii) selected sentinel animals from a litter, (iii) donor pigs after organ harvest, or (iv) any pig that dies unexpectedly. As per the Guideline, prospective organ donor pigs will be isolated for more than 3 weeks prior to organ procurement and subjected to pathogen screening during the isolation period with an expected turnaround time of 1 week for results to become available. Samples taken after entry into the final isolation (before organ harvesting) will also be tested by in vitro co-culture with human cell lines and in vivo inoculation into small mammals.

Table 8.6 Pathogens of interest in pigs proposed as sources of organs and cells for clinical xenotransplantation by Hartline et al. 2018 [17]

Target virus
Astrovirus
Bovine viral diarrhea virus
Chikungunya virus
Encephalomyocarditis virus
Hepatitis E virus
Influenza A virus
Influenza B virus
Lymphocytic choriomeningitis virus
Mycoplasma haemofelis group
Mycoplasma haemominutum group
Norovirus genogroup 2
Porcine adenovirus
Porcine cytomegalovirus
Porcine circovirus 2
Porcine circovirus 1
Porcine endogenous retrovirus C
Porcine hemagglutinating encephalomyelitis virus
Porcine lymphotropic herpesvirus 1
Porcine lymphotropic herpesvirus 2
Porcine lymphotropic herpesvirus 3
Porcine parvovirus
Porcine reproductive and respiratory syndrome virus
Rabies virus
Reovirus 1
Reovirus 2
Reovirus 3
Rotavirus
Sapovirus
Seneca valley A virus
Transmissible gastroenteritis virus
West Nile virus

Summary

In summary, the UAB off-campus xenotransplantation pig research facility is a single-purpose facility for producing a small number of genetically modified pigs for use in a pilot clinical trial. Previously, it was a large animal facility operated by the UAB Animal Resource Program that was upgraded in mid-2016 for the xenotransplantation program. Upgrades included (i) isolating the animal husbandry and procedural space within the barrier area, (ii) creating entry–exit points for workers, materials, and pigs, and (iii) upgrading to HEPA filtration for air and reverse osmosis for water. Expansions completed in 2017 provided more space for animals.

Table 8.7 Principles of production of designated pathogen-free pigs for xenotransplantation

1. All-in/all-out production
(a) Single litter production, with varying genetics
(b) To support cloned pigs *or* breeding approaches
2. Modular facility or design principles
(a) More moderate investment, replicable for scale-up
(b) Flexible for practical alterations and new techniques
(c) Facilitate parallel architectural design and construction process (design-build approach)
3. Designated pathogen-free versus germ-free
(a) Pathogen screening and response plan versus maintenance of the germ-free status
(b) Animal health status confirmed at various screening time-points
(c) Freedom from designated pathogens of interest
4. Challenging the paradigm with new technology and outside expertise
(a) Genome sequencing for product identification
(b) Rapid infectious disease screening for known and archive for unknown pathogen analysis in the future
(c) Genomic and metagenomic sequencing for pathogen detection

Comment

Quality of the "product," defined as a kidney with known genetics that is free from pathogenic agents of concern, is the primary concern in producing kidneys for use in humans. Cost and time are not insignificant concerns to the UAB program, which has an aggressive schedule and a budget that is not limitless. Past efforts elsewhere to construct and operate facilities to produce porcine tissues and cells for xenotransplantation have been expensive, and these facilities either have been closed or are operating at substantially reduced levels and high operating costs. A flexible, adaptable design is the foundation of the vision for producing clinical-quality pig kidneys, as this is the first-in-human use of genetically modified pigs for organ transplantation. Adjustments in response to progress, accomplished in a timely and efficient manner, may make the difference between a successful program and a less successful program.

The basic principles of a pragmatic xenograft production approach are highlighted in Table 8.7.

References

1. FDA, Center for Biologics Evaluation and Research. Source animal, product, preclinical, and clinical issues concerning the use of xenotransplantation products in humans. Guidance for Industry. April, 2003, updated December, 2016. www.fda.gov/downloads/BiologicsBloodVaccines/GuidanceComplianceRegulatoryInformation/Guidances/Xenotransplantation/UCM533036.pdf. Accessed 4 Sept 2017.
2. FDA, Center for Veterinary Medicine. Regulation of intentionally altered genomic DNA in animals, guidance for industry. Draft Guidance. January, 2017 (revision of earlier Guidance, January, 2009). https://www.fda.gov/downloads/AnimalVeterinary/

GuidanceComplianceEnforcement/GuidanceforIndustry/UCM113903.pdf. Accessed 4 Sept 2017.
3. U.S. Public Health Service Guideline on infectious disease issues in xenotransplantation. MMWR Recomm Rep. 2001;50:1–46.
4. Food and drugs, 21 CFR. http://www.ecfr.gov/cgi-bin/text-idx?tpl=/ecfrbrowse/Title21/21tab_02.tpl.
5. Animals and animal products, 9 CFR. http://www.ecfr.gov/cgi-bin/text-idx?tpl=/ecfrbrowse/Title09/9tab_02.tpl.
6. Public health, 42 CFR. http://www.ecfr.gov/cgi-bin/text-idx?tpl=/ecfrbrowse/Title42/42tab_02.tpl.
7. Schuurman H. Xenotransplantation: from the lab to the clinic. Clin Transpl. 2011;25:E415–21. https://doi.org/10.1111/j.1399-0012.2011.01471.x.
8. Association for Assessment and Accreditation of Laboratory Animal Care. http://www.aaalac.org/about/index.cfm. Accessed 4 Sept 2017.
9. Committee for the Update of the Guide for the Care and Use of Laboratory Animals, Institute of Laboratory Animal Research, Division on Earth and Life Studies, National Research Council. Guide for the care and use of laboratory animals. 8th ed. Washington, D.C.: The National Academies Press; 2011. https://grants.nih.gov/grants/olaw/guide-for-the-care-and-use-of-laboratory-animals.pdf. Accessed 4 Sept 2017.
10. Hawthorne WJ, Cowan PJ, Bühler LH, et al. Third WHO Global Consultation on regulatory requirements for xenotransplantation clinical trials, Changsha, Hunan, China. December 12-14, 2018. Xenotransplantation. 2019;26:e12513. https://doi.org/10.1111/xen.12513.
11. Yuan L, Jobst PM, Weiss M. Chapter 5 - Gnotobiotic pigs: from establishing facility to modeling human infectious diseases. In: Schoeb TR, Eaton KA, editors. Gnotobiotics. London: Academic Press; 2017. p. 349–68, ISBN 9780128045619. https://doi.org/10.1016/B978-0-12-804561-9.00005-0.
12. WHO. Second WHO Global Consultation on regulatory requirements for xenotransplantation clinical trials. Geneva: WHO; 2011; October 17–19.
13. Schuurman HJ. Microbiological safety of clinical xenotransplantation products: monitoring strategies and regulatory aspects. A commentary. Xenotransplantation. 2016;23:440–3.
14. Wynyard S, Nathu D, Garkavenko O, et al. Microbiological safety of the first clinical pig islet xenotransplantation trial in New Zealand. Xenotransplantation. 2014;21:309–23.
15. Fournie G, Kearsley-Fleet L, Otte J, Pfeiffer D. Trends in the emergence of swine pathogens. Animal Production and Health Commission for Asia and Pacific; 2012. 36 p.
16. Marcus-Sekura C, Richardson JC, Harston RK, Sane N, Sheets RL. Evaluation of the human host range of bovine and porcine viruses that may contaminate bovine serum and porcine trypsin used in the manufacture of biological products. Biologicals. 2011;39:359–69.
17. Hartline CB, Conner RL, James SH, et al. Xenotransplantation panel for the detection of infectious agents in pigs. Xenotransplantation. 2018;25:e12427. https://doi.org/10.1111/xen.12427.

Part IV

Antibody-Mediated Allotransplant Rejection: Lessons for Xenotransplantation

Antibody-Mediated Graft Rejection in Nonhuman Primate Models: Comparison of Sensitized Allotransplant and Xenotransplant Rejection

9

Stuart J. Knechtle, Jean Kwun, Brendon Lovasik, Alton B. Farris, A. Joseph Tector, and Andrew B. Adams

Abbreviations

AMR Antibody-mediated rejection
DSA Donor-specific alloantibody
NHP Nonhuman primate

Introduction

Antibody-mediated rejection (AMR) remains more difficult to prevent and reverse than acute cellular rejection of allotransplants. Despite advances in immunosuppressive drug management of transplant recipients, AMR, whether associated with preexisting antibody or de novo alloantibody, increases the risk of graft loss. The timing, pathology, and clinical behavior of AMR are characteristic. Banff criteria have been developed and refined for AMR in human kidney transplantation. Xenotransplants are known to also include a strong component of antibody-mediated injury, and we sought to examine similarities in a non-human primate model of kidney transplantation between the pathology of rejection of xenogeneic pig to

S. J. Knechtle (✉) · J. Kwun
Duke Transplant Center, Department of Surgery, Duke University, Durham, NC, USA
e-mail: stuart.knechtle@duke.edu

B. Lovasik · A. B. Adams
Emory Transplant Center, Department of Surgery, Emory University, Atlanta, GA, USA

A. B. Farris
Department of Pathology, Emory School of Medicine, Atlanta, GA, USA

A. J. Tector
Department of Surgery, University of Miami, Miami, FL, USA

D. K. C. Cooper, G. Byrne (eds.), *Clinical Xenotransplantation*,
https://doi.org/10.1007/978-3-030-49127-7_9

rhesus monkey kidney transplants and allogeneic pre-sensitized rhesus-to-rhesus monkey kidney transplants. Our findings suggest *considerable similarities* between the histologic findings of these two different types of renal transplants with respect to the cardinal pathologic features of AMR and/or thrombotic microangiopathy, which include glomerular capillary loop fibrin, glomerulitis, peritubular capillaritis, and interstitial hemorrhage. The time to graft loss secondary to AMR is also similar in these outbred, MHC-mismatched, large animal models. We have not compared head-to-head therapeutic strategies, but based on the similarity of pathogenesis, our results suggest that principles learned from one model may apply to the other, perhaps including therapeutic strategies.

Methods

Rhesus macaques *(Macaca mulatta)* were purchased from Alphagenesis (Yemessee, SC) through the National Institutes of Health (NIH) Non-human Primate Transplant Consortium. Animals were selected based on prescreening for MHC disparity and (SIV) seronegative status. For xeno-kidney transplantation, genetically modified donor pigs (GGTA1KO/CD55Tg) were purchased from the National Swine Resource and Research Center (University of Missouri-Columbia, Columbia, MO) [1]. All animals were cared for at a dedicated research animal facility (either the Duke Laboratory Animals for Research facility or Yerkes Primate Center), and care supervised by experienced non-human primate veterinary staff under protocols reviewed and approved by the Duke and Emory Institutional Animal Care and Use Committees (IACUCs). Humane care was guided according to the current principles of Laboratory Animal Management. Surgical procedures were conducted under general anesthesia, and pain was managed using narcotic and non-narcotic analgesics. For allosensitization, rhesus macaques were sensitized or immunized using full-thickness skin grafts procured from the dorsal skin of donor monkeys and placed on the dorsal, inter-scapular location of recipients. Two successive skin grafts were used in some monkeys to boost sensitization, placing grafts at three-week intervals. Kidney transplantation was performed by swapping left kidneys between the same pairs chosen for skin graft exchange and removing the right native kidneys such that animals depended on renal transplant function for survival. Xenotransplantation was performed in the same manner, but without placing skin grafts and rather used pigs as kidney donors to rhesus monkey recipients. Postoperative pain management was conducted according to veterinary guidelines with three times daily surveillance and drug administration as needed. Anti-rejection medications were administered daily by veterinary staff. Animals were clinically examined daily, and laboratory studies performed according to approved protocols. Kidney biopsies were done as indicated by rising serum creatinine or according to pre-determined protocols. Animals were sacrificed when they met predetermined protocol endpoints. Tissues were submitted for pathological review after sacrifice and interpreted by a single transplant pathologist (ABF).

Results

Identical induction and maintenance immunosuppressive regimen were used in both allosensitized kidney transplantation and pig-to-rhesus xenotransplantation (Fig. 9.1). Briefly, all recipients received an anti-CD4mAb (50 mg/kg) and anti-CD8mAb (25 mg/kg) prior to kidney transplantation. Posttransplant maintenance immunosuppression included Tacrolimus (target trough: 8–12 ng/ml) and mycophenolate mofetil (15 mg/kg s.c. or 30 mg/kg po BID). After skin grafting in the allosensitized model, monkeys showed elevated levels of donor-specific alloantibody (DSA, with 21.4-, 8.2-, 6.2-, 2.9-, 16.1-, and 5.9-fold increase in six monkeys). Kidneys transplanted to sensitized recipients experienced graft rejection with a mean survival time (MST) of 21.7 ± 19.0 days. For xenotransplantation, animals with preformed donor-specific IgG were excluded. Xenografts were rejected with an MST of 13.2 ± 4.8 days. Interestingly, the mean graft survival time was not significantly different between sensitized allo- and xenografts under treatment ($p = 0.53$) (Fig. 9.2). The same renal pathology specialist (ABF) assessed histology for both allosensitized- and xeno-kidney graft in a blinded fashion adopting the current human Banff scoring system [2]. The results of the AMR scoring system are summarized in Table 9.1. Similar histologic features were seen by the same pathologist in reviewing control, untreated pig-to-rhesus monkey xenografts (Fig. 9.3). The scores although variable from animal to animal were consistent with AMR and/or thrombotic microangiopathy. As shown in Fig. 9.4 and Table 9.1, both settings of transplantation showed a high incidence of thrombotic microangiopathy and

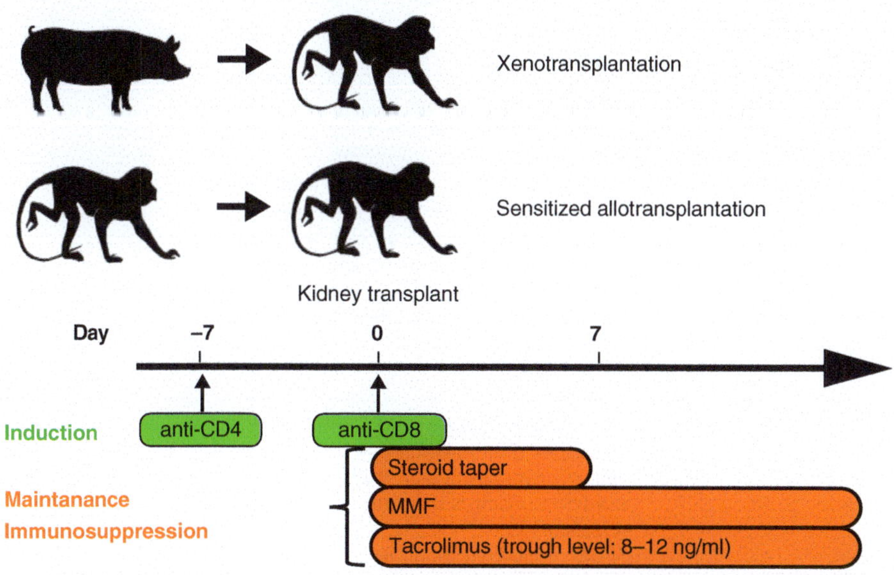

Fig. 9.1 Schematic representation of immunosuppressive regimen for sensitized-allo vs. xeno-kidney transplantation

Fig. 9.2 Comparable graft survival between sensitized allo- and xeno-kidney transplantation

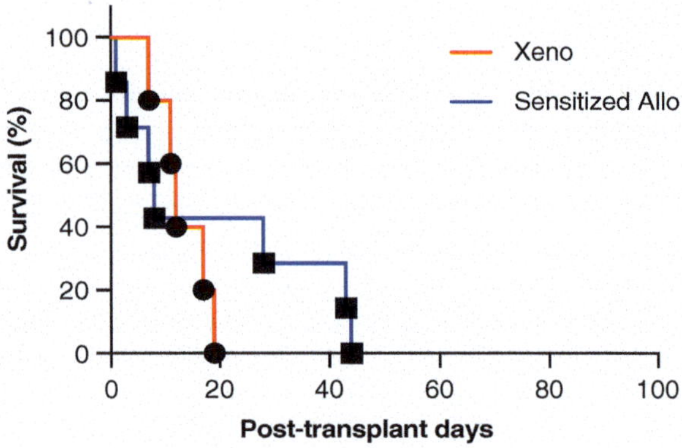

Table 9.1 Histopathologic features of xeno-kidney ($n = 5$ and sensitized allo-kidney $n = 7$) transplantations are shown, with rejection diagnoses according to the Banff criteria

ID	Graft survival (days)	Histopathological features	[a]g + ptc score (0–6)
(A) Xenotransplantation (CD4/CD8 depletion + tacrolimus/MME/steroids)			
RHU16	7	*Congestion and hemorrhage*. Negative for acute cellular rejection; cannot exclude AMR.	1
RYA16	11	Borderline changes suspicious for acute cellular rejection. *Congestion and hemorrhage*. Neutrophil casts suggest possible urinary tract infection.	5
RIU16	12	*Thrombotic microangiopathy*, suspicious for AMR. RBC congestion.	4
RHU14	17	*Thrombotic angiopathy* with fibrin in arterial walls, as well as glomerular capillary loops, peritubular capillaries, and arterioles (? AMR); subcapsular hemorrhage.	1
RPV16	19	*Thrombotic microangiopathy*, suspicious for AMR. Focal RBC congestion.	4
(B) Sensitized allotransplantation (CD4/CD8 depletion + tacrolimus/MMF/steroids)			
DW03	1	*Hyperacute rejection, thrombotic microangiopathy*. No evidence of acute cellular rejection. Acute tubular injury/necrosis. *Hemorrhage* and tubular necrosis are prominent in the medulla. Findings are possibly consistent with AMR.	4
FA6M	3	No evidence of acute cellular rejection. Acute tubular injury/necrosis.	1
GB5C	7	*Thrombotic microangiopathy*. Glomerular thrombi.	3
FE42	8	*Thrombotic microangiopathy. Glomerular congestion*. Fibrin, and glomerulitis. Apoptotic and necrotic cells in glomeruli.	4
RGm13	28	Borderline changes suspicious for acute cellular rejection.	4
GB46	43	Acute cellular rejection 2B.	5
DP78	44	Borderline changes suspicious for acute cellular rejection. Oxalate crystals, neutrophil casts suspicious of a urinary tract infection.	3

[a]*g* glomerulitis, *ptc* peritubular capillaritis

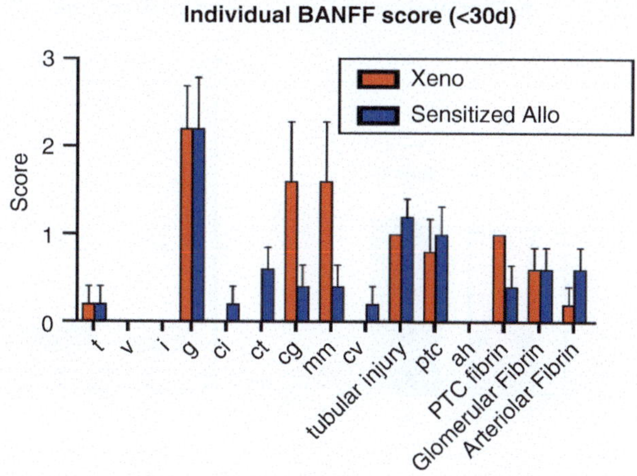

Fig. 9.3 Comparable BANFF individual gradings in sensitized allo- and xeno-kidney transplantation

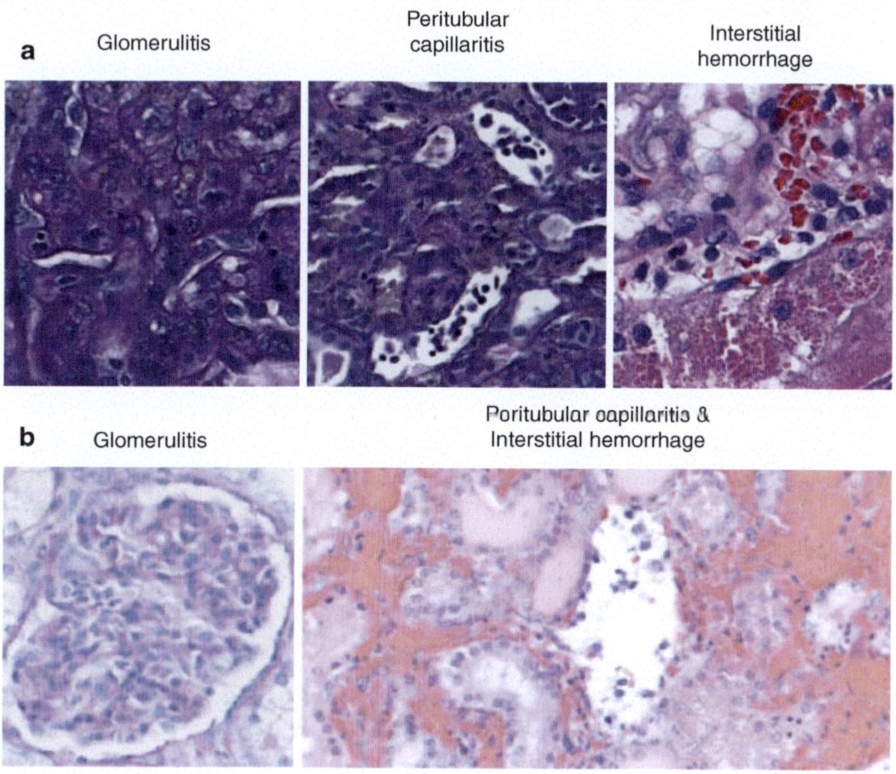

Fig. 9.4 Representative histology for common features of antibody-mediated rejection in sensitized (**a**) allo- and (**b**) xeno-kidney transplantation, including interstitial hemorrhage, peritubular capillaritis, and glomerulitis

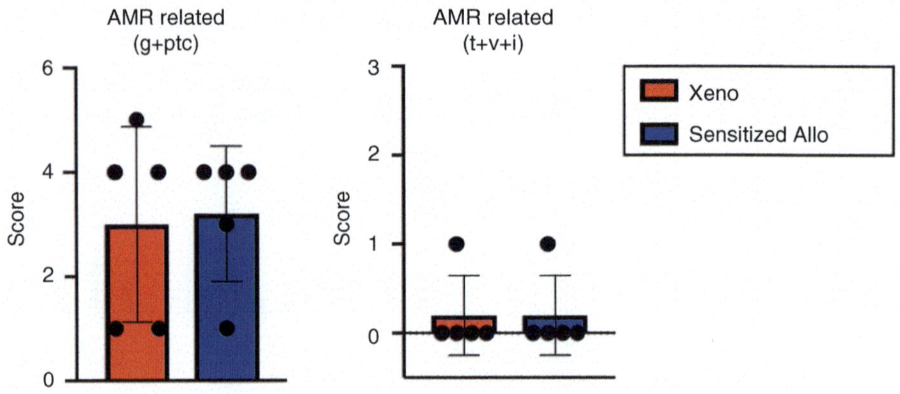

Fig. 9.5 A comparable clustered BANFF gradings for AMR and ACR in sensitized allo- and xeno-kidney transplantation. AMR-related score was calculated by combining points for tubulitis (t), intimal arteritis (v), and mononuclear infiltration (i)

interstitial hemorrhage (or glomerular fibrin) together with glomerulitis and peritubular capillaritis, which are highly associated with antibody-mediated rejection. Consequently, both groups showed high g + ptc score (Fig. 9.5) and showed high levels of c4d deposition (for allosensitized) or IgM deposition (for xenotransplants) (data not shown). These observations demonstrate striking similarities between allo- and xeno-AMR and suggest that these similarities in histologic appearance reflect mechanistic similarities with respect to immune cell activation and the effector arm of AMR.

Discussion

The contribution of antibody targeting alpha-1,3-galactosyltransferase to graft rejection in xenotransplantation is well described [3]. However, it is uncertain that antibody is the major immunologic barrier in xenotransplantation when α-gal is not present, as is the case when using transgenic pig donors lacking α-gal. Furthermore, the pathologic or histologic results of renal xenotransplantation have not been directly compared to allotransplantation. Different outbred nonhuman primate NHP recipients, induction therapy, maintenance immunosuppressive drugs, and rejection kinetics might make direct comparison difficult in a study. However, we sought to compare antibody-mediated rejection in allo- and xenotransplantation in rhesus recipients receiving very similar induction therapy and maintenance immunosuppressive agents, and who showed similar kinetics of graft rejection (Figs. 9.1 and 9.2). A head-to-head comparison of histology from allosensitized- and xenotransplantation strongly suggests that the rejection is mediated by antibodies either preformed or de novo. It is interesting that both settings lead to similar kinetics of unmodified rejection with similar histology showing features typical of antibody-mediated rejection. Allosensitized monkeys had preformed antibodies and a primed immune response while xenotransplant recipients did not show antibodies before transplantation. Further investigation is required to determine which antibodies are responsible for rejection in xeno-AMR (perhaps de novo IgM/IgG or natural antibody). Nevertheless, these similarities suggest the possibility that therapeutic approaches to pretransplant desensitization and posttransplant immunosuppression in xenotransplantation and in the sensitized allotransplantation may apply to both of these types of transplant, and that research in one area may inform the other. Clearly, transplanting the highly sensitized patient with an allograft has survival benefits compared to dialysis but also raises the risk of rejection, particularly late graft loss. The explosion of therapeutics targeting either antibody or the B cell and plasma cell offers a new opportunity to impact the outcomes of the sensitized allotransplant and the xenotransplant recipient.

The goal of our NHP studies has been to develop therapeutic strategies to address AMR in either the pre- or posttransplant period, and we have described the benefits of several approaches, including BAFF blockade [4], anti-CD40 [5–7], CTLA4-Ig [5–7], anti-CD38 [8], and proteasome inhibition [9]. We reported that presensitized recipients of allogeneic NHP kidney transplants with combined costimulation blockade (belatacept with anti-CD40mAb or belatacept alone) and plasma cell-targeted therapy (carfilzomib or bortezomib) resulted in significant lowering of the donor-specific antibody (DSA) level, prolonged survival, and less evidence of AMR by histology [6, 7]. It is quite intriguing that anti-CD154mAb (targeting the same costimulation signaling axis) in xenotransplantation showed superiority in controlling posttransplantation rejection or AMR [1].

For this reason, we suggest that increased scrutiny be given to approaches that show significant efficacy in the highly sensitized patient and that such approaches be considered for application to xenotransplantation. Evaluation of the safety and efficacy of such immunosuppressive strategies in a NHP model offers the benefits of highly controlled experiments that may appropriately precede human clinical trials. An unanswered question in xenotransplantation is whether an allosensitized patient would have an increased risk of AMR following xenotransplantation. Another question is whether xenotransplantation to humans would render the recipient allosensitized. In other words, does allosensitization cross-react with xenosensitization? These questions could be addressed in our highly allosensitized NHP model with the expectation that the principle would apply to humans as well. An answer to these questions would guide the application of xenotransplantation and tell us whether we should consider the allosensitized patient as an indication for xenotransplantation or not.

Conflict of Interest None.

References

1. Kim SC, Mathews DV, Breeden CP, et al. Long-term survival of pig-to-rhesus macaque renal xenografts is dependent on CD4 T cell depletion. Am J Transplant. 2019;19:2174; [Epub ahead of print]. https://doi.org/10.1111/ajt.15329.
2. Haas M, Sis B, Racusen LC, et al. Banff 2013 meeting report: inclusion of c4d-negative antibody-mediated rejection and antibody-associated arterial lesions. Am J Transplant. 2014;14:272–83.
3. Cooper DKC, Koren E, Oriol R. Genetically engineered pigs. Lancet. 1993;342:682–3.
4. Kwun J, Page E, Hong JJ, et al. Neutralizing BAFF/APRIL with atacicept prevents early DSA formation and AMR development in T cell depletion induced nonhuman primate AMR model. Am J Transplant. 2015;15:815–22.
5. Kim EJ, Kwun J, Gibby AC, et al. Costimulation blockade alters germinal center responses and prevents antibody-mediated rejection. Am J Transplant. 2014;14:59–69.
6. Kwun J, Burghuber C, Manook M, et al. Successful desensitization with proteasome inhibition and costimulation blockade in sensitized nonhuman primates. Blood Adv. 2017;1:2115–9.
7. Burghuber CK, Manook M, Ezekian B, et al. Dual targeting: combining costimulation blockade and bortezomib to permit kidney transplantation in sensitized recipients. Am J Transplant. 2019;19:724–36.
8. Kwun J, Matignon M, Manook M, et al. Daratumumab in sensitized kidney transplantation: potentials and limitations of experimental and clinical use. J Am Soc Nephrol. 2019;30(7):1206–19. PubMed PMID: 31227636.
9. Kwun J, Burghuber C, Manook M, et al. Humoral compensation after bortezomib treatment of allosensitized recipients. J Am Soc Nephrol. 2017;28:1991–6.

Management of Anti-HLA Antibodies and Acute Antibody-Mediated Rejection

10

Robert A. Montgomery

Abbreviations

AMR	Antibody-mediated rejection
DSA	Donor-specific antibody
HLA	Human leukocyte antigen
IdeS	IgG-degrading enzyme of streptococcus pyogenes
SLA	Swine leukocyte antigen

Introduction

In many ways, the immunologic barrier of a highly sensitized patient receiving an allograft is similar to that encountered in a pig-to-human xenotransplant. Patients who are sensitized to allo-human leukocyte antigen (HLA) have a primed humoral and cellular immune response and many will have preformed donor-specific antibody (DSA) [1]. Despite the success of editing out key glycans from the pig genome that are the targets of the bulk of human humoral xenoreactivity, there will still be preformed antibodies that will need to be managed in human xenograft trials. There are also HLA epitopes for which there appears to be cross-reactivity with antigens expressed on class I and II swine leukocyte antigens (SLA) [2]. The field of HLA typing and cross-matching is very advanced, and similar capabilities will need to be developed for xenotransplantation.

Preformed xenoreactive antibodies and anamnestic B-cell responses may produce early antibody-mediated rejection (AMR) of the kind seen in 40% of patients

R. A. Montgomery (✉)
NYU Langone Transplant Institute, NYU Langone Health, New York, NY, USA
e-mail: Robert.Montgomery@nyulangone.org

D. K. C. Cooper, G. Byrne (eds.), *Clinical Xenotransplantation*,
https://doi.org/10.1007/978-3-030-49127-7_10

transplanted across an HLA-incompatible barrier [3]. Among patients who have undergone previous transplants, we also see the development of de novo HLA and non-HLA antibodies that lead to chronic AMR and are responsible for about 40% of allograft failures [4]. These antibodies then put the patient at risk for accelerated AMR after subsequent transplants. In the xenograft setting, neoantigens are likely to trigger similar de novo antibody responses that could lead to acute or chronic xenograft injury. Thus, the AMR phenotypes encountered in allografts are likely to be recapitulated in human recipients of pig xenografts.

The field of desensitization and AMR treatment in allotransplantation has made a great progress over the last 25 years and has produced many effective treatment modalities that can be applied to xenotransplantation [5, 6]. In this review, we will discuss the diagnostic criteria of AMR, injury phenotypes, populations at risk, and established and emerging treatments. This information will inform the identification and management of immunologic responses to xenografts in humans.

Banff Diagnostic Criteria for AMR

The Banff process has continued to refine the diagnostic criteria for AMR to accommodate dizzying advancements in the technology of HLA antibody detection, microarray characterization of archetypal molecular AMR transcription signatures, and our understanding of the pathophysiology of AMR and its distinct injury phenotypes. For instance, the Banff 2007 diagnostic criteria included the requirement of the presence of diffuse C4d deposition in the peritubular capillaries. It was later discovered that anti-HLA antibody injury can result from noncomplement-activating antibody [7, 8], requiring significant changes in the Banff criteria. Banff 2017 defines two common subtypes of AMR – acute active and chronic active.

Acute active AMR has histologic evidence of acute tissue injury (glomerulitis, peritubular capillaritis [PTCitis], or vasculitis), evidence of current or recent antibody interaction with the vascular endothelium (C4d staining, microvascular inflammation, or gene transcripts of endothelial injury), and serologic evidence of DSA [9]. What differentiates chronic active AMR from acute AMR is biopsy evidence of chronic tissue injury manifest by transplant glomerulopathy, peritubular capillary basement membrane multilayering, and/or arterial intimal fibrosis. Chronic active AMR shares the diagnostic features of antibody interaction with the vascular endothelium (C4d staining, microvascular inflammation, or gene transcripts of endothelial injury) and serologic evidence of DSA, with acute AMR.

It is becoming clear that acute active AMR is generally more responsive to currently available therapeutics. If treated rapidly and effectively, acute AMR is reversible and the tissue injury can heal without progression to a chronic phenotype. Chronic active AMR, on the other hand, is probably not reversible with available treatment, leads to a truncated allograft half-life, and the best one can hope for from therapy at this point is to convert chronic active AMR to chronic injured phenotype [10, 11].

AMR Injury Phenotypes

Acute active AMR is defined by specific histologic, immunohistologic, molecular, and serologic criteria that indicate that soluble DSA is in contact with and injuring the allograft vascular endothelium. However, it does not provide information about the immunologic origin or natural history of the antibody. Acute AMR can be the result of sublethal levels of preformed DSA, a recall or anamnestic immune response in a sensitized patient or from de novo DSA formation.

In the era of routine cross-matching and sensitive solid phase assays for detecting DSA, persistent preformed antibody is often what remains after incomplete desensitization therapy. The source of this antibody is the long-lived plasma cells in the bone marrow produced in response to a previous alloantigen immunization. Because these antibodies are coming from preexisting plasma cells, this represents a rebound phenomenon or a failure of persistent DSA to disappear after transplantation. AMR from this source usually occurs in the first few months after transplantation and tends to be mild-to-moderate in severity and can be subclinical, discovered on protocol biopsies. A single center retrospective study by Orandi et al. showed improved allograft survival in patients with this phenotype who were treated with plasmapheresis compared to those who were observed [12]. Also, patients who had C4d-positive biopsies had a worse prognosis [3]. This AMR phenotype usually responds to antibody depletion therapy but can persist and cause chronic injury.

AMR from a B-cell recall or anamnestic response is generally early (within the first week after the transplant), severe, associated with a rapid rise in strong DSA, and can result in allograft loss without prompt intervention. It is the classic primed, memory immune response in a previously sensitized patient and rapidly produces high-affinity, complement-fixing IgG that causes graft dysfunction. Biopsies show the typical findings of acute active AMR, but also frequently include glomerular fibrin thrombi, interstitial hemorrhage, and scattered small infarcts associated with microvascular thrombi. If aggressive therapy is delivered rapidly (including antibody depletion, splenectomy, and complement inhibition), these organs can usually be salvaged with good function and the DSA completely eliminated [13]. However, there is a significant risk of immediate graft loss.

In general, the appearance of de novo DSA is associated with poor long-term allograft function [14]. It can occur at any time after a transplant. It may suddenly appear weeks after a cellular rejection, and de novo DSA that are detected in the first 6 months to a year usually occur in this setting. Post-one year, de novo DSA is often associated with nonadherence or immunosuppression reduction and has a mixed cellular/AMR phenotype. It may be diagnosed late, is poorly responsive to standard of care antibody reduction protocols (such as plasmaphersis or IVIg), and typically becomes chronic active AMR.

Chronic AMR, whether active or inactive, is associated with poor allograft survival. This phenotype has been recalcitrant to standard-of-care antibody-depleting therapies. If there is a response to treatment, it is usually temporary and the DSA level returns to pretreatment levels. It is most frequently associated with class II DSA, which has been shown to more commonly persist long-term after treatment and more rapidly lead to graft loss [15, 16]. There is no evidence that treatment changes the natural history of this phenotype.

Populations at Risk

Sensitized patients with detectable DSA are at the highest risk for post-transplant AMR. Most studies suggest the rate of AMR in HLA-incompatible kidney transplants is between 20% and 40% [17, 18]. Patients undergoing desensitization across a positive CDC cross-match are more likely to experience AMR than patients who have lower strength DSA. However, sensitized patients who do not have detectable DSA are at risk for anamnestic responses, although it has been difficult to measure the HLA-specific memory and assign risk, there are techniques that can allow the measurement of the frequencies of B cells of different HLA antibody specificities. There is some evidence that depleting memory B-cell pools with Rituximab (an anti-20 mAb) can reduce anamnestic responses in sensitized patients [19].

Wiebe et al. showed that the risk of developing de novo HLA antibody is greater for patients maintained on cyclosporine versus tacrolimus. Among patients treated with tacrolimus, trough levels <5 ng/ml were associated with a significantly increased risk of forming de novo DSA. Having fewer class II HLA-DR/DQ eplet mismatches also independently meant less risk of developing DSA, and a better match could positively modulate the alloimmune risk of lower tacrolimus levels [20]. This suggests that the risk of development of de novo DSA is predictable and can be modified; this information represents a major advance in this field.

Prevention and Treatment of AMR

Options Available to Highly Sensitized Patients

An unintended consequence of the historic breakthrough, complement-dependent cytotoxic (CDC) cross-match, has been the creation of an underclass of disenfranchised patients who have very limited access to transplantation [21]. The transplant rate for highly sensitized (PRA >85%) patients on the kidney waiting list has traditionally been dismal. Desensitization strategies, kidney paired donation, and deceased donor allocation changes are improving the landscape (Fig. 10.1). However, a generally poor understanding of the benefits and limitations of these options by the transplant community is resulting in missed opportunities to apply the principles of precision medicine and bring the best transplant solution to individual donor/recipient phenotypes [22].

In a single high-volume center study, patients undergoing desensitization and transplantation with a live donor have been shown to have a significant survival benefit over remaining on dialysis or waiting for a compatible deceased donor kidney [23]. These results were confirmed in a larger multicenter study involving 22 US transplant programs [24]. Matched controls for strongly sensitized patients who

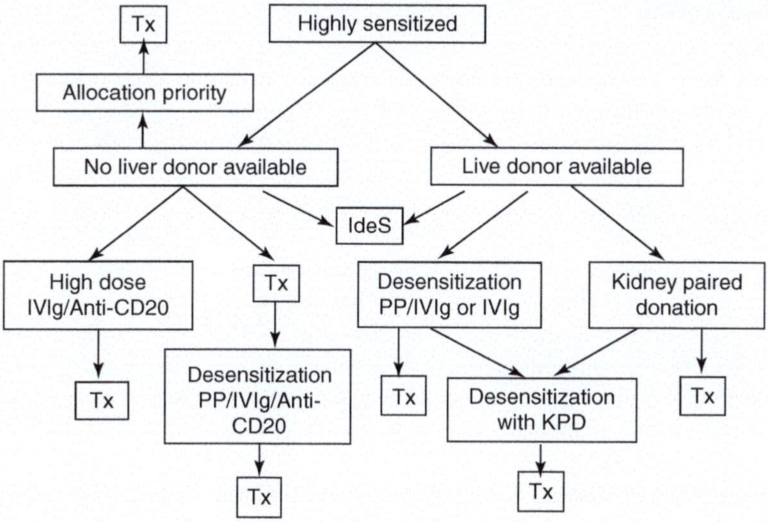

Fig. 10.1 Transplant options for highly sensitized patients. The fork in the algorithm is determined by whether or not there is a living donor available. If the answer is yes, then desensitization, kidney paired donation, and a hybrid between the two strategies can often lead to a successful transplant. Knowing which donor/recipient phenotypes benefit the most from each modality is critical to choosing the optimal transplant solution. If the patient does not have a living donor, then desensitization with high-dose IVIg or Imlifidase can be performed prior to the transplant, the patient can wait for a compatible donor enabled by allocation priority points, or they can receive post-transplant desensitization as long as the antibody strength at the time of the transplant is low

remained on dialysis during the study period had a mortality rate of 70% at 8 years, which is worse than most types of cancer. Similar data are available for sensitized patients receiving deceased donor transplants [25].

The concept of a chain of transplants initiated by an altruistic (nondirected) donor was introduced by Montgomery et al. [26] and further refined by Rees et al. to broaden chains to include nonsimultaneous extended donation in which chains would pause and be extended further at a later time [27]. Pools of incompatible pairs increase the range of HLA genotypes available to sensitized patients when compared to the small number of donors, a sensitized patient may have available to them. This increases the likelihood of finding rare genotypes that are compatible with recipients who are sensitized to common HLA antigens [28].

However, kidney paired donation has its limitations and there are certain donor/recipient phenotypes that are more or less likely to match in an incompatible pool. Clinicians need to understand matching probabilities in order to advise patients on the best option for them. For ultra-sensitized patients, paired donation pools can provide a donor for whom the recipient is not compatible, but has a lower level DSA that can be removed by desensitization [29]. Using these matching algorithms, one can determine whether paired donation, desensitization, or a combination of both provides the highest likelihood of a transplant for a particular donor/recipient phenotype [22].

Desensitization

There are two standard-of-care methods of desensitization: high-dose IVIg and plasmapheresis combined with low-dose IVIg. High-dose IVIg (2 g/kg) is usually given as 4 monthly infusions. This has been shown to increase both deceased and living donor transplant rates although it cannot be attributed directly to lowering PRA since the effect of IVIg on PRA reduction is modest [30]. There is evidence that a protocol involving two doses of IVIg and two doses of anti-CD20 improves the outcomes of desensitization and transplantation when compared to IVIg alone (Fig. 10.2a) [31, 32]. However, anti-CD20 has not been shown to reduce DSA or improve graft survival when used for the treatment of AMR [33]. Still, rituximab is utilized widely for antibody reduction despite a lack of evidence of efficacy. Mechanistically, although depletion of B cells might be predicted to prevent or dampen anamnestic responses, plasma cells do not express CD20 and are not affected by rituximab.

Plasmapheresis reduces DSA in a predictable fashion, lowering DSA by about one dilution per treatment. However, it is not an efficient way of depleting DSA since it only removes IgG from the intravascular space. The vascular space must then re-equilibrate with the larger interstitial compartment prior to the next treatment, a process that is optimized in about 48 hours. Low-dose IVIg (100 mg/kg) reduces the rebound of endogenous IgG between plasmapheresis treatments. Generally, plasmapheresis (1 volume exchange) is performed every 48 hours prior to the planned date for the transplant. The number of treatments necessary to get DSA to a safe level can be predicted based on the starting antibody strength. With increasing levels of DSA, more treatments are required. If plasmapheresis is discontinued, DSA levels will rapidly increase, and for this reason, plasmapheresis has mainly been used for patients with live donors when the date of transplant is known [34, 35].

This protocol has been modified for the treatment of AMR (Fig. 10.2b). Therapy is usually stopped when clinical, histologic, and DSA resolution is achieved. There may be some residual DSA, but it should be below the strength that would yield a positive flow cytometric cross-match. There are protocols for DSA reduction after deceased donor transplantation, but only when the initial DSA strength is at or below a level that yields a positive flow cross-match [36]. At higher antibody levels of DSA, the risk of hyperacute rejection makes these protocols unwise.

Established and Emerging Treatments

Complement Inhibition

Complement blockage has been used in both desensitization and AMR treatment regimens as "add-on" therapy to standard-of-care protocols.

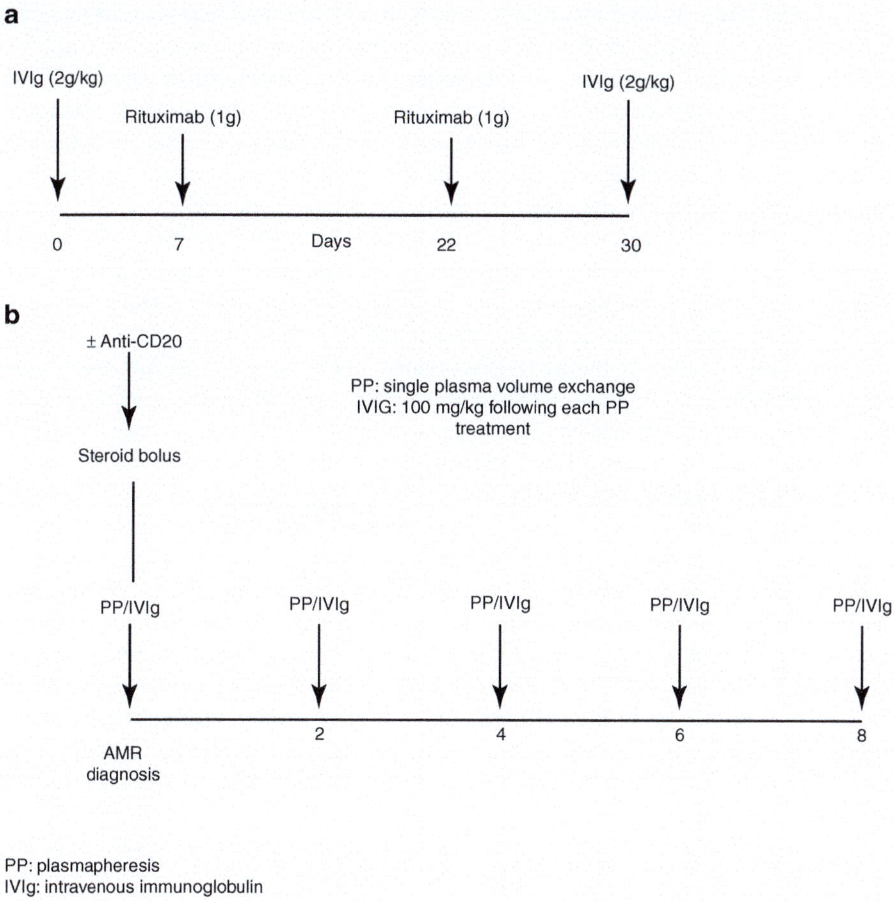

Fig. 10.2 (**a**) High-dose IVIg protocol for desensitization. Two doses of 2 g/kg and two doses of Rituximab have been shown to increase deceased donor transplant rates for highly sensitized patients and produce very good outcomes. IVIg infusions should be carefully coordinated around dialysis treatments, and products high in sucrose should be avoided. Cycles can be repeated, if necessary. (**b**) Treatment of AMR with plasmapheresis and low-dose IVIg. Once the diagnosis of AMR is confirmed by biopsy, alternate day plasmapheresis (pp) followed by 100 mg/kg of IVIg is performed. One volume exchange is replaced by albumin and normal saline. If an invasive procedure has been performed within a 24-hour period, fresh frozen plasma should make up part of the replacement fluid to reduce the risk of bleeding. If the AMR is severe, daily treatments are an option, but close monitoring for coagulopathy is advised. Therapy is usually stopped when clinical, histologic, and DSA resolution is achieved. There may be some residual DSA, but it should be below the strength that would yield a positive flow cytometric cross-match

Eculizumab was the first reported complement inhibitor used in transplant patients undergoing severe acute AMR [37]. Eculizumab prevents the cleavage of C5 to C5a and C5b, preventing the formation of the **membrane attack complex** (MAC) (C5b–C9) and cytolysis. It does not, however, prevent the upstream formation of inflammatory factors (anaphylatoxin C3a, iC3b, C3dg opsonins) or the deposition of C4d in the peritubular capillaries. The drug is FDA-approved for the treatment of paroxysmal nocturnal hemoglobinuria and atypical hemolytic uremic syndrome.

A phase II desensitization trial was initially reported by the sponsor, Alexion, as having failed to achieve a composite primary end-point which included a reduction in the rate of AMR in the first 3 months after renal transplantation. However, post hoc analysis by two study groups demonstrated that the failure of the trial was related to design flaws, including (i) discordance in the diagnosis of AMR between central and local pathologists; (ii) changes in DSA strength inclusion criteria mid-study; (iii) unconventional diagnostic criteria defining AMR; and (iv) the inclusion of patients with C1q-negative DSA. Mitigating any one of these assumptions would have resulted in a positive study [38, 39].

C1 esterase inhibitor (C1-INH) is FDA-approved for hereditary angioedema. It is an endogenous serine protease inhibitor involved in multiple biologic pathways (including coagulation, kallikrein, and complement). Nano-filtered C1-INH plasma-derived products have been used to block C1q in the classical complement pathway to prevent and treat AMR in several phase I/II placebo-controlled and uncontrolled exploratory trials [40, 41]. There were no safety signals in the transplant population. Currently, there are two phase III multicenter randomized controlled trials determining the efficacy of two formulations of C1-INH for the treatment of AMR. The Takeda/Shire trial has been discontinued due to lack of evidence of efficacy during an interim analysis.

IgG-degrading enzyme of Streptococcus pyogenes (IdeS) (Imlifidase) is an enzyme that cleaves at a specific amino acid sequence in the hinge region of human IgG, creating an Fc and F(ab')2 fragment [42]. IgG throughout the body is neutralized within 4 hours of administration. For the first 7 days after the drug is given, both soluble IgG and the B-cell receptors are undetectable. Between days 7 and 10, IgG can rebound and return to predrug levels [43]. Unfortunately, most people develop anti-IdeS neutralizing antibody after one or two doses and this prevents repeated dosing when IgG rebounds. In an initial report of sensitized patients who received a single treatment of IdeS prior to transplantation, efficacy was demonstrated in eliminating HLA antibody [44]. There was one hyperacute rejection (thought to be from IgM antibody) and 10 out of the remaining 24 patients had an episode of AMR associated with IgG rebound. All were successfully treated with standard-of-care therapy.

In a second report from Lonze et al., seven positive cross-match patients were converted to a negative cross-match and successfully transplanted with either living or deceased donors. Three of seven had DSA rebound with AMR, which resolved with standard-of-care treatment [45]. IdeS is the most rapid and potent method for eliminating HLA antibody, and is likely to play a key role in the future for the treatment of AMR [46].

There are many series and case reports of using *splenectomy* (or splenic embolization and splenic radiation) as a salvage procedure for severe early AMR, especially in highly sensitized patients [47–49]. The B-cell recall or anamnestic response phenotype seems to benefit the most from this intervention [13]. DSA rebound is profound and biopsies often demonstrate interstitial hemorrhage, glomerular fibrin thrombi, and focal necrosis. Plasmapheresis and IVIg do not effectively reduce DSA in this phenotype, and allograft loss occurs rapidly without source control of antibody production. Plasmablasts and plasma cells have been shown to traffic to the spleen under these circumstances and are accessible for elimination [50, 51]. The combination of plasmapheresis with splenectomy and complement inhibition has been shown to result in a high salvage rate and protection from transplant glomerulopathy [13].

References

1. Montgomery RA, Tatapudi VS, Leffell MS, Zachary AA. HLA in transplantation. Nat Rev Nephrol. 2018;14:558–70.
2. Martens GR, Ladowski JM, Estrada J, et al. HLA Class I-sensitized renal transplant patients have antibody binding to SLA Class I epitopes. Transplantation. 2019;103:1620–9.
3. Orandi BJ, Alachkar N, Kraus ES, et al. Presentation and outcomes of C4d-negative antibody-mediated rejection after kidney transplantation. Am J Transplant. 2016;16:213–20.
4. Lohéac C, Aubert O, Loupy A, Legendre C. Identifying the specific causes of kidney allograft loss: a population-based study. Nephrol Ther. 2018;14(Suppl 1):S39–50.
5. Montgomery RA, Loupy A, Segev DL. Antibody-mediated rejection: new approaches in prevention and management. Am J Transplant. 2018;18(Suppl 3):3–17.
6. Iyer HS, Jackson AM, Zachary AA, Montgomery RA. Transplanting the highly sensitized patient: trials and tribulations. Curr Opin Nephrol Hypertens. 2013;22:681–8.
7. Haas M, Sis B, Racusen LC, Banff Meeting Report Writing Committee, et al. Banff 2013 meeting report: inclusion of c4d-negative antibody-mediated rejection and antibody-associated arterial lesions. Am J Transplant. 2014;4:272–83.
8. Loupy A, Hill GS, Suberbielle C, et al. Significance of C4d Banff scores in early protocol biopsies of kidney transplant recipients with preformed donor-specific antibodies (DSA). Am J Transplant. 2011;11:56–65.
9. Haas M, Loupy A, Lefaucheur C, et al. The Banff 2017 Kidney Meeting Report: revised diagnostic criteria for chronic active T cell-mediated rejection, antibody-mediated rejection, and prospects for integrative endpoints for next-generation clinical trials. Am J Transplant. 2018;18:293–307.
10. Lefaucheur C, Loupy A, Hill GS, et al. Preexisting donor-specific HLA antibodies predict outcome in kidney transplantation. J Am Soc Nephrol. 2010;21:1398–406.
11. Viglietti D, Loupy A, Aubert O, et al. Dynamic prognostic score to predict kidney allograft survival in patients with antibody-mediated rejection. J Am Soc Nephrol. 2018;29:606–19.
12. Orandi BJ, Chow EH, Hsu A, et al. Quantifying renal allograft loss following early antibody-mediated rejection. Am J Transplant. 2015;15:489–98.
13. Orandi BJ, Zachary AA, Dagher NN, et al. Eculizumab and splenectomy as salvage therapy for severe antibody-mediated rejection after HLA-incompatible kidney transplantation. Transplantation. 2014;98:857–63.
14. Wiebe C, Gibson IW, Blydt-Hansen TD, et al. Evolution and clinical pathologic correlations of de novo donor-specific HLA antibody post kidney transplant. Am J Transplant. 2012;12:1157–67.

15. Zachary AA, Montgomery RA, Ratner LE, et al. Specific and durable elimination of antibody to donor HLA antigens in renal-transplant patients. Transplantation. 2003;76:1519–25.
16. Bentall A, Cornell LD, Gloor JM, et al. Five-year outcomes in living donor kidney transplants with a positive crossmatch. Am J Transplant. 2013;13:76–85.
17. Vo AA, Sinha A, Haas M, et al. Factors predicting risk for antibody-mediated rejection and graft loss in highly human leukocyte antigen sensitized patients transplanted after desensitization. Transplantation. 2015;99:1423–30.
18. Cornell LD, Schinstock CA, Gandhi MJ, Kremers WK, Stegall MD. Positive crossmatch kidney transplant recipients treated with eculizumab: outcomes beyond 1 year. Am J Transplant. 2015;15:1293–302.
19. Zachary AA, Lucas DP, Montgomery RA, Leffell MS. Rituximab prevents an anamnestic response in patients with cryptic sensitization to HLA. Transplantation. 2013;95:701–4.
20. Wiebe C, Rush DN, Nevins TE, et al. Class II eplet mismatch modulates tacrolimus trough levels required to prevent donor-specific antibody development. J Am Soc Nephrol. 2017;28:3353–62.
21. Patel R, Terasaki PI. Significance of the positive crossmatch test in kidney transplantation. N Engl J Med. 1969;280:735–9.
22. Montgomery RA. Renal transplantation across HLA and ABO antibody barriers: integrating paired donation into desensitization protocols. Am J Transplant. 2010;10:449–57.
23. Montgomery RA, Lonze BE, King KE, et al. Desensitization in HLA-incompatible kidney recipients and survival. N Engl J Med. 2011;365:318–26.
24. Orandi BJ, Luo X, Massie AB, et al. Survival benefit with kidney transplants from HLA-incompatible live donors. N Engl J Med. 2016;374:940–50.
25. Vo AA, Petrozzino J, Yeung K, et al. Efficacy, outcomes, and cost-effectiveness of desensitization using IVIG and rituximab. Transplantation. 2013;95:852–8.
26. Montgomery RA, Gentry SE, Marks WH, et al. Domino paired kidney donation: a strategy to make best use of live non-directed donation. Lancet. 2006;368:419–21.
27. Rees MA, Kopke JE, Pelletier RP, et al. A nonsimultaneous, extended, altruistic-donor chain. N Engl J Med. 2009;36011:1096–101.
28. Montgomery RA, Lonze BE, Jackson AM. Using donor exchange paradigms with desensitization to enhance transplant rates among highly sensitized patients. Curr Opin Organ Transplant. 2011;16:439–43.
29. Montgomery RA, Zachary AA, Ratner LE, et al. Clinical results from transplanting incompatible live kidney donor/recipient pairs using kidney paired donation. JAMA. 2005;294:1655–63.
30. Jordan SC, Tyan D, Stablein D, et al. Evaluation of intravenous immunoglobulin as an agent to lower allosensitization and improve transplantation in highly sensitized adult patients with end-stage renal disease: report of the NIH IG02 trial. J Am Soc Nephrol. 2004;15:3256–62.
31. Vo AA, Choi J, Cisneros K, et al. Benefits of rituximab combined with intravenous immunoglobulin for desensitization in kidney transplant recipients. Transplantation. 2014;98:312–9.
32. Vo AA, Lukovsky M, Toyoda M, et al. Rituximab and intravenous immune globulin for desensitization during renal transplantation. N Engl J Med. 2008;359:242–51.
33. Sautenet B, Blancho G, Büchler M, et al. One-year results of the effects of Rituximab on acute antibody-mediated rejection in renal transplantation: RITUX ERAH, a multicenter double-blind randomized placebo-controlled trial. Transplantation. 2016;100:391–9.
34. Montgomery RA, Zachary AA, Racusen LC, et al. Plasmapheresis and intravenous immune globulin provides effective rescue therapy for refractory humoral rejection and allows kidneys to be successfully transplanted into cross-match-positive recipients. Transplantation. 2000;70:887–95.
35. Montgomery RA, Zachary AA. Transplanting patients with a positive donor-specific crossmatch: a single center's perspective. Pediatr Transplant. 2004;8:535–42.
36. Anglicheau D, Loupy A, Suberbielle C, et al. Posttransplant prophylactic intravenous immunoglobulin in kidney transplant patients at high immunological risk: a pilot study. Am J Transplant. 2007;7:1185–92.

37. Locke JE, Magro CM, Singer AL, et al. The use of antibody to complement protein C5 for salvage treatment of severe antibody-mediated rejection. Am J Transplant. 2009;9:231–5.
38. Lefaucheur C, Viglietti D, Hidalgo LG, et al. Complement-activating anti-HLA antibodies in kidney transplantation: allograft gene expression profiling and response to treatment. J Am Soc Nephrol. 2018;29:620–35.
39. Marks WH, Mamode N, Montgomery RA, et al. Safety and efficacy of eculizumab in the prevention of antibody-mediated rejection in living-donor kidney transplant recipients requiring desensitization therapy: a randomized trial. Am J Transplant. 2019;19:2876–88.
40. Montgomery RA, Orandi BJ, Racusen L, et al. Plasma-derived C1 esterase inhibitor for acute antibody-mediated rejection following kidney transplantation: results of a randomized double-blind placebo-controlled pilot study. Am J Transplant. 2016;16:3468–78.
41. Viglietti D, Gosset C, Loupy A, et al. C1 Inhibitor in acute antibody-mediated rejection nonresponsive to conventional therapy in kidney transplant recipients: a pilot study. Am J Transplant. 2016;16:1596–603.
42. Wenig K, Chatwell L, von Pawel-Rammingen U, Björck L, Huber R, Sondermann P. Structure of the streptococcal endopeptidase IdeS, a cysteine proteinase with strict specificity for IgG. Proc Natl Acad Sci U S A. 2004;101:17371–6.
43. Järnum S, Bockermann R, Runström A, Winstedt L, Kjellman C. The bacterial enzyme ideS cleaves the IgG-type of B cell receptor (BCR), abolishes BCR-mediated cell signaling, and inhibits memory B cell activation. J Immunol. 2015;195:5592–601.
44. Jordan SC, Lorant T, Choi J, et al. IgG endopeptidase in highly sensitized patients undergoing transplantation. N Engl J Med. 2017;377:442–53.
45. Lonze BE, Tatapudi VS, Weldon EP, et al. IdeS (Imlifidase): a novel agent that cleaves human IgG and permits successful kidney transplantation across high-strength donor-specific antibody. Ann Surg. 2018;268:488–96.
46. Montgomery RA, Lonze BE, Tatapudi VS. IgG degrading enzyme of streptococcus pyogenes: an exciting new development in desensitization therapy. Transplantation. 2018;102:2–4.
47. Locke JE, Zachary AA, Haas M, et al. The utility of splenectomy as rescue treatment for severe acute antibody mediated rejection. Am J Transplant. 2007;7:842–6.
48. Roberti I, Geffner S, Vyas S. Successful rescue of refractory acute antibody-mediated renal allograft rejection with splenectomy--a case report. Pediatr Transplant. 2012;16:E49–52; Kaplan B, Gangemi A, Thielke J, et al. Successful rescue of refractory, severe antibody mediated rejection with splenectomy. Transplantation. 2007;83: 99–100.
49. Latzko M, Jasra S, Akbar S, Sun H, Palekar S. Laparoscopic splenectomy to salvage renal transplants from severe acute antibody-mediated rejection. Case Rep Transplant. 2012;2012:253173.
50. Tzvetanov I, Spaggiari M, Oberholzer J, et al. Cell population in spleens during antibody-mediated rejection: pathologic and clinical findings. Transplantation. 2012;94:255–62.
51. Jackson AM, Kuperman MB, Montgomery RA. Multiple hyperacute rejections in the absence of detectable complement activation in a patient with endothelial cell reactive antibody. Am J Transplant. 2012;12:1643–9.

Evolving Approaches to Treatment of Allosensitization and Antibody-Mediated Rejection

11

Stanley C. Jordan, Noriko Ammerman, and Ashley Vo

Abbreviations

ADCC	Antibody-dependent cellular cytotoxicity
AMR	Antibody-mediated rejection
cAMR	Chronic active antibody-mediated rejection
CDC	Complement-dependent cytotoxicity
dnDSA	De novo donor-specific antibody
DSA	Donor-specific antibody
HLA	Human leukocyte antigen
IVIg	Intravenous immunoglobulin
Tfh	T follicular helper

Introduction

Mechanisms of Allosensitization

Sensitization to alloantigen (human leukocyte antigen [HLA] class I or class II), molecules derived from the allograft, occurs through a process of antigen presentation to naïve T cells that mature to T follicular helper (Tfh) cells in the regional lymph nodes or spleen (germinal centers) of allograft recipients. In the germinal center, Tfh cell activation is critical for generation of de novo alloantibody

S. C. Jordan (✉) · N. Ammerman · A. Vo
Comprehensive Transplant Center, Transplant Immunotherapy Program, Cedars-Sinai Medical Center, Los Angeles, CA, USA
e-mail: sjordan@cshs.org

D. K. C. Cooper, G. Byrne (eds.), *Clinical Xenotransplantation*,
https://doi.org/10.1007/978-3-030-49127-7_11

production and is driven primarily by the production of interleukin (IL)-6 and IL-21. Naïve B cells subsequently become activated by Tfh cells, and the persistence of IL-6 enhances development of memory B cells and plasmablasts.

IL-6 production in plasmablasts enhances germinal center formation and terminal development of donor-specific antibody (DSA)-producing plasma cells [1, 2]. Alloantibody migrates to the graft and initiates complement-dependent cytotoxicity (CDC) and antibody-dependent cellular cytotoxicity (ADCC), with resultant antibody-mediated rejection (AMR). DSA production usually results from inadequate immunosuppression or activation of established memory responses to alloantigens in sensitized individuals. Regardless, once established, alloimmmune responses persist for the life of the allograft and beyond, creating a highly HLA-sensitized individual. For a more comprehensive overview of the etiology of allosensitization and approaches to management, several recent papers are recommended [2–4].

Antibody-Mediated Rejection (AMR)

AMR is an increasingly recognized, severe form of allograft rejection, characterized by several pathologic variants resistant to treatment with standard immunosuppressive agents. Significant advances have occurred to identify patients at risk for AMR and the pathologic features associated with this diagnosis [2–4]. The immunopathology of AMR suggests an important role for antibodies, B cells, and plasma cells. Here, intravenous immune globulin (IVIg), rituximab, and/or plasmapheresis are commonly used for the treatment of acute AMR [5–9].

Despite successes, post-transplant AMR, chronic active AMR (cAMR), and transplant glomerulopathy remain significant problems that are only modestly amenable to these therapies. Data from the Deterioration in Kidney Allograft Function (DeKAF) study show that in the current era of immunosuppression, most graft losses have evidence of cAMR with $C4d^+$ staining [10]. It is estimated that 5000 allografts are lost each year in the USA, primarily from cAMR [11]. The current treatment paradigms rely on reduction of antibody levels to prevent AMR.

AMR is often seen in patients who are noncompliant or receiving inadequate immunosuppressive therapy and those who receive HLA-incompatible transplants. Additionally, transplant glomerulopathy usually results from persistent DSA-positivity which dissipates allograft function, resulting in graft failure and return to dialysis with devastating emotional consequences for patients and their families, and financial consequences for the healthcare system [12–16]. No current therapy is approved by the US Food and Drug Administration (FDA), and patients are often treated with combination therapies making analysis of efficacy difficult. Thus, the scope of antibody-induced injury in the transplant population is significant and increasing [10]. Here, there is a large unmet clinical need.

One critical obstacle in the successful treatment of AMR is addressing the long-lived nature of plasma cells and persistent DSA, despite treatment strategies [17]. An "ideal" treatment option would eliminate circulating DSAs, inhibit CDC and ADCC, and rebound DSA generation.

Therapeutic Approaches to Treatment of AMR

The Anti-CD20 Monoclonal Antibodies

Rituximab

Rituximab is a chimeric monoclonal antibody (mAb) aimed at CD20, a cell surface antigen highly expressed on pre-B- and mature B-lymphocytes, but not differentiated plasma cells. In a single center, double-blind, placebo-controlled study by van den Hoogen et al., patients were randomized to receive induction therapy with a single dose of rituximab or placebo, with standard immunotherapy. The primary endpoint was the incidence of biopsy-proven rejection 6 months post-transplant [18]. Induction with rituximab did not result in significant reductions in allograft rejection at 6 months versus placebo. However, in high-risk patients (retransplants, DSA^{+}), a significant reduction in allograft rejection was seen with rituximab.

Our group has published a blinded, placebo-controlled study of IVIg + rituximab versus IVIg + placebo for desensitization which demonstrated an essential role of rituximab in prevention of DSA rebound, AMR/cAMR, transplant glomerulopathy, and improved post-transplant graft survival. The study was unblinded due to severe adverse events in the IVIg + placebo group, including AMR ($n = 3$) and graft loss ($n = 2$) ($P = 0.06$). Although no significant differences were seen in DSAs at transplant, rapid DSA rebound associated with severe AMR occurred within 1 month post-transplant. DSAs trended downward in those who received IVIg + rituximab, and no cases of AMR were observed on for-cause and 6-month protocol biopsies. Additionally, the IVIg + rituximab group showed significant benefits in renal function at 6 and 12 months ($P = 0.04$) [19].

Zachary et al. so elegantly demonstrated the ability of rituximab to prevent anamnestic responses to alloantigens post-transplant. In 24 patients sensitized to HLA antigens, who did not have HLA antibody before transplantation, no post-transplant antibody to HLA antigens was detected in 10 rituximab-treated patients. However, HLA antibody was detected in 13 of 16 cases without rituximab treatment ($P = 0.00006$). Thus, elimination of peripheral HLA-specific B cells in patients who are sensitized to HLA antigens, but lacking detectable antibody abrogates an anamnestic response and risk for AMR [20].

Kohei et al. showed the administration of rituximab in ABO-incompatible transplant recipients ($n = 57$) resulted in long-term prevention of *dn*DSAs and low incidence of AMR versus a cohort of ABO-compatible living donor transplants not receiving rituximab ($n = 83$). The 5-year graft survival rates were 98.1% with rituximab versus 90.3% in the ABO-compatible group. At 2 years post-transplant, the incidence of AMR in the rituximab group was 3.5% versus 22.9% in the ABO-compatible patients [21].

Recent data from a randomized, placebo-controlled clinical trial showed that anti-CD20mAb did not add benefit to treatment of AMR with plasmapheresis + IVIg [22]. However, there were many troubling issues in this trial that likely limit the validity of the results. First, rituximab was given on the same day as IVIg. This likely limits the efficacy and half-life of rituximab due to IVIg blocking of the Fc neonatal receptors (FcRn) on endothelium that are responsible for recycling IgG molecules and extending their half-life. The investigators also performed plasma exchange <24 hours after IVIg + rituximab, removing the majority of the drug [23, 24].

Another randomized, placebo-controlled study assessed the efficacy of rituximab for treatment of AMR. This multicenter trial aimed to enroll 25 patients per arm but (due to cost) enrolled 25 in total (12 rituximab, 13 placebo). The authors found no difference in outcomes at 1 year. Despite good intentions, this study was not powered to formulate the conclusions put forward.

From our work in desensitization and that of others, rituximab shows benefits in removing memory B cells and limiting antibody rebound after other treatments (i.e., IVIg, plasmapheresis, IdeS) [19, 20].

Obinutuzumab

Obinutuzumab is a humanized, type II, immunoglobulin-G1 anti-CD20mAb, FDA-approved for chronic lymphocytic leukemia. Obinutuzumab, unlike rituximab is a more powerful depleter of B cells, operating through ADCC and direct apoptosis, with little effect through CDC. This is important since the concomitant use of complement inhibitors (i.e., eculizumab) with rituximab likely diminishes efficacy of rituximab, as its primary mode of B-cell depletion is through CDC.

In vitro obinutuzumab was twofold more efficient in reducing B-cell cytotoxicity. Specifically, obinutuzumab exhibited a more potent activation of natural killer (NK) cells and neutrophils, and a more effective Fcγ receptor interaction [25–27].

A Phase Ib, open-label study of single and repeat doses of obinutuzumab to assess pharmacokinetics, pharmacodynamics, and safety in highly HLA-sensitized end-stage renal disease patients awaiting kidney transplantation did not show benefit in reducing calculated panel-reactive antibody (cPRA) values, although ~50% of patients ultimately underwent transplantation [28]. Thus, obinutuzumab, a novel type 2 anti-CD20mAb with superior B-cell depletional activity and efficacy against plasmablasts, could be of benefit in prevention and treatment of AMR.

Costimulation Blockade

CTLA4Ig (Belatacept): An Inhibitor of T and B Cells

Recent studies have highlighted the importance of costimulatory blockade in controlling B-cell-directed immune responses to allografts [29]. Post-transplant generation of *dn*DSA is recognized as a major cause for allograft failure. To date, the immunosuppressive regimen associated with low *dn*DSA development is a failure to maintain therapeutic levels of tacrolimus [30]. Recent clinical trials using the novel costimulatory-blocking IgG Fc fusion protein containing CTLA4 (CTLA4-Ig) show that kidney transplant recipients treated with belatacept have better graft survival, graft function, and a lower proportion of *dn*DSAs versus cyclosporine [31]. Chen et al. showed that CTLA4-Ig treatment of allosensitized mice resulted in significant suppression of B memory cell responses [32].

Our group conducted dosing experiments in a mouse model of allogeneic sensitization to evaluate the efficacy of CTLA4-Ig treatment in DSA suppression. We found that CTLA4-Ig significantly inhibited *dn*DSA IgM and IgG production in mice sensitized to HLA-A2$^+$ skin grafts. In longitudinal experiments, we found that CTLA4-Ig administered during T-cell priming 90 days after primary skin graft exposure had a long-lasting effect in reducing DSA IgG memory responses to HLA-A2$^+$ skin grafts. The inhibitory effect of CTLA4-Ig in suppressing DSA memory responses was significantly enhanced by the addition of an anti-IL-6R antibody. We also demonstrated that dnDSA suppression by CTLA4-Ig is due to inhibition of T-dependent B-cell activation secondary to Tfh cell inhibition. In in vitro experiments with alloreactive plasma cells, we found that CTLA4-Ig inhibited plasma cell proliferation and Ig production. This suggests that plasma cells may depend on costimulation through CD28/B7 as a mechanism of activation [33].

Leibler et al. recently investigated the mechanisms involved in the control of humoral responses by analyzing the effect of belatacept on different steps of the B-cell-mediated response in humans. In vitro belatacept reduced plasmablast differentiation, Ig production, and the expression of the major transcription factor involved in plasma cell function, Blimp-1, in a *T-cell-independent* manner. Belatacept reduced the expression of CD86 on antigen-presenting cells (APCs). Additionally, belatacept blocked CD28-mediated activation of Tfh in an autologous Tfh-memory B-cell model.

In kidney transplant recipients treated with belatacept, investigators demonstrated that patients treated with belatacept had reduced effector B cells and activated Tfh cells compared with calcineurin inhibitor-treated patients. They concluded that belatacept modulates B-cell function directly at the level of B cell–Tfh interaction and these interactions are likely responsible for the modulation of humoral immunity seen in belatacept-treated patients [34]. This paper is of great interest since belatacept may emerge as an important agent for prevention and treatment of AMR.

Anti-plasma Cell Therapies

Daratumumab

Daratumumab is an anti-CD38mAb which induces potent CDC and ADCC against CD38$^+$ cells in patients with multiple myeloma. Daratumumab is the first anti-CD38mAb, and received FDA breakthrough status in 2015 [35]. CD38 is a type II transmembrane glycoprotein heavily involved in intracellular signaling via cell adhesion, calcium-dependent signal cascade, activation of NK cells, and IgG1 production from B- and T-cell signal transmission [36]. CD38 is more highly expressed on malignant cells and is present on the surface of short- and long-lived plasma cells. In a dose escalation study for multiple myeloma, daratumumab significantly reduced bone marrow plasma cells. CD38$^+$ T regulatory cells (Tregs) and CD38$^+$ B regulatory cells (Bregs) were decreased by daratumumab administration. This mechanism may provide a potential treatment of plasma cell-induced AMR. Overall, the safety profile of daratumumab is acceptable [37]. However, the impact on depletion of Tregs/Bregs by daratumumab may be a concern for induction of cell-mediated rejection in HLA-sensitized patients.

Proteasome Inhibitors

Bortezomib

Targeting antibody-producing plasma cells was felt to be a superior strategy to use of anti-CD20mAb in treating AMR. Several centers promoted the use of bortezomib, a proteasome inhibitor approved for the treatment of multiple myeloma. Data supporting the use of bortezomib were limited to single centers, and evidence on efficacy and safety from a larger cohort of patients with AMR was lacking. However, two recent papers have addressed this issue.

First, Eskandary et al. reported results of the first prospective, randomized, placebo-controlled trial of bortezomib in patients with late active AMR (BORTEJECT Trial) [38]. Of 744 patients, 44 met study criteria and were randomized to two cycles of bortezomib ($n = 21$) or placebo ($n = 23$). In direct contradiction to previous reports, these investigators found that bortezomib had no effect on outcomes over 2 years of follow-up (GFR slope −4.7 versus −5.2 ml/min per 1.73 m^2 per year). Proteinuria, DSA and AMR histology, and molecular microscopic analysis did not differ between the two groups. Bortezomib therapy was associated with more drug-related side effects [38].

Moreno Gonzales et al. recently reported that 32 doses of bortezomib for desensitization were not well tolerated and had only a modest impact on anti-HLA antibodies [39]. Thus, the role of bortezomib as a future therapeutic agent in treatment and prevention of AMR is questionable, and may be more effective when combined with other therapies.

Carfilzomib

Ensor et al. reported on carfilzomib-based therapy for treatment of AMR in lung transplant recipients. Carfilzomib is a second-generation proteasome inhibitor that irreversibly binds the 26s proteasome and permanently inhibits activity. These investigators found that treatment with carfilzomib resulted in significant reductions in DSAs and improvement in lung allograft function during the treatment period in 10 of 14 patients. However, seven deaths occurred in carfilzomib responders due to allograft failure. The authors suggest that severe AMR may not be amenable to intermittent carfilzomib therapy [40]. It is also likely that rebound DSA responses after cessation of carfilzomib accelerated the decline in allograft function. Carfilzomib + CTLA4-Ig may offer an excellent therapeutic approach for prevention and treatment of AMR [41].

IL-6 and IL-6R Inhibitors

IL-6 was first described as a multifunctional cytokine that directed the development and maturation of B cells to plasma cells and sustained antibody production [1, 42–44]. In this regard, the role of IL-6 in induction of Tfh cells is critical for initiation of adaptive immune responses, progression of naive B cells to plasma cells, and production of high-affinity antibodies [1, 44, 45]. Additionally, persistence of IL-6/

IL-6R (IL-6 receptor) signaling inhibits Treg cell development, thus enhancing Tfh and Th17 pathogenic antibody and inflammatory functions. Importantly, anti-IL-6/IL-6R therapies are known to have effects in reducing Th17 and Tfh cells, which block autoimmunity and reduce pathogenic antibody production. The ability of anti-IL-6/IL-6R therapy to inhibit Tfh activity and reduce alloreactive B cells, plasmablasts, and DSA production is a significant consideration in the prevention and treatment of alloantibody-induced injury.

IL-6 is a growth factor critical for B cells and plasma cells and is produced by plasmablasts, resulting in new germinal center formation. IL-6 inhibition significantly reduces Th17 and Tfh cells, plasmablasts, and upregulates Treg cells. Clinical trials of anti-IL-6/IL-6R therapies have been completed or are now underway in kidney transplantation for the treatment and prevention of AMR [46–50]. Data on the two most important IL-6 inhibitors are below.

Tocilizumab

Tocilizumab is a first-in-class mAb directed at the IL-6R. Tocilizumab was FDA-approved for treatment of rheumatoid arthritis and juvenile idiopathic arthritis in 2011. Tocilizumab resulted in reductions in peripheral pre- and postswitch memory B cells, IgG^+ and IgA^+ B cells, and significantly reduced B-cell hyper-reactivity.

Our group conducted a single-center phase I/II open-label study in HLA-sensitized patients with end-stage renal disease who had failed desensitization with IVIg + rituximab ± plasmapheresis. All patients received IVIg 10% 2 g/kg on days 1 and 30, and tocilizumab 8 mg/kg on day 15, then monthly for 6 months. Of the ten patients, five received transplants (two were withdrawn due to noncompliance with protocol pretransplant). Mean time to transplant from first desensitization decreased from 25 ± 10.5 to 8 ± 5.4 months with tocilizumab treatment. Reductions in immunodominant DSAs were seen in all transplanted patients at transplant ($p = 0.024$) and most significantly at 12 months ($p = 0.0003$). All patients received six tocilizumab doses post-transplant and 6-month protocol biopsies showed no evidence of rejection. The estimated glomerular filtration rate (eGFR) at 12 months in transplanted patients was 60 ± 25 ml/min [51].

In another single center, open-label study conducted by our group, highly sensitized patients with DSA^+ cAMR, who failed IVIg + rituximab ± plasmapheresis therapy, received tocilizumab 8 mg/kg monthly for 6–25 months. A total of 36 patients were assessed (between 2011 and 2016) for allograft loss, patient survival, DSA reduction, and improvement in biopsy results. They were compared with a historical cohort treated with standard-of-care therapy (IVIG ± rituximab ± plasmapheresis) ($n = 39$). The median follow-up was 3.26 years (maximum 7 years). Four of 36 tocilizumab-treated patients had graft failure (11.1%). Repeat biopsies in tocilizumab-treated patients showed significant reductions in glomerulitis + peritubular capillaritis and $C4d^+$ scores. This contrasted to 21 of 39 graft losses (54%) in patients treated with standard-of-care therapy. At 6 years post-cAMR diagnosis, the tocilizumab-treated patients had a graft and patient survival probability of 80% and 91%, respectively [52].

Clazakizumab

Clazakizumab is an immunoglobulin G1 (IgG1) mAb aimed at the IL-6 ligand. Clazakizumab has been evaluated extensively in patients with rheumatoid arthritis, but is not FDA-approved for any condition [53]. Our center has recently initiated two phase I/II, open-label, single-arm exploratory studies.

First, an AMR study ($n = 10$) will examine the safety and tolerability of clazakizumab 25 mg subcutaneously every 4 weeks in DSA$^+$ patients with cAMR transplant glomerulopathy on biopsy [NCT03380377] [48]. Second, a desensitization study will assess 10 highly sensitized patients with cPRA >50% awaiting either a living donor or deceased donor kidney transplant. Eligible patients will receive plasmapheresis + IVIg followed by 6 months of clazakizumab 25 mg subcutaneously until transplantation. If transplanted, patients will receive six additional doses [NCT03444103] [47]. An additional placebo-controlled study assessing the utility of clazakizumab for treatment of cABMR is being conducted in Vienna and Berlin [46]. A large blinded, placebo-controlled multicenter study in cAMR is set to start in 2019.

IgG-Degrading Enzyme of *Streptococcus pyogenes* (*IdeS, imlifidase*)

Imlifidase is a novel drug that is being developed for desensitization and treatment of AMR in kidney transplant patients. Imlifidase is an immunomodulating enzyme that cleaves all four IgG antibody subclasses into F(ab')2 and Fc fragments at the lower hinge region with high specificity. Other immunoglobulins, including IgA, IgM, IgE, and IgD, are not affected by the administration of imlifidase [54]. A critical observation in early assessments of IdeS in vitro is the inability of F(ab)'2 fragments to mediate CDC and ADCC [55].

In 2015, a single-arm, single-center, Phase II dose-finding study was completed in Sweden using Imlifidase in HLA-sensitized patients awaiting kidney transplantation to evaluate safety, tolerability, pharmacokinetics, and efficacy of imlifidase. The results demonstrated that imlifidase treatment eliminated all HLA antibodies detected by Luminex single antigen bead assays 6 hours after infusion and, more importantly, eliminated all complement-activating (C1q$^+$) HLA antibodies 1 hour postinfusion [56].

We have also shown that IdeS is a potent inhibitor of ADCC [limiting NK-cell γ-IFN (interferon) release induced by anti-HLA antibodies binding to target endothelial cells] [57]. These initial observations led to the development of an open-label, Phase I/II study of imlifidase for desensitization in 25 highly sensitized patients with DSAs undergoing living and deceased donor kidney transplantation in Sweden and USA. All patients received imlifidase infusion prior to transplantation. The objective of the study was to assess the ability of imlifidase to eliminate DSAs in patients who were DSA$^+$, with a positive crossmatch at time of transplantation. The patient group at Cedars-Sinai received desensitization therapy with IVIg 2 g/kg + rituximab, while the patients in Uppsala did not receive desensitization.

Of the 25 patients who received IdeS, 24 were successfully transplanted. AMR occurred at a mean of 2 weeks post-transplant in three patients in the Swedish arm due to rebound DSA, with C4d-positivity on for-cause biopsies. At Cedars-Sinai, AMR occurred in two patients at 2 and 5 months post-transplant, which correlated with an increase in DSA intensity and resolved with treatment. The differences in rebound times likely reflect the post-transplant use of IVIg + rituximab in the US patients. Long-term outcomes for these patients have been good [58]. A trial of imlifidase for treating AMR is now underway. In summary, imlifidase may represent an important breakthrough in prevention and treatment of AMR.

Important Considerations When Administering Biologic Agents for AMR

IdeS (imlifidase)

Imlifidase cleaves human and rabbit IgG at the hinge region creating F(ab)'2 and Fc fragments of all IgG molecules in the body in 4–6 hours after administration. The half-life of IdeS is approximately 8–12 hours, but IgG cleaving capacity may last for up to 4 days. This poses a problem for induction therapy post-transplant. Prior to performing the first clinical desensitization trials, we found that alemtuzumab was rapidly cleaved by imlifidase-treated sera up to 4 days post-transplant [58]. We altered our induction protocol to use high-dose steroids on days 1–4 and alemtuzumab on day 4. This gave similar T-cell depletion as seen in non-imlifidase-treated patients. Since thymoglobulin is also cleaved by IdeS, one should use a similar approach or consider horse anti-thymocyte globulin. Additionally, the post-transplant administration of IVIg + rituximab should be delayed for 4–5 days after IdeS administration.

Neonatal Fc Receptors and Half-Life of IgG

A common characteristic of IgG molecules is their long serum half-life of 3–4 weeks. This is related to IgG's interaction with the FcRn. The function of FcRn is twofold. First, FcRn binds to serum IgG that has been endocytosed into lysosomes by endothelial cells or myeloid cells under low pH conditions. Here, the IgG is recycled back to the cell surface and, under neutral pH conditions, released back into the serum. Second, FcRn are expressed on the villi of placentas that are exposed to maternal blood and, after the 28th week of gestation, are responsible for endocytosing and transporting IgG molecules at a high rate from mother to neonate [23, 24]. In animal models of FcRn-knockouts, there is a dramatic reduction in IgG half-life from weeks to a few days.

It is also known that IVIg can occupy FcRn and thus accelerate the turnover of pathogenic antibodies. This may explain why large amounts of IVIg are needed for therapeutic activity in autoimmune and alloimmune disorders. Here, the antibodies in IVIg preparations compete with pathological autoantibodies and alloantibodies for FcRn binding. This explanation is supported by the observations that IVIg resulted in a reduction of approximately 50% in autoantibody half-life in a rat model of immune thrombocytopenic purpura (ITP) and in a neonatal mouse model of bullous pemphigoid [24].

Although FcRn saturation by IVIg is likely important in accelerating the clearance of pathogenic antibodies, it could also result in the rapid clearance of therapeutic antibodies. Little information is available, although there is one report of accelerated clearance of eculizumab when given with high-dose IVIg for treatment of multiple sclerosis [59]. In this regard, it is critical to avoid dosing therapeutic mAbs in temporal proximity to high-dose IVIg. This is likely to rapidly diminish the therapeutic efficacy of the monoclonal antibodies. This is important since reports assessing the efficacy of rituximab in the treatment of AMR in a placebo-controlled trial concluded that rituximab added no benefit to IVIg and plasmapheresis. Here, rituximab was given immediately after IVIg, which likely diminished the half-life and efficacy of rituximab [25]. In our practice, we wait ~7 days after high-dose IVIg administration to proceed with any chimeric or humanized mAbs. We have recently completed a trial to more thoroughly assess the impact of concomitant IVIg treatment on half-life of humanized mAb (anti-C5) therapy [NCT02878616] [60].

Summary

Modification of alloimmunity and alloantibodies will have relevance to all solid organ allotransplantation and to xenotransplantation, where xenoantibodies present a formidable obstacle. The ease of administration of biologic agents will likely change our views of immunosuppression to one of immune modulation that will ultimately result in better, more effective, and less toxic allograft-sustaining therapies, and also increase patient compliance. It is critical to advance transplant therapeutics to the next level where biologic agents are likely to play important roles in addressing the persisting barriers to successful transplantation created by alloimmunity.

Acknowledgments The authors acknowledge the team members of the Kidney Transplant Program, Transplant Immunology Laboratory, and HLA Laboratory at Cedars-Sinai Medical Center for their commitment to improving lives through transplantation. The authors also want to acknowledge the patients who kindly participated in clinical trials to evaluate these potentially important drugs.

References

1. Chavele KM, Merry E, Ehrenstein MR. Cutting edge: circulating plasmablasts induce the differentiation of human T follicular helper cells via IL-6 production. J Immunol. 2015;194:2482–5.
2. Jordan SC, Ammerman N, Choi J, et al. Novel therapeutic approaches to allosensitization and antibody-mediated rejection. Transplantation. 2019;103:262–72.

3. Loupy A, Lefaucheur C. Antibody-mediated rejection of solid-organ allografts. N Engl J Med. 2018;379:1150–60.
4. Valenzuela NM, Reed EF. Antibody-mediated rejection across solid organ transplants: manifestations, mechanisms, and therapies. J Clin Invest. 2017;127:2492–504.
5. Jordan SC, Tyan D, Stablein D, et al. Evaluation of intravenous immunoglobulin as an agent to lower allosensitization and improve transplantation in highly sensitized adult patients with end-stage renal disease: report of the NIH IGO2 trial. J Am Soc Nephrol. 2004;15:3256–62.
6. Marfo K, Lu A, Ling M, Akalin E. Desensitization protocols and their outcome. Clin J Am Soc Nephrol. 2011;6:922–36.
7. Montgomery RA, Lonze BE, King KE, et al. Desensitization in HLA-incompatible kidney recipients and survival. N Engl J Med. 2011;365:318–26.
8. Vo AA, Lukovsky M, Toyoda M, et al. Rituximab and intravenous immune globulin for desensitization during renal transplantation. N Engl J Med. 2008;359:242–51.
9. Vo AA, Petrozzino J, Yeung K, et al. Efficacy, outcomes, and cost-effectiveness of desensitization using IVIG and rituximab. Transplantation. 2013;95:852–8.
10. Sellarés J, DeFreitas DG, Mengel M, et al. Understanding the causes of kidney transplant failure: the dominant role of antibody-mediated rejection and nonadherence. Am J Transplant. 2012;12:388–99.
11. Loupy A, Hill GS, Jordan SC. The impact of donor-specific anti-HLA antibodies on late kidney allograft failure. Nat Rev Nephrol. 2012;8:348–57.
12. Solez K, Colvin RB, Racusen LC, et al. Banff '05 meeting report: differential diagnosis of chronic allograft injury and elimination of chronic allograft nephropathy ('CAN'). Am J Transplant. 2007;7:518–26.
13. Gloor J, Cosio F, Lager DJ, et al. The spectrum of antibody-mediated renal allograft injury: implications for treatment. Am J Transplant. 2008;8:1367–73.
14. Lefaucheur C, Loupy A, Vernerey D, et al. Antibody-mediated vascular rejection of kidney allografts: a population-based study. Lancet. 2013;381:313–9.
15. Jordan SC, Reinsmoen N, Peng A, et al. Advances in diagnosing and managing antibody-mediated rejection. Pediatr Nephrol. 2010;25:2035–45.
16. Haas M, Sis B, Racusen LC, et al. Banff 2013 meeting report: inclusion of c4d-negative antibody-mediated rejection and antibody-associated arterial lesions. Am J Transplant. 2014;14:272–83.
17. Schrezenmeier E, Jayne D, Dorner T. Targeting B cells and plasma cells in glomerular diseases: translational perspectives. J Am Soc Nephrol. 2018;29:741–58.
18. van den Hoogen MW, Kamburova EG, Baas MC, et al. Rituximab as induction therapy after renal transplantation: a randomized, double-blind, placebo-controlled study of efficacy and safety. Am J Transplant. 2015;15:407–16.
19. Vo AA, Choi J, Cisneros K, et al. Benefits of rituximab combined with intravenous immunoglobulin for desensitization in kidney transplant recipients. Transplantation. 2001;98:312–9.
20. Zachary AA, Lucas DP, Montgomery RA, et al. Rituximab prevents an anamnestic response in patients with cryptic sensitization to HLA. Transplantation. 2013;95:701–4.
21. Kohei N, Hirai T, Omoto K, et al. Chronic antibody-mediated rejection is reduced by targeting B-cell immunity during an introductory period. Am J Transplant. 2012;12:469–76.
22. Moreso F, Crespo M, Ruiz JC, et al. Treatment of chronic antibody mediated rejection with intravenous immunoglobulins and rituximab: a multicenter, prospective, randomized, double-blind clinical trial. Am J Transplant. 2018;18:927–35.
23. Roopenian DC, Christianson GJ, Sproule TJ, et al. The MHC class I-like IgG receptor controls perinatal IgG transport, IgG homeostasis, and fate of IgG-Fc-coupled drugs. J Immunol. 2003;170:3528–33.
24. Gelfand EW. Intravenous immune globulin in autoimmune and inflammatory diseases. N Engl J Med. 2012;367:2015–25.
25. Reddy V, Klein C, Isenberg DA, et al. Obinutuzumab induces superior B-cell cytotoxicity to rituximab in rheumatoid arthritis and systemic lupus erythematosus patient samples. Rheumatology. 2017;56:1227–37.

26. Obinutuzumab (Gazyva) [package insert]. Genentech, Inc., San Francisco; 2017.
27. Jain P, Kanagal-Shamanna R, Wierda W, et al. Membranoproliferative glomerulonephritis and acute renal failure in a patient with chronic lymphocytic leukemia: response to obinutuzumab. Hematol Oncol Stem Cell Ther. 2017;10:151–4.
28. A study of obinutuzumab to evaluate safety and tolerability in hypersensitized adult participants with end stage renal disease awaiting transplantation. Available from: https://ClinicalTrials.gov/show/NCT02586051.
29. Leibler C, Thiolat A, Elsner RA, El Karoui K, Samson C, Grimbert P. Costimulatory blockade molecules and B-cell-mediated immune response: current knowledge and perspectives. Kidney Int. 2019;95:774–86.
30. Davis S, Gralla J, Klem P, et al. Lower tacrolimus exposure and time in therapeutic range increase the risk of de novo donor-specific antibodies in the first year of kidney transplantation. Am J Transplant. 2018;18:907–15.
31. Vincenti F. Belatacept and long-term outcomes in kidney transplantation. N Engl J Med. 2016;374:2600–1.
32. Chen J, Wang Q, Yin D, et al. Cutting edge: CTLA-4Ig inhibits memory B cell responses and promotes allograft survival in sensitized recipients. J Immunol. 2015;195:4069–73.
33. Kim I, Wu G, Chai NN, et al. Immunological characterization of de novo and recall alloantibody suppression by CTLA4Ig in a mouse model of allosensitization. Transpl Immunol. 2016;38:84–92.
34. Leibler C, Thiolat A, Hénique C, et al. Control of humoral response in renal transplantation by belatacept depends on a direct effect on B cells and impaired T follicular helper-B cell crosstalk. J Am Soc Nephrol. 2018;29:1049–62.
35. Dolgin E. Cancer's true breakthroughs. Nat Med. 2013;19:660–3.
36. de Weers M, Tai YT, Van Der Veer MS, et al. Daratumumab, a novel therapeutic human CD38 monoclonal antibody, induces killing of multiple myeloma and other hematological tumors. J Immunol. 2011;186:1840–8.
37. Daratumumab (Darzalex) [package insert]. Janssen Biotech, Inc, Horsham; 2017.
38. Eskandary F, Regele H, Baumann L, et al. A randomized trial of bortezomib in late antibody-mediated kidney transplant rejection. J Am Soc Nephrol. 2018;29:591–605.
39. Gonzales MAM, Gandhi MJ, Schinstock CA, et al. 32 doses of bortezomib for desensitization is not well tolerated and is associated with only modest reductions in anti-HLA antibody. Transplantation. 2017;101:1222–7.
40. Ensor CR, Yousem SA, Marrari M, et al. Proteasome inhibitor carfilzomib-based therapy for antibody-mediated rejection of the pulmonary allograft: use and short-term findings. Am J Transplant. 2017;17:1380–8.
41. Kwun J, Burghuber C, Manook M, et al. Successful desensitization with proteasome inhibition and costimulation blockade in sensitized nonhuman primates. Blood Adv. 2017;1:2115–9.
42. Hunter CA, Jones SA. IL-6 as a keystone cytokine in health and disease. Nat Immunol. 2015;16:448–57.
43. Liu X, Jones GW, Choy EH, et al. The biology behind interleukin-6 targeted interventions. Curr Opin Rheumatol. 2016;2:152–60.
44. Tanaka T, Kishimoto T. The biology and medical implications of interleukin-6. Cancer Immunol Res. 2014;2:288–94.
45. Wood KJ, Bushell A, Hester J. Regulatory immune cells in transplantation. Nat Rev Immunol. 2012;12:417–30.
46. A pilot trial of clazakizumab in late ABMR. Available from: https://ClinicalTrials.gov/show/NCT03444103.
47. Clazakizumab in highly-HLA sensitized patients awaiting renal transplant. Available from: https://ClinicalTrials.gov/show/NCT03380962.
48. Clazakizumab for chronic and active Antibody mediated rejection post-kidney transplant. Available from: https://ClinicalTrials.gov/show/NCT03380377.
49. A safety study of tocilizumab to Improve transplant rates in highly sensitized patients awaiting kidney transplantation. Available from: https://ClinicalTrials.gov/show/NCT01594424.

50. Cedars-Sinai Medical C. A safety study of tocilizumab to improve transplant rates in highly sensitized patients awaiting kidney transplantation. 2014 November. Available from: https://ClinicalTrials.gov/show/NCT01594424.
51. Vo AA, Choi J, Kim I, et al. A phase I/II trial of the interleukin-6 receptor-specific humanized monoclonal (tocilizumab) + intravenous immunoglobulin in difficult to desensitize patients. Transplantation. 2015;99:2356–63.
52. Choi J, Aubert O, Vo AA, et al. Assessment of tocilizumab (anti-interleukin-6 receptor monoclonal) as a potential treatment for chronic antibody-mediated rejection and transplant glomerulopathy in HLA-sensitized renal allograft recipients. Am J Transplant. 2017;17:2381–9.
53. Weinblatt ME, Mease P, Mysler E, et al. The efficacy and safety of subcutaneous clazakizumab in patients with moderate-to-severe rheumatoid arthritis and an inadequate response to methotrexate: results from a multinational, phase IIb, randomized, double-blind, placebo/active-controlled, dose-ranging study. Arthritis Rheumatol. 2015;67:2591–600.
54. von Pawel-Rammingen U, Johansson BP, Bjorck L. IdeS, a novel streptococcal cysteine proteinase with unique specificity for immunoglobulin. EMBO J. 2002;21:1607–15.
55. Tradtrantip L, Asavapanumas N, Verkman AS. Therapeutic cleavage of anti-aquaporin-4 autoantibody in neuromyelitis optica by an IgG-selective proteinase. Mol Pharmacol. 2013;83:1268–75.
56. Phase II Study, evaluation of safety and efficacy of IdeS in chronic kidney disease. Available from: https://ClinicalTrials.gov/show/NCT02224820.
57. Toyoda M, Shin BH, Ge S, et al. Impact of desensitization on antiviral immunity in HLA-sensitized kidney transplant recipients. J Immunol Res. 2017;2017:5672523.
58. Jordan SC, Lorant T, Choi J, et al. IgG endopeptidase in highly sensitized patients undergoing transplantation. N Engl J Med. 2017;377:442–53.
59. Fitzpatrick AM, Mann CA, Barry S, et al. An open label clinical trial of complement inhibition in multifocal motor neuropathy. J Peripher Nerv Syst. 2011;16:84–91.
60. Study in end-stage renal disease patients awaiting kidney transplant to investigate the potential effect of IVIG treatment on the pharmacokinetics and pharmacodynamics of LFG316. Available from: https://ClinicalTrials.gov/show/NCT02878616.

Lessons from ABO-Incompatible Cardiac Allotransplantation in the Newborn

12

Jeffrey L. Platt and Marilia Cascalho

Abbreviations

Gal Galactose-α1,3-galactose
HLA Human leukocyte antigen

Introduction

Management of severe cardiac failure in newborn infants represents the most agonizing impact of the shortage of human hearts for transplantation. As we recently discussed [1], most newborn infants with severe cardiac failure owing to the hypoplastic left heart syndrome and related univentricular defects are offered palliative procedures rather than transplantation. This is because the few hearts of a size suitable for newborn infants are directed to patients with congenital cardiomyopathy, which is not amenable to palliation.

The shortage of human hearts also motivated the deliberate use of ABO-incompatible hearts for transplantation in young infants [2]. Similarities between the natural antibodies that recognize human blood group A and B antigens and natural xenoreactive antibodies, and similarities in the pathogenic consequences of binding of these antibodies to organ transplants, suggested to others and to us that ABO-incompatible organ transplants could model some aspects of the immune barrier to xenotransplantation [3–7].

J. L. Platt (✉) · M. Cascalho
Departments of Surgery and Microbiology & Immunology, University of Michigan, Ann Arbor, MI, USA
e-mail: plattjl@umich.edu

D. K. C. Cooper, G. Byrne (eds.), *Clinical Xenotransplantation*, https://doi.org/10.1007/978-3-030-49127-7_12

The past 20 years has brought much more experience transplanting organs across blood group A and B barriers, especially in newborn recipients, and new insights into the barriers to xenotransplantation in nonhuman primate recipients. Here, we discuss how eventually lessons drawn from cardiac transplantation across ABO blood group barriers in newborn recipients might impact the clinical application of xenotransplantation.

The Advent of ABO-Incompatible Transplantation

ABO-incompatible kidney transplants performed during the first decades of the era of clinical transplantation often suffered immediate nonfunction (ischemia-reperfusion injury and/or hyperacute rejection) or early severe rejection and graft loss [8–11]. For example, Wildbrandt et al. [11] described the course of 12 ABO-incompatible kidney transplants – seven underwent severe rejection or thrombosis within 2 weeks, and ultimately all but one failed. Despite occasional success of individual transplants, the generally dismal outcome of ABO-incompatible kidney transplants led to the widely accepted view that ABO incompatibility of organ allografts should be avoided [12, 13].

The outcomes of ABO-incompatible kidney transplants improved dramatically in the 1980s when procedures, such as plasmapheresis, that remove natural antibodies from graft recipients were used in conjunction with splenectomy [14]. This improvement was observed before [14] and after cyclosporine was available [15]. These manipulations together prevented hyperacute rejection and nearly averted early antibody-mediated rejection that previously had been found to cause the loss of most ABO-incompatible kidney transplants. Similar success was observed for transplantation of kidneys of blood group A2 into immunosuppressed, but otherwise unmanipulated, recipients of blood group O, in which setting the antibody-antigen reaction is much diminished [16]. These observations together provided compelling evidence that early vulnerability of ABO-incompatible transplants resulted from the reaction of isohemagglutinins of the recipient with blood group antigen in the graft, a conclusion confirmed by analysis of the specificity-bound antibody [17, 18].

The early experience with ABO-incompatible transplants revealed a phenomenon potentially pertinent to xenotransplantation. When early rejection of ABO-incompatible kidney transplants was averted, antibodies against the blood group antigens expressed in the graft sometimes returned in the circulation without inducing rejection and graft loss [18, 19]. Of various potential explanations for the absence of acute rejection, including (i) a change in graft antigen, or (ii) in the properties of the antibodies, or (iii) an acquired resistance of the graft to injury [5], acquired resistance emerged as the preferred and most broadly applicable mechanism [7, 20, 21]. The phenomenon was named "accommodation" [5]. Accommodation was also found to occur in porcine organ xenografts in nonhuman primates treated with similar regimens [22–24].

However, after accommodation began, ABO-incompatible kidney transplants and organ xenografts exhibited dramatic differences in outcome. ABO-incompatible transplants in which accommodation occurred could endure the intermittent or continuous presence of anti-blood group antibodies, i.e., accommodation persisted, and overall success approached the success of ABO-compatible transplants [25–27]. Organ xenografts, however, with accommodation might survive for days or weeks, but ultimately the xenografts developed the pathologic lesions characteristic of antibody-mediated rejection and failed [6, 28]. Thus, while ABO-incompatible transplants could model the inital immune barrier to xenotransplantation, by the 1990s it became apparent that ABO-incompatible transplants drammatically differed from organ xenografts in long term outcome and this difference revealed the existence of distinct immunological or biological barriers to xenotransplantation.

Immunity Versus Biochemical Incompatibility in the Failure of Organ Grafts

Why do xenografts fail under conditions in which ABO-incompatible transplants succeed? One or both of two explanations are commonly proposed. One explanation is that xenografts may elicit more intense and destructive immune reactions than allografts. The other explanation is that biochemical incompatibility between the xenograft and the recipient causes xenografts to fail under conditions that allografts thrive. Addressing this question could have important implications for the clinical application of xenotransplantation.

If immunity poses the main barrier today, then more robust immunosuppression, and possibly tolerance, is probably needed to make the outcome of xenografts approach the outcome of ABO-incompatible allografts. In the 1960s and 1970s, when immunosuppressive regimens were far less effective than today, efforts to improve the outcome of transplants by administration of agents that interfere with effector pathways led to no clinical advances. In contrast, better immunosuppressive agents or regimens eventuated in progressive improvement in transplant outcomes (see [29] as one example).

On the other hand, we showed years ago that partial, and even transient, correction of incompatibility between the complement system of the recipient and the complement-regulatory proteins expressed in a xenograft can have significant impact on the outcome of xenografts [30], well beyond what can be achieved with incremental improvement in immunosuppressive regimens.

As we discussed previously [5, 7], the greater diversity and abundance of foreign antigens in xenografts could fuel more powerful immune responses. On the other hand, incompatibilities between the co-receptors and cytokines of the graft and recipient could limit the ability of T cells to recognize xenogeneic cells and in this way suppress the intensity of immunity [31]. Still, regardless of whether molecular incompatibilities limit the initiation of immune responses to xenotransplantation, the immune responses that do occur have a greater impact on xenogeneic targets because of incompatibilities in the control of complement and coagulation between species. Thus, the effector function systems responding to xenografts, allografts, and infectious agents recruit complement and coagulation among other inflammatory pathways to modify the physiology of blood vessels. However, incompatibility between the recipient and a xenograft limits the control of effector pathways, thereby amplifying the impact of effector pathways [32, 33].

Since current knowledge does not provide the full range of insights needed to engineer human system compatibility into pigs [34], the best possibility for weighing the importance of immunity and biochemical compatibility would involve induction of immunological tolerance to test residual incompatibility or is by countering incompatibilities in ways that allow analysis of the intensity of immunity to a xenograft [7].

ABO-Incompatible Cardiac Transplantation in Infancy: An Experiment of Nature

For reasons that may be obvious, ABO-incompatible hearts are only occasionally used for transplantation in adults. Still, the outcomes of ABO-incompatible cardiac transplants appear to mirror the outcomes (described above) for ABO-incompatible kidney transplants. That is, notwithstanding a few cases of hyperacute and early acute antibody-mediated rejection in a relatively small numbers of cases, ABO-incompatible cardiac transplants in adult recipients exhibited the same frequency of cumulative graft failure as ABO-compatible transplants [35]. It is important to consider that all or nearly all ABO-incompatible cardiac transplants are performed across human leukocyte antigen (HLA) barriers, that the circumstances leading to ABO-incompatible transplantation could sensitize some potential recipients to HLA, and that it is difficult or impossible to determine whether antibody-mediated rejection is caused by antibodies against allogeneic blood group antigens or HLA.

Although far fewer cardiac transplants are performed in infants and young children, the experience with ABO-incompatible cardiac transplantation has been significant and instructive [1, 36]. ABO-incompatible cardiac transplants in young recipients generate outcomes similar to those observed for ABO-compatible cardiac transplants in the same age group – similar overall survival, similar freedom from rejection, and absence of antibody-mediated rejection [36–38].

The unique characteristics of the immune system of newborn infants make the responses to ABO-incompatible transplants in this group of recipients especially illuminating. Most newborn infants lack or have only low concentrations of antibodies against blood groups A and B and against galactose-α1,3-galactose (Gal) [39, 40]. Newborn infants are also less likely than mature individuals to be sensitized to HLA or cross-reactive antigens. Therefore, the early rejection once characteristic of ABO-incompatible transplants generally does not occur in ABO-incompatible cardiac transplants in newborn recipients. Consistent with these concepts and despite unique technical hurdles, ABO-incompatible and ABO-compatible cardiac allografts in newborn infants undergo fewer episodes of acute rejection and achieve the same or better long-term success than cardiac allografts in older individuals [41–44].

Consistent with immunological immaturity, newborn recipients of ABO-incompatible cardiac allografts generally continue to lack detectable antibodies against donor-specific blood groups and may be less apt to produce antibodies against donor HLA [45] and to develop chronic vasculopathy than recipients that underwent transplantation in childhood or maturity [46, 47].

In addition, newborn infants with ABO-incompatible cardiac allografts can become tolerant to the allogeneic blood group antigens expressed in the graft [48]. The absence over time of antibodies directed against allogeneic blood groups expressed in cardiac allografts could reflect absorption of the antibodies to the graft or decreased production. However, ELISPOT analysis designed to enumerate B cells specific for blood group antigens reveals that at least some newborn recipients of ABO-incompatible cardiac transplants continue to lack B cells specific for the blood group antigen in the graft but to have B cells specific for allogeneic blood group antigens *not* present in the graft [48]. In short, these recipients developed B cell tolerance.

Given the similarities between antibodies specific for blood groups A and B and antibodies specific for Gal and some other xenogeneic saccharides, and in the delayed ontogeny of B cells specific for those saccharides, there is reason to think that newborn recipients of grafts expressing Gal and other xenogeneic saccharides could develop tolerance to these antigens [49]. Although the recipients cannot be expected to develop tolerance to MHC-encoded antigens or other polypeptides with conventional regimens of immunosuppression [1], production of such "elicited antibodies" might be limited, as observed in recipients of ABO-incompatible transplants [45].

Further, to the extent that newborn recipients have naive repertoires of B cells and T cells, the immunosuppressive regimen needs only to prevent primary B cell and T cell responses for which the threshold for response is relatively high. In contrast, mature recipients have immune memory for antigens that might cross-react with HLA [50] – the threshold for recruitment of which is much lower. Accordingly, newborn recipients may offer an ideal model in which to weigh the importance of the immune barrier to xenotransplantation and the barrier posed by biochemical incompatibility.

Accommodation in Allografts and Xenografts

A puzzling observation regarding ABO-incompatible transplants led to the discovery of accommodation. Frequent analysis of blood from the recipients of ABO-incompatible transplants revealed absolutely no relationship between the levels of antibodies against allogeneic blood group antigens in the graft and graft function [15, 18, 19]. In light of the clear dose-response relationship observed when varying concentrations of anti-blood group A or B antibodies are combined with standard numbers of erythrocytes plus complement, this inconsistency was striking. The range of responses of ABO-incompatible kidney transplants and organ xenografts to the presence of graft-specific antibodies led to the idea that, unlike erythrocytes, transplanted organs might play an active role in determining whether, and to which extent, anti-graft antibodies induce injury. The observations also led to an initial operative definition of accommodation as a condition in which an organ graft continues to function without acute rejection despite the presence in the recipient of antibodies directed against the graft [5]. Based on this definition, nearly all ABO-incompatible kidney transplants in mature individuals, and most kidney transplants in pre-sensitized individuals, demonstrate accommodation at one time or another, if not persistently.

The need for accommodation to sustain graft integrity and function was supported by investigation of the response of cultured cells to binding by antibodies, activation of complement, and interaction with leukocytes. These lines of inquiry showed that very small amounts of bound antibodies, or interaction with only a few activated leukocytes, induce (in minutes to hours) the pathophysiologic changes in endothelium thought to underlie acute rejection [51–58]. Clearly, transplanted organs, and particularly the endothelial lining of blood vessels in transplanted organs, are protected from acute responses in ways cultured cells are not.

One mechanism of protection involves the diversion of reactants in the blood away from portions of the vascular network [59]. This mechanism explains some phenomena associated with accommodation [60], but the vessels into which reactants flow are still susceptible, and hence cellular changes that increase resistance to injury must still occur.

Another mechanism that is surely essential for accommodation is acquired resistance to cytotoxicity [61]. Binding of antibodies that induce expression of cytoprotective genes is an appealing example. The products of these genes protect against complement-mediated cytotoxicity, and, indeed, expression is essential to avoid severe ischemia-reperfusion injury. However, this explanation does not suffice. As we review elsewhere [21], heightened expression of cytoprotective genes does not necessarily prevent injury in transplants over time, and therefore more enduring processes are also needed. We shall soon offer a more comprehensive model, but for the present, we shall emphasize one pertinent feature.

Under physiologically normal conditions, cultured cells and living organs can take up and metabolize substantial amounts of antibody [62]. The ability of foreign organs to quantitatively remove antibodies from blood was observed more than 50 years ago when clinicians and investigators began to explore the mechanisms of rejection (see [63, 64] as examples). Consequently, antibodies against graft antigens may not appear in the blood of a transplant recipient until rejection is advanced and the cellular injury compromises uptake. We have postulated further that one cellular change associated with accommodation is an increase in ability of cells to process bound antibody. Our preliminary work supports this concept.

The capacity of organ transplants to absorb antibodies from the blood has practical significance beyond the understanding of accommodation [65, 66]. To the extent that transplants deplete antibodies specific for the graft, the steady-state concentration of these antibodies in blood will be lower, and assays used to assess donor-specific antibodies before transplantation could miss or underestimate the levels of antibodies after transplantation. Still more important is the impact absorption may have on conclusions regarding antigen specificity of donor-specific antibodies. After transplantation, antibodies of the highest affinity and/or against the most abundant antigens will be preferentially depleted.

To address these problems, and particularly to enable us to estimate the full scope of accommodation, we devised approaches to enumerate and characterize donor-specific B cells in transplant recipients [67]. We expect soon to report pertinent lessons in detail, but for the moment, we would suggest that accommodation might be more common than generally thought.

Accommodation of the Newborn Heart

We would hypothesize that the heart in newborn infants is "naturally" accommodated and, as such, might be poised to resist and/or repair ischemic injury that would irreversibly damage a mature heart. Although cardiac transplants in newborn recipients are potentially beset by technical and physiologic challenges, we suspect the "accommodation," or greater ability than mature hearts to repair and regenerate after injury, may help to explain the better long-term outcomes discussed above.

Development of the heart with expansion and remodeling of myocardium, parturition, changes in availability of oxygen, and the rapid change in cardiac circulation after birth impose significant stresses and demand for repair on the heart. How the fetal and newborn heart meets those stresses, resists and repairs damage, and regenerates damaged myocardium is incompletely understood [68–71]. Particularly important may be the capacity of newborn myocardium to regenerate by replication of progenitor cells [71]. Older hearts, having less capacity for self-renewal, may instead regenerate by cell fusion. Cell fusion causes DNA breaks that can resolve by recombination [72] or can induce apoptosis. The importance of cell fusion in the regeneration of the heart is suggested by the finding that the hearts of newborn mice contain relatively frequent mutations associated with DNA recombination (a consequence of cell fusion) and recombination increases with age; by contrast, small intestine accumulates numerous point mutations (a consequence of mitosis), but fewer mutations owed to recombination [73]. Weighing these mechanisms in large animals or humans without labeling would be difficult or impossible. However, the fusion of cells of disparate species is readily detected and the implications of recombination potentially deduced [74, 75]. Therefore, cardiac xenografts using newborn hearts might, in addition to treating cardiac insufficiency, offer the possibility of resolving important questions regarding repair and regeneration of myocardium.

Concluding Remarks

The compelling challenge of developing physiologically optimal approaches to treatment of severe heart failure in newborn infants has fueled interest in xenotransplantation for decades [1, 49, 76]. Yet, at this juncture, it may be difficult to know when, or even if, xenotransplantation of the heart will be applied in newborn infants. It is not difficult, however, to envision the advances in knowledge that xenotransplantation would bring and to begin to glimpse that knowledge from the experience of ABO-incompatible transplantation in the newborn.

Because ABO-incompatible transplantation expands, if only modestly, the availability of transplantation for infants with severe cardiac failure, such transplants hint at the potential benefit xenotransplantation could add. We recently discussed this subject in another communication [21].

Above, we suggest that cardiac xenotransplantation in newborn infants could help resolve questions about the relative contribution of immunity and incompatibility to the biological barrier to xenotransplantation. Although biochemical incompatibility has been considered mainly with respect to xenografts, research in our laboratories and others in recent years indicates that differences in biochemical networks likely influence the recovery and function of all transplants [77–79], although not to the extent observed in xenografts. The answers to those questions have profound, but parochial, application.

Xenotransplantation in the newborn may also expand knowledge regarding the mechanisms of accommodation and or repair and regeneration of the heart. This knowledge has obvious application in transplantation, but it also has broad relevance to health and disease. Thus, if addressing the urgent need to provide a physiologically normal option for the occasional newborn infant with severe cardiac failure fuels interest in xenotransplantation, the benefits from successful application of xenotransplantation extend much further in fields of biology and medicine.

Acknowledgment Supported by grants from the National Institutes of Health (AI123262, AI122369, OD023138)

References

1. Platt JL, West LJ, Chinnock RE, Cascalho M. Toward a solution for cardiac failure in the newborn. Xenotransplantation. 2018;25:e12479.
2. West LJ, Pollock-Barziv SM, Dipchand AI, et al. ABO-incompatible heart transplantation in infants. N Engl J Med. 2001;344:793–800.
3. Hammer C, Chaussy CH, Brendel W. Preformed natural antibodies in animals and man: outlook on xenotransplantation. Eur Surg Res. 1973;5:162166.
4. Hammer C. Isohemagglutinins and preformed natural antibodies in xenogeneic organ transplantation. Transplant Proc. 1987;19:4443–7.
5. Platt JL, Vercellotti GM, Dalmasso AP, et al. Transplantation of discordant xenografts: a review of progress. Immunol Today. 1990;11:450–6.
6. Parker W, Saadi S, Lin SS, Holzknecht ZE, Bustos M, Platt JL. Transplantation of discordant xenografts: a challenge revisited. Immunol Today. 1996;17:373–8.
7. Cascalho M, Platt JL. The immunological barrier to xenotransplantation. Immunity. 2001;14:437–46.
8. Hume DM, Merrill JP, Miller BF, Thorn GW. Experiences with renal homotransplantation in the human: report of nine cases. J Clin Investig. 1955;34:327–82.
9. Starzl TE, Marchioro TL, Holmes JH, et al. Renal homografts in patients with major donor-recipient blood group incompatibilities. Surgery. 1964;55:195–209.

10. Gleason RE, Murray JE. Report from kidney transplant registry: analysis of variables in the function of human kidney transplants. Transplantation. 1967;5:343–59.
11. Wilbrandt R, Tung KSK, Deodhar SD, Nakamoto S, Kolff WJ. ABO blood group incompatibility in human renal homotransplantation. Am J Clin Pathol. 1969;51:15–23.
12. Moore FD, Burch GE, Harken DE, Swan HJ, Murray JE, Lillihei CW. Cardiac and other organ transplantation. In the setting of transplant science as a national effort. JAMA. 1968;206:2489–500.
13. Rapaport FT, Dausset J. Activity of the ABO antigen system as a determinant of histocompatibility in human transplantation. Transplant Proc. 1987;19:4487–91.
14. Alexandre GP, De Bruyere M, Squifflet JP, Moriau M, Latinne D, Pirson Y. Human ABO-incompatible living donor renal homografts. Neth J Med. 1985;28:231–4.
15. Alexandre GPJ, Squifflet JP, De Bruyere M, et al. Present experiences in a series of 26 ABO-incompatible living donor renal allografts. Transplant Proc. 1987;19:4538–42.
16. Breimer ME, Brynger H, Le Pendu J, et al. Blood group ABO-incompatible kidney transplantation biochemical and immunochemical studies of blood group A glycolipid antigens in human kidney and characterization of the antibody response (antigen specificity and antibody class) in O recipients receiving A2 grafts. Transplant Proc. 1987;19:226–30.
17. Paul LC, van Es LA, de la Reviere GB, Eernisse G, de Graeff J. Blood group B antigen on renal endothelium as the target for rejection in an ABO-incompatible recipient. Transplantation. 1978;26:268–71.
18. Chopek MW, Simmons RL, Platt JL. ABO-incompatible renal transplantation: initial immunopathologic evaluation. Transplant Proc. 1987;19:4553–7.
19. Bannett AD, McAlack RF, Morris M, Chopek M, Platt JL. ABO incompatible renal transplantation: a qualitative analysis of native endothelial tissue ABO antigens after transplant. Transplant Proc. 1989;21:783–5.
20. Koch CA, Khalpey ZI, Platt JL. Accommodation: preventing injury in transplantation and disease. J Immunol. 2004;172:5143–8.
21. Garcia de Mattos Barbosa M, Cascalho M, Platt JL. Accommodation in ABO-incompatible organ transplants. Xenotransplantation. 2018;25:e12418.
22. Platt JL, Fischel RJ, Matas AJ, Reif SA, Bolman RM, Bach FH. Immunopathology of hyperacute xenograft rejection in a swine-to-primate model. Transplantation. 1991;52:214–20.
23. Platt JL. Xenotransplantation: immunological hurdles and prospects for future success. J Nephrol. 1993;6:182–92.
24. Lin SS, Weidner BC, Byrne GW, et al. The role of antibodies in acute vascular rejection of pig-to-baboon cardiac transplants. J Clin Investig. 1998;101:1745–56.
25. Opelz G, Morath C, Susal C, Tran TH, Zeier M, Dohler B. Three-year outcomes following 1420 ABO-incompatible living-donor kidney transplants performed after ABO antibody reduction: results from 101 centers. Transplantation. 2015;99:400–4.
26. Montgomery JR, Berger JC, Warren DS, James NT, Montgomery RA, Segev DL. Outcomes of ABO-incompatible kidney transplantation in the United States. Transplantation. 2012;93:603–9.
27. Tanabe K, Ishida H, Inui M, et al. ABO-incompatible kidney transplantation: long-term outcomes. Clin Transpl. 2013:307–12.
28. Bustos M, Platt JL. The pathology of cardiac xenografts. J Card Surg. 2001;16:357–62.
29. Chinnakotla S, Verghese P, Chavers B, et al. Outcomes and risk factors for graft loss: lessons learned from 1,056 pediatric kidney transplants at the University of Minnesota. J Am Coll Surg. 2017;224:473–86.
30. McCurry KR, Kooyman DL, Alvarado CG, et al. Human complement regulatory proteins protect swine-to-primate cardiac xenografts from humoral injury. Nat Med. 1995;1:423–7.
31. Pierson RN, Winn HJ, Russell PS, Auchincloss H. Xenogeneic skin graft rejection is especially dependent on CD4+ T cells. J Exp Med. 1989;170:991–6.
32. Lawson JH, Platt JL. Molecular barriers to xenotransplantation. Transplantation. 1996;62:303–10.

33. Hammer C. Physiological obstacles after xenotransplantation. Ann N Y Acad Sci. 1998;862:19–27.
34. Platt JL, Cascalho M, Piedrahita JA. Xenotransplantation: progress along paths uncertain from models to application. ILAR J. 2018;59:286–308.
35. Bergenfeldt H, Andersson B, Bucin D, et al. Outcomes after ABO-incompatible heart transplantation in adults: a registry study. J Heart Lung Transplant. 2015;34:892–8.
36. Urschel S, West LJ. ABO-incompatible heart transplantation. Curr Opin Pediatr. 2016;28:613–9.
37. Dipchand AI, Pollock-Barziv SM, Manlhiot C, West LJ, Vandervliet M, McCrindle BW. Equivalent outcomes for pediatric heart transplantation recipients: ABO-blood group incompatible versus ABO-compatible. Am J Transplant. 2009;10:389–97.
38. Henderson HT, Canter CE, Mahle WT, et al. ABO-incompatible heart transplantation: analysis of the Pediatric Heart Transplant Study (PHTS) database. J Heart Lung Transplant. 2012;31:173–9.
39. Parker W, Bruno D, Platt JL. Xenoreactive natural antibodies in the world of natural antibodies: typical or unique? Transpl Immunol. 1995;3:181–91.
40. Parker W, Lundberg-Swanson K, et al. Isohemagglutinins and xenoreactive antibodies are members of a distinct family of natural antibodies. Hum Immunol. 1996;45:94–104.
41. Chinnock RE, Bailey LL. Heart transplantation for congenital heart disease in the first year of life. Curr Cardiol Rev. 2011;7:72–84.
42. Copeland H, Razzouk A, Chinnock R, Deming D, Hasaniya N, Bailey L. Pediatric recipient survival beyond 15 post-heart transplant years: a single-center experience. Ann Thorac Surg. 2014;98:2145–50; discussion 50–51.
43. John M, Bailey LL. Neonatal heart transplantation. Ann Cardiothorac Surg. 2018;7:118–25.
44. Dipchand AI. Current state of pediatric cardiac transplantation. Ann Cardiothorac Surg. 2018;7:31–55.
45. Urschel S, Meyer SR, Larsen IM, et al. Absence of donor-specific anti-HLA antibodies after ABO-incompatible heart transplantation in infancy - altered immunity or age? Am J Transplant. 2010;10:149–56.
46. Kobayashi D, Du W, L'Ecuyer TJ. Predictors of cardiac allograft vasculopathy in pediatric heart transplant recipients. Pediatr Transplant. 2013;17:436–40.
47. Chen CY, Warner P, Albers EL, et al. Donor-specific anti-HLA antibody production following pediatric ABO-incompatible heart transplantation. Pediatr Transplant. 2019;23:e13332.
48. Fan X, Ang A, Pollock-BarZiv SM, et al. Donor-specific B-cell tolerance after ABO-incompatible infant heart transplantation. Nat Med. 2004;10:1227–33.
49. Cooper DKC, Hara H, Iwase H, Banks CA, Cleveland DC. An approach to induction of tolerance to pig cardiac xenografts in neonates. Xenotransplantation. 2018;25(6):e12454.
50. Lombardi G, Sidhu S, Daly M, Batchelor JR, Makgoba W, Lechler RI. Are primary alloresponses truly primary? Int Immunol. 1990;2:9–13.
51. Saadi S, Holzknecht RA, Patte CP, Stern DM, Platt JL. Complement-mediated regulation of tissue factor activity in endothelium. J Exp Med. 1995;182:1807–14.
52. Saadi S, Platt JL. Transient perturbation of endothelial integrity induced by natural antibodies and complement. J Exp Med. 1995;181:21–31.
53. Malyguine AM, Saadi S, Platt JL, Dawson JR. Human natural killer cells induce morphologic changes in porcine endothelial cell monolayers. Transplantation. 1996;61:161–4.
54. Malyguine AM, Saadi S, Holzknecht RA, et al. Induction of procoagulant function in porcine endothelial cells by human NK cells. J Immunol. 1997;159:4659–64.
55. Bustos M, Coffman TM, Saadi S, Platt JL. Modulation of eicosanoid metabolism in endothelial cells in a xenograft model: role of cyclooxygenase-2. J Clin Investig. 1997;100:1150–8.
56. Nagayasu T, Saadi S, Holzknecht RA, Patte CA, Plummer TB, Platt JL. Expression of tissue factor mRNA in cardiac xenografts: clues to the pathogenesis of acute vascular rejection. Transplantation. 2000;69:475–82.
57. Bustos M, Saadi S, Platt JL. Platelet-mediated activation of endothelial cells: implications for the pathogenesis of transplant rejection. Transplantation. 2001;72:509–15.

58. Parker W, Lin SS, Platt JL. Antigen expression in xenotransplantation: how low must it go? Transplantation. 2001;71:313–9.
59. Saadi S, Wrenshall LE, Platt JL. Regional manifestations and control of the immune system. FASEB J. 2002;16:849–56.
60. Williams JM, Holzknecht ZE, Plummer TB, Lin SS, Brunn GJ, Platt JL. Acute vascular rejection and accommodation: divergent outcomes of the humoral response to organ transplantation. Transplantation. 2004;78:1471–8.
61. Bach FH, Ferran C, Hechenleitner P, et al. Accommodation of vascularized xenografts: expression of "protective genes" by donor endothelial cells in a host Th2 cytokine environment. Nat Med. 1997;3:196–204.
62. Parker W, Holzknecht ZE, Song A, et al. Fate of antigen in xenotransplantation: implications for acute vascular rejection and accommodation. Am J Pathol. 1998;152:829–39.
63. Altman B. Tissue transplantation: circulating antibody in the homotransplantation of kidney and skin. Ann R Coll Surg Engl. 1963;33:79–104.
64. Milgrom F, Litvak BI, Kano K, Witebsky E. Humoral antibodies in renal homograft. JAMA. 1966;198:226–30.
65. Platt JL, Cascalho M. Donor specific antibodies after transplantation. Pediatr Transplant. 2011;15:686–90.
66. Cascalho MI, Chen BJ, Kain M, Platt JL. The paradoxical functions of B cells and antibodies in transplantation. J Immunol. 2013;190:875–9.
67. Lynch RJ, Silva IA, Chen BJ, Punch JD, Cascalho M, Platt JL. Cryptic B cell response to renal transplantation. Am J Transplant. 2013;13:1713–23.
68. Ali SR, Hippenmeyer S, Saadat LV, Luo L, Weissman IL, Ardehali R. Existing cardiomyocytes generate cardiomyocytes at a low rate after birth in mice. Proc Natl Acad Sci U S A. 2014;111:8850–5.
69. van Berlo JH, Molkentin JD. An emerging consensus on cardiac regeneration. Nat Med. 2014;20:1386–93.
70. Zhang Y, Mignone J, MacLellan WR. Cardiac regeneration and stem cells. Physiol Rev. 2015;95:1189–204.
71. Sereti KI, Nguyen NB, Kamran P, et al. Analysis of cardiomyocyte clonal expansion during mouse heart development and injury. Nat Commun. 2018;9:754.
72. Zhou X, Merchak K, Lee W, Grande JP, Cascalho M, Platt JL. Cell fusion connects oncogenesis with tumor evolution. Am J Pathol. 2015;185:2049–460.
73. Dolle ME, Snyder WK, Gossen JA, Lohman PH, Vijg J. Distinct spectra of somatic mutations accumulated with age in mouse heart and small intestine. Proc Natl Acad Sci U S A. 2000;97:8403–8.
74. Ogle BM, Butters KB, Plummer TB, et al. Spontaneous fusion of cells between species yields transdifferentiation and retroviral in vivo. FASEB J. 2004;18:548–50.
75. Ogle BM, Cascalho M, Platt JL. Biological implications of cell fusion. Nat Rev Mol Cell Biol. 2005;6:567–75.
76. Bailey LL, Gundry SR. Survival following orthotopic cardiac xenotransplantation between juvenile baboon recipients and concordant and discordant donor species: foundation for clinical trials. World J Surg. 1997;21:943–50.
77. Platt JL, Nath KA. Heme oxygenase: protective gene or Trojan horse. Nat Med. 1998;4:1364–5.
78. Platt JL, Cascalho M. IgM in the kidney: a multiple personality disorder. Kidney Int. 2015;88:439–41.
79. Platt JL, Kaufman CL, Garcia de Mattos Barbosa M, Cascalho M. Accommodation and related conditions in vascularized composite allografts. Curr Opin Organ Transplant. 2017;22:470–6.

Part V

Patient Evaluation and Selection for First Clinical Trials of Kidney or Heart Xenotransplantation

Defining an "Acceptable Risk Threshold": Who Should Be the First Kidney Xenotransplant Recipient?

13

Jayme E. Locke

Abbreviation

ESRD End-stage renal disease

Introduction

Allotransplantation has been established as the gold standard for the treatment of end-stage renal disease (ESRD), in part because the survival benefit of transplantation compared to remaining on dialysis exists independent of donor quality (e.g., donor age, comorbid disease) and donor-recipient matching (e.g., human leukocyte antigen [HLA] and/or blood group incompatibility) [1–3]. Not surprisingly, the number of patients seeking this lifesaving therapy continues to grow, with more than 90,000 ESRD patients waiting for a kidney transplant in the USA alone [4]. This extraordinary medical and surgical feat requires a source of donor kidneys for transplantation. Donation is thus the cornerstone of transplantation. Despite global efforts to increase the number of donors, and therefore kidneys for transplant, deceased donation supplies <10% of the need, and thousands of patients die each year while waiting [4]. While efforts to increase living kidney donations have increased transplantation rates, this growth has been modest and falls short of the global need [4].

Similar to past generations, like that of Nobel Laureate Joseph Murray, who performed the first successful kidney transplant (between identical twins) in 1954,

J. E. Locke (✉)
Division of Transplantation, Department of Surgery, University of Alabama at Birmingham, Birmingham, AL, USA
e-mail: jlocke@uabmc.edu

D. K. C. Cooper, G. Byrne (eds.), *Clinical Xenotransplantation*,
https://doi.org/10.1007/978-3-030-49127-7_13

the current generation of scientists find new and unique ways to innovate and solve the unimaginable while rewriting history in their own right [5]. This scientific evolution, or some might say revolution, now stands on the threshold of solving the organ shortage through the development of xenotransplantation, in which modified/humanized pig xenografts are utilized for transplantation into humans. No longer simply a grandiose idea, recent data suggest xenotransplantation is within arm's reach of solving a global crisis.

The First Xenotransplantation Patient?

So then, the next logical step is "who should be the first xenotransplant recipient?" The answer is elusive, as medical practitioners must weigh the risks, benefits, and alternatives to such an experiment – a task that on first glance seems larger than life and begs the cycle of "what if" questioning. What if the first recipient dies? What if the first recipient contracts a zoonosis (porcine-borne disease)? What if the xenograft fails? The burden is seemingly so overwhelming that our paternalistic approach to medicine may result in a hard and emphatic "NO!" But should not the ESRD patient have some autonomy in decision-making? Imagine an ESRD patient's cycle of "what if" questioning. What if I don't live to see my children grow up? What if I can't provide for my family? Shouldn't these "what if" questions matter too? A balance between our paternalistic need to protect our patients and a patient's autonomy in calculated risk-taking must be struck. Perhaps an "acceptable risk threshold" can be found among one of these three patient scenarios [6].

Old Transplant Candidate

A 68-year-old male, blood group O, cPRA 3%, and ESRD secondary to polycystic kidney disease. He has no potential living kidney donor and is listed at a transplant center with a waiting time of greater than 10 years for blood group O candidates. His "what if" is "what if I am no longer medically suitable for transplant at 78 years of age?"

Recurrent Disease Candidate

A 27-year-old female, blood group A, cPRA 87%, and ESRD secondary to focal sclerosing glomerulosclerosis (FSGS). She received her first transplant at the age of 15 years, a living kidney from her mother, which failed 3 years later from recurrent FSGS. She received her second transplant at the age of 21 from her sister, which failed 2 years later from recurrent FSGS. She has since run out of dialysis access and undergoes dialysis through a transhepatic catheter. She has 100% risk of recurrent disease with re-transplant. Her "what if" questions are "what if I run out of dialysis access? What if just maybe my disease can't recur in a pig kidney?"

No Access to Dialysis Candidate

A 25-year-old female, blood group B, cPRA 0%, and chronic kidney disease stage IV secondary to human immunodeficiency virus-associated nephropathy (HIVAN), living in a low-to-middle income country with limited dialysis facilities and HIV an absolute contraindication to offering a dialysis slot. Her "what if" question is "what if I die before my daughter is old enough to care for herself?"

Comment

Balancing medical paternalism and patient autonomy to define an acceptable risk threshold will likely differ among key stakeholders – US Food and Drug Administration (FDA), transplant centers, payers, xeno-companies – but nonetheless it is time to begin the conversation in anticipation of the first xeno kidney transplant. After all, our patients in need are asking "what if they never try?"

References

1. Merion RM, Ashby VB, Wolfe RA, et al. Deceased-donor characteristics and the survival benefit of kidney transplantation. J Am Med Assoc. 2005;294:2767–33.
2. Montgomery RA, Lonze BE, King KE, et al. Desensitization in HLA-incompatible kidney recipients and survival. N Engl J Med. 2011;365:318–26.
3. Montgomery RA, Locke JE, King KE, et al. ABO incompatible renal transplantation: a paradigm ready for broad implementation. Transplantation. 2009;87:1246–55.
4. Hobeika MJ, Miller CM, Pruetet TL, et al. PROviding better access to organs: a comprehensive overview of organ-access initiatives from the ASTS PROACTOR Task Force. Am J Transplant. 2017;17:2546–58.
5. Murray JE, Merrill JP, Harrison JH. Kidney transplantation between seven pairs of identical twins. Ann Surg. 1958;148:343–59.
6. Lentine KL, Kasiske BL, Levey AS, et al. Summary of kidney disease: improving global outcomes (KDIGO) clinical practice guideline on the evaluation and care of living kidney donors. Transplantation. 2017;101:1783–92.

Selection of Patients for the Initial Clinical Trials of Kidney Xenotransplantation

14

David K. C. Cooper, Abhijit Jagdale, Roslynn B. Mannon, Vineeta Kumar, Robert Gaston, Michael J. Hanaway, Douglas J. Anderson, and Devin E. Eckhoff

Abbreviations

ESRD	End-stage renal disease
HLA	Human leukocyte antigen
NHPs	Nonhuman primates
PERVs	Porcine endogenous retroviruses
SLA	Swine leukocyte antigen

Introduction

The incidence of end-stage renal disease (ESRD) is increasing worldwide. Old age, diabetes, hypertension, obesity, and cardiovascular disease all contribute to the development of chronic kidney disease [1]. Since the mid-1970s, hemodialysis has been life-sustaining for millions of patients with ESRD. Although it may be life-supporting, chronic dialysis is a suboptimal form of therapy for many patients with

D. K. C. Cooper (✉) · A. Jagdale · D. J. Anderson · D. E. Eckhoff
Xenotransplantation Program, Department of Surgery,
University of Alabama at Birmingham (UAB), Birmingham, AL, USA
e-mail: dkcooper@uabmc.edu

R. B. Mannon · V. Kumar · R. Gaston
Division of Nephrology, Department of Medicine,
University of Alabama at Birmingham (UAB), Birmingham, AL, USA

M. J. Hanaway
Division of Transplantation, Department of Surgery,
University of Alabama at Birmingham (UAB), Birmingham, AL, USA

D. K. C. Cooper, G. Byrne (eds.), *Clinical Xenotransplantation*,
https://doi.org/10.1007/978-3-030-49127-7_14

ESRD, and their quality of life remains poor (reviewed in [2]). Renal transplantation offers a better quality of life [3–6]. However, there is a critical and continuing shortage of deceased human donor kidneys.

As a consequence of comorbidities or socioeconomic factors, most dialysis patients are not considered candidates for a kidney transplant, though the absolute number is debatable [7]. The benefit of a kidney transplant is most evident when the outcome is compared to that of dialysis in wait-listed patients thought reasonable candidates who did not receive an allograft. In comparison with approximately 500,000 patients on chronic dialysis, there are currently >80,000 patients wait-listed for kidney transplantation in the USA [8, 9]. Therefore, less than 20% of the dialysis population is currently on transplant waiting lists.

The overall survival of patients on chronic hemodialysis in the USA is 78% at 1 year, 57% at 3 years, and only 42% at 5 years (Fig. 14.1) [10]. Deaths are mainly related to comorbidities, e.g., congestive heart failure, malnutrition, or cancer,

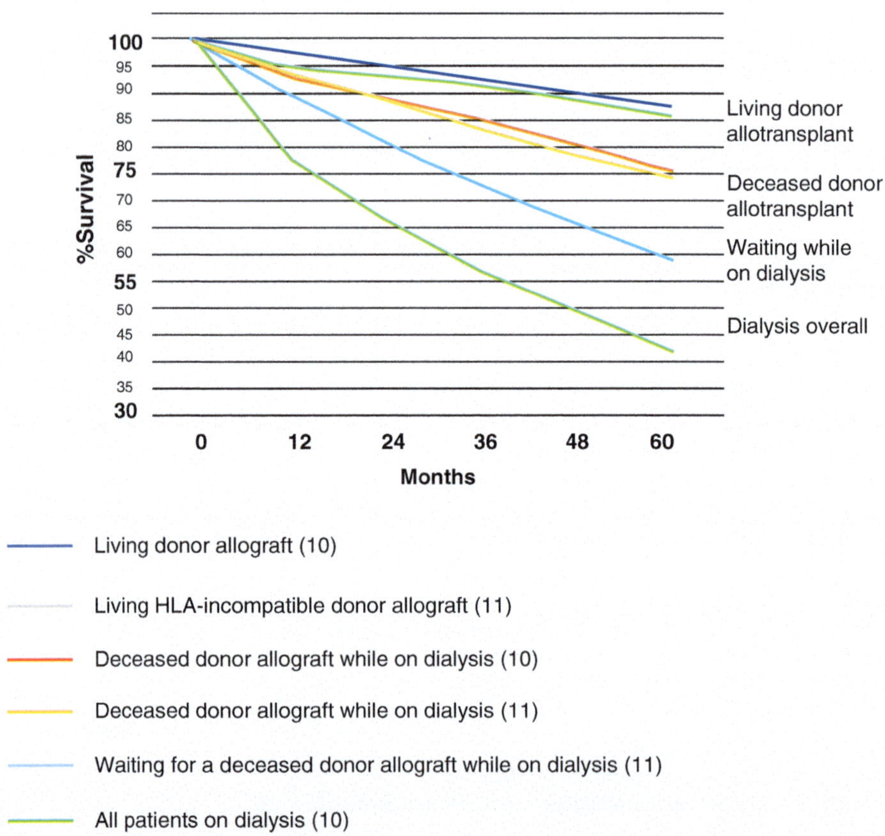

Fig. 14.1 Percentage survival of ESRD patients by treatment modality in 2010. (Modified from Jagdale et al. [2] and based on data from two sources: (i) USRDS 2017 [10] and (ii) Orandi et al. [11])

accompanied by kidney failure. Mortality among patients >65 years of age is higher than in younger patients. Remarkably, patients with ESRD on dialysis have poorer survival and fewer remaining years of life than many with cancer, diabetes, and cardiovascular disease [10]. These mortality rates, however, largely reflect outcomes among those who are *not* candidates for renal transplantation. Nevertheless, any procedure that might (i) delay the need for dialysis or (ii) reduce the period during which the patient is on dialysis while awaiting an allograft would be worthwhile.

On average, more than 20 patients are removed from the kidney wait-list each day, either because they die or because they become too sick to tolerate the transplant procedure [8]. Forty percent of *wait-listed* patients are likely to die within 5 years [10, 11] (Fig. 14.1). There is a great deal of geographic disparity, with some centers demonstrating greater wait-list mortality than transplant rates. Importantly, in one analysis, most of those who died on the wait-list had been excellent candidates at the time of listing [12].

The very poor quality of life of some of these patients is illustrated by the fact that a significant percentage of them choose withdrawal of dialysis [13]. In the USA in 2014, approximately 13–17% did so. Remarkably, in New England, almost every third patient chooses to withdraw from treatment [10, 14]. Patients who withdraw from dialysis usually die within 10 days (median, 8 days) [14, 15]. However, it should again be emphasized that these patients are mainly *not* those who are on the waiting list for a kidney transplant.

Kidney allotransplantation can ameliorate many of the problems associated with ESRD and chronic dialysis [16–25], addressing both survival and quality of life issues. Survival of patients with renal allotransplants from deceased donors is 85% after 3 years and rises to 93% when a living donor kidney has been transplanted. At 5 years, survival in these two groups is 76% and 88%, respectively [10]. However, if the patient is not carefully selected, renal allotransplantation can be associated with serious complications that impact the patient's quality of life. This could be even more so in those undergoing the initial renal xenotransplants, and so very careful patient selection will be essential.

The lack of deceased human donor kidneys could be resolved if kidneys from genetically engineered pigs offered an alternative with an acceptable clinical outcome, e.g., a good quality of life in the absence of major morbidity. The potential advantages of xenotransplantation are several (Table 14.1) [26]. The results of genetically engineered pig kidney transplantation in nonhuman primates (NHPs) have improved significantly in recent years (Chap. 5), sufficiently to encourage consideration for initial clinical trials. It is reasonable to presume that clinical trials of kidney xenotransplantation will at least offer an outcome competitive with chronic dialysis.

Selection of Patients for the First Clinical Trial

This topic has been considered and patients with various conditions have been discussed (Table 14.2) [2, 27]. An important point to bear in mind is that, for the simple reason that xenotransplantation will be offered to patients who are unable to receive

Table 14.1 Advantages of xenotransplantation over allotransplantation

1. Unlimited supply of "donor" organs. (This will be particularly important to the millions of patients with diabetes [+/− ESRD] in whom pig islet transplantation may control glycemia.)
2. Organs available electively. (Patients with ESRD will no longer require chronic dialysis. Patients with acute failure of the liver or heart will no longer need prolonged intensive care or mechanical support.)
3. Avoids the detrimental effects on the donor organs of brain death.
4. The "donors" will be infection-free.
5. "Borderline" transplant candidates, i.e., those with health problems that may be detrimental to prolonged patient survival, e.g., poorly controlled diabetes, peripheral or cerebrovascular disease, will be more acceptable (as they will no longer be competing for scarce organs with other potential transplant candidates).
6. Avoids the "cultural" barriers to deceased human organ donation, e.g., in countries such as Japan.

Table 14.2 Potential conditions for which initial clinical trials of pig kidney xenotransplantation may be justified*

Elderly patients without significant concomitant disease
Patients of blood group B or O often wait for >5 years for a suitable donor. The mortality of wait-list patients is 40% at 5 years, and so many of these patients, particularly in the age range 55–65 years, will not survive until a deceased human kidney becomes available
Recurrent kidney disease
Recurrent focal segmental glomerulosclerosis (FSGS)
Recurrence can be very rapid in some patients. If recurrence occurs rapidly in the pig kidney, this may not be a valid test of xenotransplantation
Other potentially recurrent diseases
Recurrence is slower in several other disease states, e.g., immunoglobulin A (IgA) nephropathy, membranoproliferative glomerulonephritis (MPGN) type II, and so these patients might possibly be candidates for a trial of pig kidney transplantation
High sensitization to HLA
There is evidence for some cross-reactivity between anti-HLA antibodies and SLA, suggesting that patients sensitized to HLA should be excluded from the first clinical trials
Loss of vascular access for dialysis
These patients have often been on dialysis for some time (years rather than months) and may have diseased blood vessels making kidney transplantation technically difficult. They are frequently less than ideal candidates even for allotransplantation

*Modified from Cooper et al. [27]

a timely allograft, the results of the initial pig kidney xenotransplants should be compared with those for comparable patients maintained on chronic dialysis, but not with those for patients receiving kidney allografts.

A second major point is that the patient's general physical state should make him or her an acceptable candidate for allotransplantation. To select patients who are unlikely to survive after receiving an allograft, e.g., from general frailty, chronic infection, previous or current neoplasia, or other comorbidities, would not prove to be an adequate trial of xenotransplantation, as the patient would be equally unlikely to survive.

We suggest it would be ethical to offer a pig kidney transplant to selected patients whose life expectancy is less than the time it will take for them to obtain a deceased human donor organ. The median waiting period for a patient with ESRD to obtain a deceased human donor kidney is 3.9 years [10], by which time approximately 35% of transplant candidates may have died or been removed from the wait-list (Fig. 14.1). Approximately 50% of the patients therefore wait for a period significantly longer than 4 years. Those of blood group B or O may experience a wait of 7 or more years [10], with an *average* waiting time of almost 5 years, even when the patient has no antibodies directed to human leukocyte antigens (HLA) (Table 14.3).

Many of the patients might choose to receive a timely and life-supporting pig kidney as long as a reasonable period of graft function could be anticipated without the need for excessive immunosuppressive therapy (compared with that required to maintain an allograft). Other parameters that we believe may be important are listed in Table 14.4. Importantly, the patients undergoing xenotransplantation would remain on the wait-list for an allograft and so would not be penalized in this respect. For many, the period of support by the xenograft would be a welcome relief from the restrictions imposed by chronic dialysis, even if only for a year or so.

Table 14.3 Average waiting time (in years) for a deceased human kidney transplant by blood type and percentage panel-reactive antibodies (PRA), 1998–2010 [10]

	Blood group			
PRA (%)	A	B	AB	O
0	2.9	4.8	2.0	4.7
>0 < 20	2.5	4.8	1.4	4.7
>20 < 80	2.5	3.7	1.4	4.3
>80 < 98	3.7	[a]	3.2	4.8
>98 < 100	[a]	5.9	[a]	[a]

[a]As the estimated time to the transplant probability had not reached 50% (median) at the end of follow-up, the median waiting time could not be calculated

Table 14.4 Factors considered in the selection of patients for the first clinical trials of pig kidney xenotransplantation

1. Age 55–65 years. (As the anticipated period of pig graft survival remains uncertain, younger patients, who are more likely to survive until a suitable allograft becomes available, should perhaps be excluded from the initial trials.)
2. No significant health problems except ESRD. (A patient with a pig xenograft may possibly (i) require more immunosuppressive therapy than one with an allograft and (ii) may need to return to chronic dialysis, which may be associated with a higher morbidity than initially. It is therefore important that the patient should have no other health problems except ESRD. A patient with isolated polycystic kidney disease might be a preferred candidate.)
3. Blood type B or O, as these patients spend longer on the waiting list for a deceased human donor kidney.
4. No anti-HLA antibodies (to avoid any risk of cross-reactivity of anti-HLA antibodies with SLA).
5. Supported by dialysis, but for less than 12 months. (Initiation of dialysis will confirm to the patient and his/her family that ESRD has advanced sufficiently to warrant kidney transplantation, but the period of dialysis has not been so long to be associated with complications or general debility.)
6. Fulfill all other criteria for allotransplantation, e.g., absence of potentially life-threatening infections or malignant disease.
7. Vascular access may be problematic or is likely to become limited.

Our understanding is that, at present, the FDA recommends that a patient should only be considered for a pig organ transplant if his/her life expectancy is anticipated to be less than 2 years. It may be difficult to predict the exact survival of a patient, and the guideline, if followed, might rule out many of the patients we have suggested above. Furthermore, some patients who are unlikely to survive for 2 years have already been supported by dialysis for several years and have developed comorbidities and are no longer ideal transplant candidates. In this case, patients in whom vascular access for dialysis is becoming difficult could be considered, but again many of these have been receiving dialysis for a prolonged period of time and, for this and other reasons, may not be suitable candidates for inclusion in a clinical trial. Selection of the initial patients, therefore, will require very careful consideration.

A second group of patients who could be considered as possible candidates for xenotransplantation is those with an underlying disease that has recurred in a second or even third allograft (Table 14.2) [27]. However, some of these patients will likely have developed antibodies to HLA and, therefore, if we follow our current criteria (see below), would be excluded. It is unknown whether any of these diseases will recur in a pig kidney, but, even if they do, the xenograft will have allowed for the allocation of an allograft to another recipient in whom it may have been better utilized. For the first clinical trial, however, we suggest that these patients, particularly those in whom the disease might recur rapidly, e.g., focal segmental glomerulosclerosis (FSGS), may not be ideal candidates.

At present, we would exclude patients with any sensitization to HLA. Although our own studies have repeatedly indicated that sensitization to HLA would not be detrimental to survival of a pig kidney graft [28–31], several other studies have indicated that there can be cross-reactivity between anti-HLA antibodies and some swine leukocyte antigens (SLA) ([31], reviewed in [32, 33]) (Chaps. 3, 4, and 18). Furthermore, HLA sensitization might be associated with T cell activation that might be detrimental to the survival of a pig graft. Methods are being developed to delete or replace specific SLA against which there might be cross-reactivity [34, 35] (Chap. 18). Although HLA-sensitized patients may be those who ultimately benefit most from pig kidney xenotransplantation, we submit that no risks in this respect should be taken in the first clinical trial.

In addition, the kidney allocation system (KAS) introduced in the USA in December 2014 prioritizes allocation of donor kidneys to HLA highly sensitized recipients if there are no preformed antibodies against that specific donor. This has resulted in increased rates of transplantation for highly sensitized patients, some of whom may now have a shorter wait-list time than less-sensitized patients [36]. There has been a 21% reduction in access, with greater dependence on lower-quality kidneys (i.e., those with a high kidney donor profile index [KDPI]), for wait-list candidates >65 years of age [37]. If they remain free of significant comorbidities, these patients might therefore benefit from renal xenotransplantation.

The results of preemptive kidney transplantation are superior to those of transplantation *after* dialysis is underway [38, 39]. More than 50% of patients return to some form of paid work after preemptive kidney transplantation, whereas only approximately 25% rejoin the workforce if hemodialysis precedes transplantation [40]. All pig kidney xenotransplants could be preemptive. Although a preemptive transplant might be most beneficial, for the first clinical trial, we suggest that recipient selection should be limited to those already on dialysis, as this removes the additional variable of native renal function from the interpretation of the study results. Furthermore, we suggest that, if the patient is already undergoing dialysis, there will be no doubt in the mind of the prospective patient or his/her family that ESRD is advanced enough to warrant a kidney transplant. However, we would recommend that the patients considered for the first clinical trial of pig kidney xenotransplantation should not have been on dialysis for more than a few months, as this will reduce the risks of comorbidities.

There were (and possibly still are) some hospitals that refused to undertake kidney allotransplantation in patients >65 years of age, citing the adverse effect on survival of frailty and comorbidities. However, survival of patients in this age group on dialysis has been reported to be 81% at 1 year, but only 30% at 5 years, and 15% at 7 years, whereas after renal allotransplantation, survival was 93%, 70%, and 46% at the same time intervals [41, 42]. More recent data continue to indicate survival of patients >65 years of age after renal allotransplantation of approximately >90%, 80–90%, and 70–80% at 1, 3, and 5 years, respectively [43, 44]. In a highly relevant study, Heldal et al. compared survival after renal allotransplantation in patients >70 years of age with that of similar-aged patients on the waiting list for an allograft [45]. At 1, 3, and 5 years, survival was 89% (after transplantation) vs. 98% (on the wait-list), 74% vs. 56%, and 64% vs. 33%, suggesting that, if the patient survived the initial posttransplant period, transplantation offered better long-term outcome. Advancing age, therefore, should not be an absolute contraindication to kidney xenotransplantation.

The average remaining lifespan for a patient aged 65–69 years on dialysis is only 4.6 years [46], whereas it is 11.4 years after a kidney transplant, approximately 5 years less than the general population [10]. Once again, however, the patients on dialysis include many who have comorbidities rendering them unsuitable for transplantation, and therefore they cannot be compared directly with those with functioning renal allografts.

In summary, therefore, because they are at greater risk of dying before a suitable renal allograft becomes available, we would suggest that older non-HLA-sensitized patients, e.g., 55 or older, particularly if of blood group B or O, who have recently begun chronic dialysis but who remain free of significant comorbidities, would be candidates who could be considered for the initial trials of pig kidney transplantation. It is these patients who might benefit most from undergoing pig kidney xenotransplantation, even if only to delay the need for chronic dialysis for a significant period of time while they await allotransplantation.

Although there are additional risks associated with retransplantation with allograft survival ranging from approximately 60–90% at 5 years [47–51], in our opinion an initial pig renal transplant could be justified because, in its absence, an older patient may not survive until an allograft becomes available.

Importantly, the current (limited) evidence is that a failed pig xenograft, even if anti-pig antibodies developed, would not be detrimental to a subsequent allograft [32] (Chap. 3).

Expenditure Related to Renal Replacement Therapy

According to data from the USRDS, in the USA in 2015, expenditure on patients with ESRD was $33.8 billion [10] and accounted for 7.1% of the overall costs of Medicare. In the same year, the total cost of care of patients with chronic kidney disease or ESRD was $98 billion. The cost of hemodialysis was $88,750 per person per year and of peritoneal dialysis was $75,140. In contrast, the cost after kidney allotransplantation was $34,084 per person per year [46]. In a detailed analysis of costs, Held and colleagues provided slightly different data [52]. From their data, it could be concluded that, if a patient with a functioning renal allograft survives for longer than 2 years, the procedure has been cost-effective.

Whether the costs of pig kidney xenotransplantation will be comparable to, or greater than, those associated with allotransplantation remains unknown. Many factors have to be taken into consideration. If the immunosuppressive therapy required is comparable (which is not yet certain), then the only major difference in costs may be the acquisition of the kidney. In the USA, the costs associated with retrieval of a single deceased human kidney are considerable [53] and generally vary from approximately $25,000–$40,000 (mean $33,000) but can be much greater, which is passed on to the recipient. When genetically engineered pig organs become commercially available, it could well be that a pig kidney will be priced significantly higher than this, in part to defray the very considerable research and development expenditure that has been incurred during the past 20–30 years [54]. However, against this expense, the considerable savings in the cost of caring for a patient with kidney failure, including chronic dialysis, convenience of ready availability of the organ, etc., will need to be considered.

The Future

The transplantation of organs, tissues, and cells from genetically engineered pigs has immense clinical therapeutic potential, not only with regard to whole-organ transplantation but also to pancreatic islet [55], corneal [56], neuronal cell [57], and skin [58] transplantation and as a source or red blood cells for transfusion [59].

The genetic manipulations that have been introduced in the pigs to date have largely been directed to overcome the innate immune response, for which effective drug therapy is very limited. In the future, however, the pigs will also be manipulated to control the adaptive immune response, thus enabling exogenous immunosuppressive therapy to be significantly reduced or, indeed, ultimately unnecessary.

There are numerous other genetic manipulations in pigs that have been, or are currently being, explored, many of which may be beneficial to long-term pig graft survival. These include inactivation of porcine endogenous retroviruses (PERVs) [60, 61] or deletion of PERVs [62], though present opinion is that this will be unnecessary. Breeding and housing in a biosecure, designated pathogen-free environment should rid the pigs of all other potentially pathogenic microorganisms [63] (Chaps. 8 and 17).

The early rapid growth of pig kidneys after transplantation into NHPs has been described (Chap. 5) but is unlikely to be problematic in clinical kidney xenotransplantation, particularly in adult recipients. To date, pig kidney function in a NHP has not been comprehensively investigated, but the data available have indicated no definitive problems that would preclude a successful outcome.

The ultimate goal of both allotransplantation and xenotransplantation is the induction of a state of immunological tolerance to the graft, in which the recipient no longer attempts to reject the graft, even in the absence of exogenous immunosuppressive therapy. Although efforts in this respect in xenotransplantation have to date been unsuccessful, in view of the potential offered by genetic engineering of the donor, it would seem it is more likely to be achieved in xenotransplantation than in allotransplantation.

The clinical potential of xenotransplantation is enormous, and we suggest that eventually progress in the field will render allotransplantation to be of historic interest only.

Acknowledgment Work on xenotransplantation at the University of Alabama at Birmingham is supported in part by NIH NIAID U19 grant AI090959 and in part by a grant from United Therapeutics, Silver Spring, MD (the parent company of Revivicor, Blacksburg, VA, that provides UAB with genetically engineered pigs).

Conflict of Interest The authors declare no conflicts of interest.

References

1. Levey AS, Coresh J. Chronic kidney disease. Lancet. 2012;379:165–80.
2. Jagdale A, Cooper DKC, Iwase H, Gaston RS. Chronic dialysis in patients with end-stage renal disease: relevance to kidney xenotransplantation. Xenotransplantation. 2018:e12471. https://doi.org/10.1111/xen.12471.
3. Matas AJ, Halbert R, Barr ML, et al. Life satisfaction and adverse effects in renal transplant recipients: a longitudinal analysis. Clin Transpl. 2002;16:113–21.
4. Ogutmen B, Yildirim A, Sever MS, et al. Health-related quality of life after kidney transplantation in comparison intermittent hemodialysis, peritoneal dialysis, and normal controls. Transplant Proc. 2006;38:419–21.
5. Liem YS, Bosch JL, Arends LR, Heijenbrok-Kal MH, Hunink M. Quality of life assessed with the medical outcomes study short form 36-item health survey of patients on renal replacement therapy: a systematic review and meta-analysis. Value Health. 2007;10:390–7.

6. Dabrowska-Bender M, Dykowska G, Żuk W, Milewska M, Staniszewska A. The impact on quality of life of dialysis patients with renal insufficiency. Patient Prefer Adherence. 2018;12:577–83.
7. Schold JD, Meier-Kriesche HU. Which renal transplant candidates should accept marginal kidneys in exchange for a shorter waiting time on dialysis? Clin J Am Soc Nephrol. 2006;1:532–8.
8. SRTR. Annual Data Report 2017 [cited 2019 4/08/2019]. Available from: https://www.srtr.org/.
9. USRDS. Annual data report 2018 [cited 2019 4/08/2019]. Available from: https://www.usrds.org/adr.aspx.
10. USRDS. Annual data report 2017 [cited 2019 4/08/2019]. Available from: https://www.usrds.org/adr.aspx.
11. Orandi BJ, Luo X, Massie AB, et al. Survival benefit with kidney transplants from HLA-incompatible live donors. New Engl J Med. 2016;374:940–50.
12. Casingal V, Glumac E, Tan M, Sturdevant M, Nguyen T, Matas A. Death on the kidney waiting list—good candidates or not? Am J Transplant. 2006;6:1953–6.
13. Khan IH. Comorbidity: the major challenge for survival and quality of life in endstage renal disease. Nephrol Dial Transplant. 1998;13(1):76–9.
14. Germain MJ, Cohen LM, Davison SN. Withholding and withdrawal from dialysis: what we know about how our patients die. Semin Dial. 2007;20:195–9.
15. Chater S, Davison S, Germain M, Cohen L. Withdrawal from dialysis: a palliative care perspective. Clin Nephrol. 2006;66:364–72.
16. Kawamura M, Fijimoto S, Hisanaga S, Yamamoto Y, Eto T. Incidence, outcome, and risk factors of cerebrovascular events in patients undergoing maintenance hemodialysis. Am J Kidney Dis. 1998;31:991–6.
17. La Rocca E, Fiorina P, Di Carlo V, et al. Cardiovascular outcomes after kidney–pancreas and kidney–alone transplantation. Kidney Int. 2001;60:1964–71.
18. Oliveras A, Roquer J, Puig JM, et al. Stroke in renal transplant recipients: epidemiology, predictive risk factors and outcome. Clin Transpl. 2003;17:1–8.
19. Seliger SL, Gillen DL, Tirschwell D, Wasse H, Kestenbaum BR, Stehman-Breen CO. Risk factors for incident stroke among patients with end-stage renal disease. J Am Soc Nephrol. 2003;14:2623–31.
20. Snyder JJ, Kasiske BL, Maclean R. Peripheral arterial disease and renal transplantation. J Am Soc Nephrol. 2006;17:2056–68.
21. Aull-Watschinger S, Konstantin H, Demetriou D, et al. Pre-transplant predictors of cerebrovascular events after kidney transplantation. Nephrol Dial Transpl. 2007;23:1429–35.
22. Kunzendorf U, Krämer BK, Arns W, et al. Bone disease after renal transplantation. Nephrol Dial Transplant. 2008;23:450–8.
23. Lentine KL, Rey LAR, Kolli S, et al. Variations in the risk for cerebrovascular events after kidney transplant compared with experience on the waiting list and after graft failure. Clin J Am Soc Nephrol. 2008;3:1090–101.
24. Zolty R, Hynes P, Vittorio T. Severe left ventricular systolic dysfunction may reverse with renal transplantation: uremic cardiomyopathy and cardiorenal syndrome. Am J Transplant. 2008;8:2219–24.
25. Willicombe M, Kumar N, Goodall D, et al. Incidence, risk factors, and outcomes of stroke post-transplantation in patients receiving a steroid sparing immunosuppression protocol. Clin Transpl. 2015;29:18–25.
26. Cooper DKC, Gollackner B, Sachs DH. Will the pig solve the transplantation backlog? Annu Rev Med. 2002;53:133–47.
27. Cooper DKC, Wijkstrom M, Hariharan S, et al. Selection of patients for initial clinical trials of solid organ xenotransplantation. Transplantation. 2016. [Epub ahead of print]. 2017; 101:1551–1558.
28. Cooper DKC, Tseng Y, Saidman S. Alloantibody and xenoantibody cross-reactivity in transplantation. Transplantation. 2004;77:1–5.
29. Hara H, Ezzelarab M, Rood PP, et al. Allosensitized humans are at no greater risk of humoral rejection of GT-KO pig organs than other humans. Xenotransplantation. 2006;13:357–65.

30. Wong BS, Yamada K, Okumi M, et al. Allosensitization does not increase the risk of xenoreactivity to α1,3-Galactosyltransferase gene-knockout miniature swine in patients on transplantation waiting lists. Transplantation. 2006;82:314–9.
31. Zhang Z, Hara Z, Long C, et al. Immune responses of HLA-highly-sensitized and non-sensitized patients to genetically engineered pig cells. Transplantation. 2018;102:e195–204.
32. Li Q, Hara H, Breimer ME, Wang Y, Cooper DKC. Is sensitization to pig antigens detrimental to subsequent allotransplantation? Xenotransplantation. 2018;25:e12393. https://doi.org/10.1111/xen.12393.
33. Byrne GW. Does human leukocyte antigens sensitization matter for xenotransplantation? Xenotransplantation. 2018;25(3):e12411.
34. Martens GR, Reyes LM, Butler JR, et al. Humoral reactivity of renal transplant-waitlisted patients to cells from GGTA1/CMAH/B4GalNT2, and SLA class I knockout pigs. Transplantation. 2017;101:e86–92.
35. Ladowski JM, Reyes LM, Martens GR, et al. Swine leukocyte antigen class II is a xenoantigen. Xenotransplantation. 2018;102:249–54.
36. Jackson KR, Covarrubias K, Holscher CM, et al. The national landscape of deceased donor kidney transplantation: transplant rates, waitlist mortality, and posttransplant survival under KAS. Am J Transplant. 2019;19:1129–38.
37. Stewart D, Kucheryavaya A, Klassen D, Turgeon N, Formica R, Aeder M. Changes in deceased donor kidney transplantation one year after KAS implementation. Am J Transplant. 2016;16:1834–47.
38. Kasiske BL, Snyder JJ, Matas AJ, Ellison MD, Gill JS, Kausz AT. Preemptive kidney transplantation: the advantage and the advantaged. J Am Soc Nephrol. 2002;13:1358–64.
39. Innocenti GR, Wadei HM, Prieto M, et al. Preemptive living donor kidney transplantation: do the benefits extend to all recipients? Transplantation. 2007;83:144–9.
40. Abecassis M, Bartlett ST, Collins AJ, et al. Kidney transplantation as primary therapy for end-stage renal disease: a National Kidney Foundation/Kidney Disease Outcomes Quality Initiative (NKF/KDOQI™) conference. Clin J Am Soc Nephrol. 2008;3:471–80.
41. Mallick N, El Marasi A. Dialysis in the elderly, to treat or not to treat? Nephrol Dial Transplant. 1999;14:37–9.
42. Oniscu GC, Brown H, Forsythe JL. How great is the survival advantage of transplantation over dialysis in elderly patients? Nephrol Dial Transplant. 2004;19:945–51.
43. Rao PS, Merion RM, Ashby VB, Port FK, Wolfe RA, Kayler LK. Renal transplantation in elderly patients older than 70 years of age: results from the Scientific Registry of Transplant Recipients. Transplantation. 2007;83:1069–74.
44. Huang E, Poommipanit N, Sampaio MS, et al. Intermediate-term outcomes associated with kidney transplantation in recipients 80 years and older: an analysis of the OPTN/UNOS database. Transplantation. 2010;90:974–9.
45. Heldal K, Hartmann A, Grootendorst DC, et al. Benefit of kidney transplantation beyond 70 years of age. Nephrol Dial Transplant. 2010;25:1680–7.
46. USRDS. Annual Data Report 2015 [cited 2018 3/30/2018]. Available from: https://www.usrds.org/adr.aspx.
47. Coupel S, Giral-Classe M, Karam G, et al. Ten-year survival of second kidney transplants: impact of immunologic factors and renal function at 12 months. Kidney Int. 2003;64:674–80.
48. Pour-Reza-Gholi F, Nafar M, Saeedinia A, et al. Kidney retransplantation in comparison with first kidney transplantation. Transplant Proc. 2005;37:2962–4.
49. Surga N, Viart L, Wetzstein M, et al. Impact of renal graft nephrectomy on second kidney transplant survival. Int Urol Nephrol. 2013;45:87–92.
50. Dinis P, Nunes P, Marconi L, et al. Kidney retransplantation: removal or persistence of the previous failed allograft? Transplant Proc. 2014;46:1730–4.
51. Fadli SE, Pernin V, Nogue E, et al. Impact of graft nephrectomy on outcomes of second kidney transplantation. Int J Urol. 2014;21:797–802.
52. Held PJ, McCormick F, Ojo A, Roberts JP. A cost-benefit analysis of government compensation of kidney donors. Am J Transplant. 2016;16:677–85.

53. Saari R, Cooper DKC. Financial aspects of organ procurement from deceased donors in the USA – relevance to xenotransplantation. Xenotransplantation. 2017;24(4). https://doi.org/10.1111/xen.12322.
54. Cooper DKC. Financial support for xenotransplantation research (Letter). Xenotransplantation. 2019;26(3):e12483. https://doi.org/10.1111/xen.12483.
55. Liu Z, Hu W, He T, et al. Pig-to-primate islet xenotransplantation – past, present, and future. Cell Transplant. 2017;26:925–47.
56. Kim MK, Hara H. Current status of corneal xenotransplantation. Int J Surg. 2015;23:255–60.
57. Vadori M, Aron Badin R, Hantraye P, Cozzi E. Current status of neuronal cell xenotransplantation. Int J Surg. 2015;23:267–72.
58. Yamamoto T, Iwase H, King TW, Hara H, Cooper DKC. Skin xenotransplantation: historical review and clinical potential. Burns. 2018;44:1738–49. https://doi.org/10.1016/j.burns.2018.02.029. Epub 2018 Mar 27.
59. Smood B, Hara H, Schoel LJ, Cooper DKC. Genetically-engineered pigs as sources for clinical red blood cell transfusion: what pathobiological barriers need to be overcome? Blood Rev. 2019. pii: S0268-960X(18)30095-X; https://doi.org/10.1016/j.bire.2019.01.003.
60. Dieckhoff B, Petersen B, Kues WA, Kurth R, Niemann H, Denner J. Knockdown of porcine endogenous retrovirus (PERV) expression by PERV-specific shRNA in transgenic pigs. Xenotransplantation. 2008;15:36–45.
61. Ramsoondar J, Vaught T, Ball S, et al. Production of transgenic pigs that express porcine endogenous retrovirus small interfering RNAs. Xenotransplantation. 2009;16:164–80.
62. Niu D, Wei HJ, Lin L, et al. Inactivation of porcine endogenous retrovirus in pigs using CRISPR-Cas9. Science. 2017;357:1303–7.
63. Fishman JA. Infectious disease risks in xenotransplantation. Am J Transplant. 2018;18:1857–64.

Selection of Patients for the Initial Clinical Trials of Cardiac Xenotransplantation

15

Richard N. Pierson III, Lars Burdorf, and David A. D'Alessandro

Abbreviations

MCSD	Mechanical circulatory support device
PRA	Panel-reactive antibodies
TAH	Total artificial heart
VA ECMO	Venoarterial extracorporeal membrane oxygenator
VAD	Ventricular assist device

Introduction

A major limiting factor constraining the demonstrated efficacy of cardiac allotransplantation remains the availability of organs from human donors. Each year, over 150 of the ~4000 patients on the United Network for Organ Sharing (UNOS) waiting list die resulting from the unavailability of a suitable human donor heart [1, 2]. Many additional patients are removed from consideration because of clinical deterioration while waiting.

More than half of the 3500 North American patients fortunate enough to receive a heart transplant each year have previously required "bridging" with a mechanical circulatory support device (MCSD) to stabilize them until a suitable donor can be identified. In addition to the substantial added costs associated with MCSD, many device-treated patients experience major morbidities that complicate or prevent

R. N. Pierson III (✉) · L. Burdorf
Division of Cardiac Surgery and Center for Transplantation Sciences, Department of Surgery, Massachusetts General Hospital and Harvard University, Boston, MA, USA
e-mail: rpierson@mgh.harvard.edu

D. A. D'Alessandro
Division of Cardiac Surgery, Department of Surgery, Massachusetts General Hospital and Harvard University, Boston, MA, USA

D. K. C. Cooper, G. Byrne (eds.), *Clinical Xenotransplantation*,
https://doi.org/10.1007/978-3-030-49127-7_15

subsequent transplantation. Use of hearts from "expanded criteria" donors or resuscitated after "donation after cardiac death" is being investigated in clinical trials but appears unlikely to fill the gap between supply and demand [3]. If predictably healthy organs were dependably available when needed, many patients who are currently not offered heart transplantation might also benefit if cardiac xenotransplantation is proven effective. These considerations justify efforts to develop pig cardiac xenotransplantation as a readily available alternative source of hearts for patients with life-threatening cardiac diseases [1].

Survival of nonhuman primate recipients of life-supporting porcine heart grafts for 6 months strongly suggests that clinical application of heart xenografts is likely to be successful [4]. For the purposes of argument, we presume that a combination of pig phenotype and clinically acceptable drug regimen will be defined that is successful in the preclinical life-supporting orthotopic heart xenograft model [5–8]. If renal xenograft clinical trials advance more rapidly to the clinic, successful approaches and lessons learned will be useful to the implementation of the first heart efforts.

Background

The use of pigs as a source of hearts for use as "xenografts" in humans (cardiac xenotransplantation) could potentially address the current donor heart shortfall. Substantial prolongation in survival of genetically modified pig organ grants in nonhuman primates has recently been reported from a number of groups, with consistent prevention of "delayed xenograft rejection" ("thrombotic microangiopathy" and "consumptive coagulopathy") [6, 9]. The mechanistic causes of these phenomena are addressed by a combination of innovations in genetic engineering of "donor" pigs [10–12] and development of several effective immunosuppressive strategies, primarily based on monoclonal antibodies that block key costimulation pathway molecules [7, 13].

For example, some recipients of life-supporting kidney xenografts have survived for over 1 year [13–15]. Nonlife-supporting heart xenografts continue to function in healthy recipients, with preserved myocardial histology, for >2 years as long as immunosuppression is continued [7]. Using a highly similar regimen, baboon recipients of life-supporting orthotopic pig xenografts have survived to elective termination at 6 months [4]. Although the immunosuppressive regimens used in these examples are currently not clinically approved for use in humans, extensive preclinical experience in our hands and by others suggests that they would likely have a favorable clinical safety profile. Based on these reports, we believe that it is reasonable to assume that similar or equivalent biologics will become available or that further genetic engineering of the organ-source pig will allow conventional immunosuppressive therapy to prevent the adaptive immune response.

Various currently available alternative approaches to mitigate the numerical disparity between potential recipients and deceased donors are discussed below, including the use of extended criteria donors and ventricular device (VAD) implantation either as a bridge to allotransplantation or as definitive ("destination") therapy. For completeness, we recognize that repopulation of decellularized heart tissue scaffolds with autologous recipient cells has yielded organs that survive for short periods after implantation, but cannot yet generate clinically useful cardiac performance or durability [16, 17]. Growth of whole organs from stem cells is also being explored but is at a very early stage [18, 19]. Thus, despite important recent progress in each of these areas, these "tissue engineering" and "regenerative medicine" approaches are unlikely to resolve the ongoing gap between supply and demand for allogeneic hearts in the foreseeable future.

Selection of Initial Patients for Heart Xenotransplantation

Ethical concerns regarding design and conduct of clinical trials of xenotransplantation are substantial and have been discussed by others [20–22]. Cardiac xenograft recipient selection must be considered in the context of the several available MCSDs, including ventricular assist devices (LVADs, RVADs), total artificial heart (TAH), and venoarterial extracorporeal membrane oxygenator (VA ECMO) technology. These technologies are currently used effectively as alternative "destination" or "bridge-to-transplant" options for many patients with acute or chronic heart failure. However, a substantial number of heart failure patients are poorly served by current "device therapy" alternatives, because those approaches are either not feasible or have predictably poor results. Here we describe candidates from among those facing a grim prognosis with currently available options for whom ethical equipoise with respect to cardiac xenografting may be present.

It is important to remember that, even if proposed benchmarks for preclinical results are achieved [21], the predictive power of the pig-to-baboon model has never been tested. This fact should be considered in developing the informed consent process for the initial clinical attempt.

The US Food and Drug Administration (FDA) has recommended that xenotransplantation should first be attempted in "patients with serious or life-threatening diseases for whom adequately safe and effective alternative therapies are not available" and in subjects "who have potential for a clinically significant improvement with increased quality of life following the procedure" [23]. Initial subjects should be very unlikely to be allocated a human organ, but medically eligible for heart replacement treatment, and reasonably expected to benefit if a heart xenograft works well. Of note, minors should probably be excluded from the initial clinical xenotransplantation trials because international guidelines generally discourage inclusion of children in "first-in-human" trials unless testing in adults is not feasible, particularly for high-risk studies.

Initial cardiac xenografting candidates should meet institutional heart transplant candidacy "listing" guidelines [24–26]. Accordingly, evaluation should include a current negative screen for active malignancy and for infectious diseases likely to be aggravated by immunosuppression, such as untreated or latent mycobacterial infection, untreated viral hepatitis, and chronic or recurrent lung, biliary, skin, or urinary tract infections. The reversibility of systemic impairments such as renal insufficiency or hepatic congestion that might be attributable to the underlying cardiac disease should be plausible if the proposed procedure is technically successful and the associated treatment regimen is well-tolerated. Psychosocial factors and general condition should not be overlooked, because psychiatric disease, poor compliance, or inadequate family supports might not only interfere with determining the procedures efficacy but also prevent the subject from realizing the benefits of a technically successful operation [26]. Public confidence would be quickly eroded if poor initial outcomes were associated with deviations from time-tested heart transplant candidate selection criteria.

Cardiac transplantation remains the most effective, durable therapeutic option for treatment of end-stage heart failure and has become firmly established as the standard-of-care for medically suitable patients. However, many patients are excluded from consideration (never referred) and succumb to acute cardiac decompensation before referral to a transplant center can be accomplished. Emergence of VA ECMO and temporary percutaneous heart support technologies as effective and increasingly available rescue measures to stabilize patients in cardiogenic shock might be more broadly applied, and yield improved outcomes, if healthy pig organs provided a readily available 'exit strategy'. However, these patients are critically ill, often with incipient or established failure of multiple other organ systems. Applying heart xenotransplantation in emergent circumstances is likely to fail for reasons independent of heart xenograft performance. Further, multisystem derangements typical of desperate clinical circumstances would complicate efforts to understand and address residual barriers to cardiac xenotransplantation that may not be revealed in preclinical models.

Patient selection criteria for initial xenotransplantation clinical trials were developed by experts convened under the auspices of the International Xenotransplantation Association (IXA) and World Health Organization (WHO) [21]. Their ethical guidelines include informing not only the initial trial participants but also their close contacts regarding known and possible unforeseen risks and doing so deliberately, alongside therapeutic alternatives, under noncoercive circumstances. When facing critical illness, and after exhausting currently available conventional treatment options, patients and their families are unlikely to perceive that they have a real choice. Thus, cardiac xenotransplantation should not be considered in emergent circumstances until proven feasible and reasonably successful in circumstances where existing ethical guidelines for obtaining informed consent can be followed. Similarly, candidates who are deemed ineligible based on medical conditions that would significantly reduce quality of life or predictably limit survival should probably not be included initially. On the other hand, those patients might be among those who most benefit from xenotransplantation of hearts, alone, or with other organs, once the field has been established.

Results with "destination therapy" VADs now approach those achieved following heart transplant subpopulations of carefully selected patients [27–31], and VADs are also commonly used as a "bridge-to-transplant." Indeed, over 50% of recent heart transplant recipients are previous VAD recipients [1, 2]. Among the population of patients bridged with a VAD, some have (or develop) high levels of "panel-reactive antibodies" (PRA). High PRA levels are closely associated with long waiting times and poor outcomes after heart transplantation [32, 33]. Some of these patients might be eager to participate in a heart xenograft trial, particularly if they lack antibodies reactive with a possible pig donor and were otherwise facing risks associated with a different high-risk procedure (pump exchange for thrombosis) or treatment alternative (alloantibody desensitization protocol).

BiVAD or TAH Candidates

Several categories of patients exhibit relative or absolute contraindications to VAD implantation (listed in Table 15.1). Some of these patients are likely to perceive that participating in the initial cardiac xenotransplantation trial offers the significant advantage relative to their current available options, particularly those who are likely to acquire a temporary or permanent RVAD support. BiVAD recipients, particularly those who are highly sensitized to alloantigens, have a high mortality rate on the heart transplant waiting list. Current US Food and Drug Administration (FDA)-approved VADs that are deployed as RVADs are extracorporeal and not designed (or approved) for out-of-hospital use. Many BiVAD patients typically miss "windows of opportunity" between implantation and succumbing to device-related complications because a donor does not become available. During recuperation after BiVAD surgery, education regarding a xenograft trial could be accomplished and the transplant procedure performed electively in a fully informed,

Table 15.1 Criteria for initial pig heart xenotransplantation trial candidacy

1. *High immunologic risk, rapidly progressive failure of prior heart allograft*
A. Rapid progression of cardiac allograft vasculopathy or myocardial fibrosis
B. Broadly reactive, high-titer antibody against HLA antigens
(i) Allosensitization by transfusion, prior pregnancy, or allograft implant
2. *Absolute or relative contraindications to VAD implantation (rapidly deteriorating)*
A. Aortic valve insufficiency; aortic root aneurysmal disease
B. Left ventricular thrombus
C. Left-sided mechanical prosthetic valves
D. Postinfarction ventricular septal defect
E. Restrictive or hypertrophic cardiomyopathy
F. Acute myocardial necrosis of the left ventricular apex
G. Congenital or acquired single-ventricle physiology
3. *Adult congenital heart disease (rapidly deteriorating)*
A. Single-ventricle physiology
B. Declining reversibility of pulmonary vascular resistance (PVR) elevation (falling left-to-right shunt)
4. *Severe biventricular failure without established end-organ failure*
A. Severe right ventricular failure in the context of left ventricular failure (BiVAD candidate)
B. TAH candidate

Coexistence of multiple criteria is common in heart allograft candidates and would identify patients most likely to have a favorable risk/benefit profile from a novel, high-risk experimental intervention medically stabilized patient. Biventricular support using currently available TAH devices is constrained by sizing issues and is associated with a high complication rate and inferior bridge-to-transplant outcomes [30, 31]. These considerations identify BiVAD and TAH candidates as candidates for consideration in initial enrollment in a clinical cardiac xenograft trial.

Chronic Rejection After Prior Heart Transplant

Chronic rejection is prevalent among heart transplant recipients [2]. A minority receive a second (or third) allograft, in competition with those awaiting their first transplant. Particularly those with high PRA have substantial mortality while waiting for a "negative crossmatch" and inferior survival after transplant, even in the absence of "donor-specific" antibody [2, 34]. Patients with chronic rejection of their prior cardiac allograft, particularly those without preformed antibodies against the donor pig [35–40], might be willing participants in a heart xenotransplantation trial.

Congenital Heart Disease

Despite successful palliative procedures, adult patients with complex congenital heart disease remain vulnerable to late myocardial dysfunction and a variety of other complications [41, 42]. These patients are relatively disadvantaged by current heart allograft allocation criteria [2, 41, 42]. In addition, many have been sensitized to alloantigens by blood transfusion or tissue homograft exposures [43]. VAD implantation is frequently not feasible or effective in patients with Fontan failure where the development of complications, such as protein-losing enteropathy and desaturation secondary to intrapulmonary venoarterial shunts, occurs even in the setting of normal systolic function [42]. For some highly sensitized adult congenital patients with progressive heart failure or worsening arrhythmias, xenotransplantation could be an attractive alternative.

Strategic and Technical Considerations

Xenograft as a Bridge to Allotransplantation

The use of a heart xenograft as a temporary bridge to allotransplantation has been suggested, but would not ameliorate the donor organ shortage. Because immunosuppression would almost certainly be necessary to protect the heart xenograft, the patient would face higher risks of infection at the time of subsequent allotransplantation. Clinical translation of heart xenografts is unlikely to be considered until durable life-supporting cardiac xenograft performance and a favorable safety profile are predicted by results in baboons. In this context, it might prove difficult to justify removal of a well-functioning heart xenograft until and unless quality and length of life with the xenograft was predictably inferior to the added risks associated with another major transplant operation. Several investigators have shown that antibodies against human leukocyte antigens (HLA) are not usually elicited after exposure of humans to pig cells or tissues [44, 45]. However, design of a clinical heart xenograft trial should take the eventuality of incidental cross-sensitization to alloantigens into account, particularly if bridging to an allograft is anticipated, either as a planned strategy or a "bailout" option.

Heterotopic Heart Transplantation

The use of the heterotopic heart transplant technique would reduce the recipients' dependence on robust initial heart xenograft function, either initially after implantation or in case of later immune injury. Because this heterotopic technique is difficult and was associated with respiratory and embolic complications, this approach should probably be reserved for unusual cases (high pulmonary vascular resistance) where it would also be used for an available allograft.

Facing "Unknown Unknowns"

The patient selection, education, and consenting process will need to be deliberate, rigorous, and transparent, acknowledging the many "unknown unknowns" facing trial participants and investigators and the implications of predictable as well as unforeseen complications not only for each trial participant but also for their family or close contacts.

Multi-organ Xenotransplantation, Including the Heart

Patients who would otherwise die because they are not acceptable as candidates for heart-only allotransplantation, due to potentially reversible liver or kidney dysfunction, could be considered for heart xenotransplantation, using the pig heart to evaluate whether these problems are reversible, with restoration of normal cardiac function (bridge-to-end-organ recovery). Adequate temporary support may provide time for sufficient recovery of native liver and/or kidney function to enable recuperation to proceed to heart allotransplantation and might be facilitated by co-transplantation of a kidney xenograft or an auxiliary liver xenograft from the same pig. However, as noted above, this strategy adds significant complexity to designing and evaluating results from an initial clinical trial and is likely to result in a high incidence of adverse outcomes, particularly if the liver or kidney dysfunction proves irreversible. Because of these considerations, we do *not* favor this "rescue" or "bridge-to-decision" approach for the initial clinical studies of cardiac xenotransplantation.

Combined Transplantation of the Heart and Lungs

Experimental pig lung xenotransplantation studies in ex vivo human blood perfusion models and in vivo in nonhuman primates have revealed numerous obstacles to lung xenograft survival that do not usually limit graft or recipient survival after pig kidney or heart xenotransplantation [15, 46]. Although considerable progress has recently been reported [47], further progress will be required before a clinical trial of heart-lung or lung xenotransplantation will be seriously contemplated. Nevertheless, we believe it is valuable to consider what form a clinical trial would take once justified by consistently improved preclinical results.

Initial heart-lung clinical trials might include otherwise qualified heart transplant candidates excluded solely on fixed high pulmonary vascular resistance (PVR), particularly among adult congenital heart patients. Multiple patients with end-stage lung disease (e.g., chronic obstructive pulmonary disease [COPD], idiopathic pulmonary fibrosis [IPF], sarcoidosis) are found to have severe atherosclerotic, valvular, or myopathic heart disease that currently disqualifies them from lung transplant candidacy. Lung and heart-lung patients with rapidly progressing chronic rejection (bronchiolitis obliterans syndrome) fare poorly after lung re-transplantation, particularly in the setting of the restrictive bronchiolitis obliterans syndrome (BOS) phenotype and in the setting of high PRA. Once safety and efficacy have been established in non-emergent conditions, heart-lung xenotransplantation could be offered for patients who require prolonged VA ECMO support for disorders affecting both the heart and the lungs.

Current United Network for Organ Sharing (UNOS) policies are single organ-centered and intended to optimize efficacy of a scarce resource on a societal basis. In addition, those who might benefit from combined transplantation of the heart-lung block along with another organ, such as a kidney, are not prioritized by current donor allocation rules due to concerns regarding futility and organ scarcity. If multi-organ xenografts became available, not only might the allo- (and auto-)immune barrier be avoided and access to high-quality organs be facilitated, but the obstacles to organ access based on current allocation policies that adversely impact patients who are afflicted with simultaneous failure of multiple organs would become irrelevant.

Conclusion

Recent preclinical results predict that pig hearts with a limited set of targeted genetic modifications will be protected from immune injury in humans using a clinically applicable calcineurin inhibitor-free immunosuppressive regimen, setting the stage for design of clinical pig-to-human heart xenotransplantation trials. Because of the broad variety among patients with end-stage heart failure and the availability of VADs, TAHs, and ECMO as temporizing or even durable therapeutic alternatives, selection of patients for the initial clinical application of heart xenotransplantation is not straightforward.

Here we argue that clinical heart xenotransplantation should *not* be first attempted under emergent circumstances or in patients for whom allotransplantation is contraindicated. Rather, we identify multiple categories of patients who are currently awaiting heart transplantation, but whose timely access to a human heart is severely constrained by their clinical circumstances, and for whom participating in an initial exploratory study of heart xenotransplantation would satisfy ethical equipoise. Consistent demonstration of sustained life-supporting xenograft function in a preclinical model, as stipulated by the International Society for Heart and Lung Transplantation (ISHLT) and WHO-convened international experts, should be sufficient to establish this equipoise and to win necessary approval from institutional review boards and national regulatory authorities, affirming prior [48] and contemporary [21] consensus statements.

Once established as safe and effective in such carefully selected patients, we believe that pig heart xenotransplantation, alone and in conjunction with other organ xenografts, will be among the options offered to all patients with terminal heart failure. The recent accelerating rate of progress in pig-to-nonhuman primate kidney and heart transplantation gives us optimism that clinical heart xenotransplantation trials will be undertaken within the next 5 years.

Acknowledgment The authors wish to thank Dr. Paul S. Russell, Jr., Dr. Joren C. Madsen, and Dr. A. Benedict Cosimi for their careful reading and constructive comments in response to an early draft.

Disclosures The authors declare no conflicts of interest.

Funding Work on xenotransplantation in the Center for Transplantation Sciences at MGH is supported by grants from NIH (U19 AI090959), by contributed research materials from Revivicor, Blacksburg, VA, and by funding and contributed research materials from eGenesis, Cambridge, MA. Prior work at the University of Maryland was supported by grants from NIH (U19 AI090959) and by gifts, sponsored research agreement funding, and contributed research materials from Revivicor, Blacksburg, VA, Lung Biotechnologies PC, and United Therapeutics, Silver Spring, MD.

References

1. http://optn.transplant.hrsa.gov.
2. Colvin M, Smith JM, Hadley N, et al. OPTN/SRTR 2018 annual data report: heart. Am J Transplant. 2020;20:340–426.
3. Israni AK, Zaun D, Hadley N, et al. OPTN/SRTR 2018 annual data report: deceased organ donation. Am J Transplant. 2020;20:509–41.
4. Längin M, Mayr T, Reichart B, et al. Consistent success in life-supporting porcine cardiac xenotransplantation. Nature. 2018;564:430–3.
5. Pierson RN III. Current status of xenotransplantation. J Am Med Assoc. 2009;301:967–9.
6. Pierson RN III, Dorling A, Ayares D, et al. Current status of xenotransplantation and prospects for clinical application. Xenotransplantation. 2009;16:263–80.

7. Mohiuddin M, Singh AK, Corcoran PC, et al. Chimeric 2C10R4 anti-CD40 antibody therapy is critical for long-term survival of GTKO.hCD46.hTBM pig-to-primate cardiac xenografts. Nat Commun. 2016;7:11138.
8. Pierson RN III. A major advance toward clinical cardiac xenotransplantation. J Thorac Cardiovasc Surg. 2019. pii: S0022-5223(19)31024-4.; https://doi.org/10.1016/j.jtcvs.2019.04.087.
9. Cooper DKC, Ezzelarab MB, Hara H, et al. The pathobiology of pig-to-primate xenotransplantation: a historical review. Xenotransplantation. 2016;23:83–105.
10. Cooper DKC, Ekser B, Ramsoondar J, Phelps C, Ayares D. The role of genetically-engineered pigs in xenotransplantation research. J Pathol. 2016;238:288–99.
11. Cooper DKC, Hara H, Iwase H, et al. Justification of specific genetic modifications in pigs for clinical organ xenotransplantation. Xenotransplantation. 2019;26:e12516.
12. Niu D, Wei HJ, Lin L, et al. Inactivation of porcine endogenous retrovirus in pigs using CRISPR-Cas9. Science. 2017;357:1303–7.
13. Kim SC, Mathews DV, Breeden CP, et al. Long-term survival of pig-to-rhesus macaque renal xenografts is dependent on CD4 T cell depletion. Am J Transplant. 2019;19:2174–85.
14. Higginbotham L, Mathews D, Breeden CA, et al. Pre-transplant antibody screening and anti-CD154 costimulation blockade promote long-term xenograft survival in a pig-to-primate kidney transplant model. Xenotransplantation. 2015;22:221–30.
15. Cooper DKC, Satyananda V, Ekser B, et al. Progress in pig-to-nonhuman primate transplantation models (1998–2013): a comprehensive review of the literature. Xenotransplantation. 2014;21:397–419.
16. Guyette JP, Charest JM, Mills RW, et al. Bioengineering human myocardium on native extracellular matrix. Circ Res. 2016;118:56–72.
17. Tong C, Li C, Xie B, et al. Generation of bioartificial hearts using decellularized scaffolds and mixed cells. Biomed Eng Online. 2019;18:71.
18. Gao L, Kupfer ME, Jung JP, et al. Myocardial tissue engineering with cells derived from human-induced pluripotent stem cells and a native-like, high-resolution, 3-dimensionally printed scaffold. Circ Res. 2017;120:1318–25.
19. Li RA, Keung W, Cashman TJ, et al. Bioengineering an electro-mechanically functional miniature ventricular heart chamber from human pluripotent stem cells. Biomaterials. 2018;163:116–27.
20. Smetanka C, Cooper DKC. The ethics debate in relation to xenotransplantation. Rev Sci Tech. 2005;24:335–42.
21. Hawthorne WJ, Cowan PJ, Bühler LH, et al. Third WHO global consultation on regulatory requirements for xenotransplantation clinical trials, Changsha, Hunan, China December 12–14, 2018: "the 2018 Changsha Communiqué" the 10-year anniversary of the international consultation on xenotransplantation. Xenotransplantation. 2019;26:e12513. https://doi.org/10.1111/xen.12513.
22. Paris W, Seidler RJH, FitzGerald K, et al. Jewish, Christian and Muslim theological perspectives about xenotransplantation. Xenotransplantation. 2018;25:e12400. https://doi.org/10.1111/xen.12400.
23. http://www.fda.gov/cber/guidelines.htm.
24. Cooper DKC, Wijkstrom M, Hariharan S, et al. Selection of patients for initial clinical trials of solid organ xenotransplantation. Transplantation. 2017;101:1551–8.
25. Ponikowski P, Voors AA, Anker SD, ESC Scientific Document Group, et al. 2016 ESC guidelines for the diagnosis and treatment of acute and chronic heart failure. Eur Heart J. 2016;37:2129–200. Erratum in: *Eur Heart J*. 2016 Dec 30.
26. Dew MA, DiMartini AF, Dobbels F, et al. The 2018 ISHLT/APM/AST/ICCAC/STSW recommendations for the psychosocial evaluation of adult cardiothoracic transplant candidates and candidates for long-term mechanical circulatory support. Psychosomatics. 2018;59:415–40.

27. Kirklin JK. Terminal heart failure: who should be transplanted and who should have mechanical circulatory support? Curr Opin Organ Transplant. 2014;19:486–93.
28. Kirklin JK, Naftel DC, Pagani FD, et al. Sixth INTERMACS annual report: a 10,000-patient database. J Heart Lung Transplant. 2014;33:555–64.
29. Kanwar MK, Bailey S, Murali S. Challenges and future directions in left ventricular assist device therapy. Crit Care Clin. 2018;34:479–92.
30. Melton N, Soleimani B, Dowling R. Current role of the total artificial heart in the management of advanced heart failure. Curr Cardiol Rep. 2019;21:142.
31. Beaupré RA, Frazier OH, Morgan JA. Total artificial heart implantation as a bridge to transplantation: a viable model for the future? Expert Rev Med Devices. 2018;15:701–6.
32. Kidambi S, Mohamedali B, Bhat G. Clinical outcomes in sensitized heart transplant patients bridged with ventricular assist devices. Clin Transpl. 2015;29:499–505.
33. Chiu P, Schaffer JM, Oyer PE, et al. Influence of durable mechanical circulatory support and allosensitization on mortality after heart transplantation. J Heart Lung Transplant. 2016;35:731–42.
34. Goerler H, Simon A, Gohrbandt B, et al. Cardiac retransplantation: is it justified in times of critical donor organ shortage? Long-term single-center experience. Eur J Cardiothorac Surg. 2008;34:1185–90.
35. Wong BS, Yamada K, Okumi M, et al. Allosensitization does not increase the risk of xenoreactivity to alpha1,3-galactosyltransferase gene-knockout miniature swine in patients on transplantation waiting lists. Transplantation. 2006;82:314–9.
36. Hara H, Ezzelarab M, Rood PP, et al. Allosensitized humans are at no greater risk of humoral rejection of GT-KO pig organs than other humans. Xenotransplantation. 2006;13:357–65.
37. Naziruddin B, Durriya S, Phelan D, et al. HLA antibodies present in the sera of sensitized patients awaiting renal transplant are also reactive to swine leukocyte antigens. Transplantation. 1998;66:1074–80.
38. Martens GR, Reyes LM, Li P, et al. Humoral reactivity of renal transplant-waitlisted patients to cells from GGTA1/CMAH/B4GalNT2, and SLA class I knockout pigs. Transplantation. 2017;101(4):e86–92. https://doi.org/10.1097/TP.0000000000001646. Erratum in: *Transplantation*. 2018; 102(2): e88.
39. Ladowski JM, Reyes LM, Martens GR, et al. Swine leukocyte antigen class II is a xenoantigen. Transplantation. 2018;102:249–54.
40. Ladowski JM, Martens GR, Reyes LM, et al. Examining the biosynthesis and xenoantigenicity of class II swine leukocyte antigen proteins. J Immunol. 2018;200:2957–64.
41. Dipchand AI, Kirk R, Edwards LB, et al. The registry of the International Society for Heart and Lung Transplantation: sixteenth official pediatric heart transplantation report – 2013; focus theme: age. J Heart Lung Transplant. 2013;32:979–88.
42. Gelow JM, Song HK, Weiss JB, Mudd JO, Broberg CS. Organ allocation in adults with congenital heart disease listed for heart transplant: impact of ventricular assist devices. J Heart Lung Transplant. 2013;32:1059–64.
43. Jayakumar KA, Addonizio LJ, Kichuk-Chrisant MR, et al. Cardiac transplantation after the Fontan or Glenn procedure. J Am Coll Cardiol. 2004;44:2065–72.
44. Key T, Schuurman HJ, Taylor CJ. Does exposure to swine leukocyte antigens after pig-to-nonhuman primate xenotransplantation provoke antibodies that cross-react with human leukocyte antigens? Xenotransplantation. 2004;11:452–6.
45. Li Q, Hara H, Zhang Z, Breimer ME, Wang Y, Cooper DKC. Is sensitization to pig antigens detrimental to subsequent allotransplantation? Xenotransplantation. 2018;25:e12393. https://doi.org/10.1111/xen.12393.
46. Kubicki N, Laird C, Burdorf L, Pierson RN III, Azimzadeh AM. Current status of pig lung xenotransplantation. Int J Surg. 2015;23:247–54.

47. Burdorf L, Azimzadeh AM, Pierson RN III. Progress and challenges in lung xenotransplantation: an update. Curr Opin Organ Transplant. 2018; https://doi.org/10.1097/MOT.0000000000000582.
48. Cooper DKC, Keogh AM, Brink J, et al. Report of the xenotransplantation advisory Committee of the International Society for heart and lung transplantation: the present status of xenotransplantation and its potential role in the treatment of end-stage cardiac and pulmonary diseases. J Heart Lung Transplant. 2000;19:1125–65.

Selection of Pediatric Patients for the Initial Clinical Trials of Cardiac Xenotransplantation

16

James K. Kirklin and David C. Cleveland

Abbreviations

ECMO	Extracorporeal membrane oxygenator
HLHS	Hypoplastic left heart syndrome
MCS	Mechanical circulatory support

Hierarchy

In December 2017, members of the transplant community gathered in Cape Town, South Africa, to celebrate the 50th anniversary of the first human-to-human heart transplant. On December 3, 1967, Christiaan Barnard electrified the world when he transplanted the heart of Denise Darvall, the victim of a hit-and-run motor vehicle accident, into Louis Washkansky, a former pugilist, dying from the ravages of multiple heart attacks. Three days later, Adrian Kantrowitz transplanted the heart of an anencephalic infant into a baby dying from Ebstein's anomaly, only to see the transplanted heart fail several hours after the operation.

The next truly seminal event in pediatric heart transplantation, even though unsuccessful, occurred 17 years later when Leonard Bailey and his team at Loma Linda transplanted the heart of a baboon into Baby Fae, a newborn baby with

J. K. Kirklin (✉)
Division of Cardiothoracic Surgery, Kirklin Institute for Research in Surgical Outcomes (KIRSO), Department of Surgery, University of Alabama at Birmingham (UAB), Birmingham, AL, USA
e-mail: jkirklin@uabmc.edu

D. C. Cleveland
Division of Cardiothoracic Surgery, Department of Surgery, University of Alabama at Birmingham, Children's Hospital of Alabama, Birmingham, AL, USA

D. K. C. Cooper, G. Byrne (eds.), *Clinical Xenotransplantation*,
https://doi.org/10.1007/978-3-030-49127-7_16

hypoplastic left heart syndrome (HLHS). Even though the baby died after 2 weeks, the absence of rejection offered hope to the transplant community regarding the possibilities of xenotransplantation. However, public outcry about the use of primates as organ donors was intense and personal, and Bailey never again ventured into the arena of xenotransplantation. But, he and his colleagues ushered in the era of infant heart transplantation [1], and Loma Linda became the world leader in transplanting hearts in small babies. His group demonstrated convincingly superior outcomes with neonatal heart transplantation, reporting 15-year survival of nearly 70% [2].

Today, with over two decades of orderly scientific progress in xenotransplantation in general and within the specific domain of cardiac xenotransplantation, investigators appear on the verge of optimizing genetically engineered triple-knockout (TKO) pigs suitable for heart transplantation in humans (Fig. 16.1) [3].

So, who would be the first candidates for pediatric heart transplantation? By virtue of being "the first," short- or longer-term outcomes data with genetically engineered pig-to-human heart transplantation will not be available. Most likely, an initial experience with pig-to-human kidney transplantation will have emerged, so the early immunologic issues will have been recognized. The most logical first step in applying xenotransplantation to the first infant or child would be to consider patient groups for which heart transplantation would be desirable but is currently rarely achievable due to lack of available organs in a suitable time frame. This primarily includes patients whose hemodynamic instability (and limited time window available before death) mandates a donor heart sooner than is currently feasible. In older children and adults, mechanical circulatory support (MCS) is generally an effective means of "bridging" to transplantation when hemodynamics are compromised. So, we must also examine pediatric conditions for which no reliable MCS system is currently available. In the following discussion, if consistent 6 months'

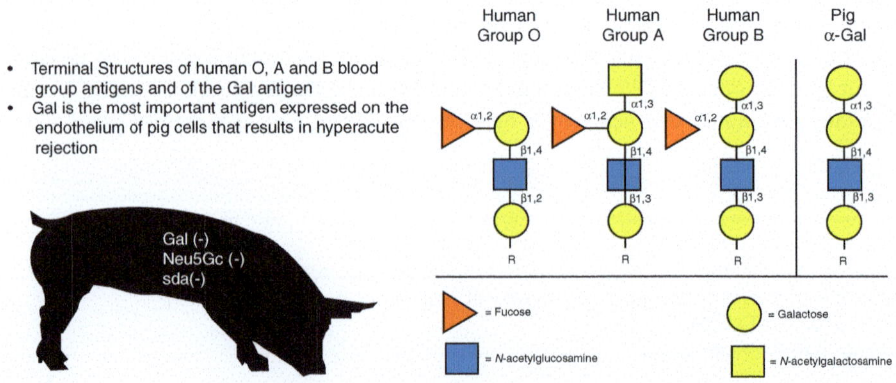

Fig. 16.1 Terminal structures of the human O, A, and B blood group antigens and of the Gal antigen. (Reproduced with permission from reference [3])

survival in a genetically engineered pig-to-primate model could be demonstrated, a trial using xenotransplantation initially in a "bridge-to-allotransplantation" mode would be most likely.

Hypoplastic Left Heart Syndrome: Heart Transplantation as Primary Therapy

During the late 1980s and much of the 1990s, heart transplantation was the favored therapy for HLHS in a number of institutions, despite the challenges of access to donors before deterioration occurred on prostaglandin support (to maintain ductal patency). However, as initial palliation with the Norwood operation showed improved early survival, most centers abandoned transplantation as the preferred therapy given the high mortality and morbidity while on the waiting list (Fig. 16.2) [4].

But what about the long-term outcomes in those babies with HLHS who did receive a heart transplant? A long-term study from Bailey's group, as well as a multi-institutional study from the Pediatric Heart Transplant Society (PHTS), reported 15-year survival of 70–75% after primary heart transplantation for HLHS [2], which is clearly superior to survival after staged palliation resulting in the Fontan procedure for HLHS. Unpublished data from the Congenital Heart Surgeons' Society also

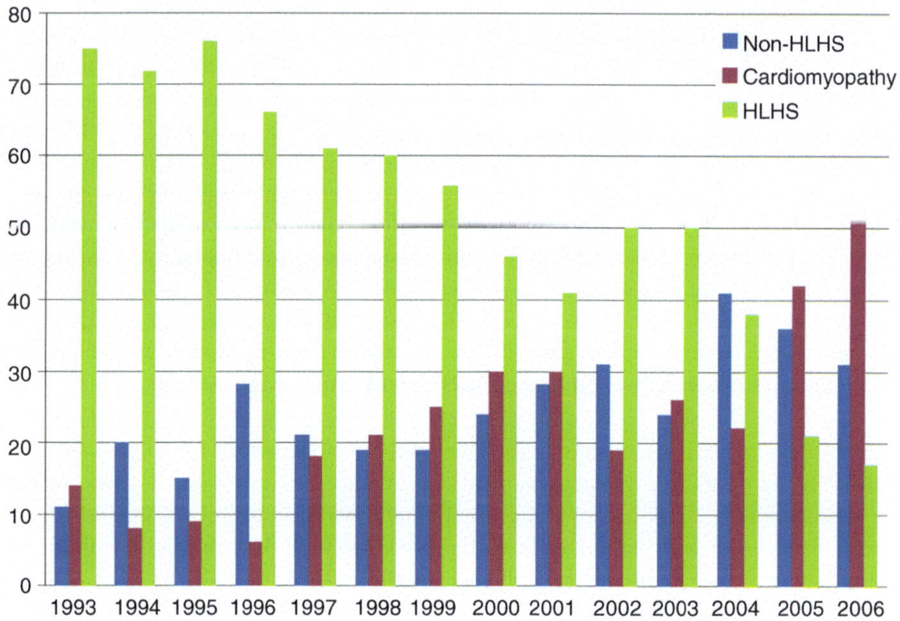

Fig. 16.2 The distribution of number of patients listed for heart transplantation in the Pediatric Heart Transplant Study in each year from 1993 to 2006 with cardiomyopathy, hypoplastic left heart syndrome (HLHS), and other congenital heart disease (non-HLHS), without previous surgery. (Reproduced with permission from reference [4])

showed a higher longer-term hazard of mortality after the Fontan operation for HLHS than after heart transplantation (personal communication, Stackhouse et al., 2019).

However, the argument for primary cardiac allotransplantation for HLHS (before even considering xenotransplantation) is not without controversy. In fact, ethicists have argued that, given the current more favorable outcomes with three-stage palliation (even if inferior to primary heart transplant), a policy of primary transplant for HLHS is inappropriate.... "where issues of social justice must be considered, such allocation of a limited resource .... should be given to infants who cannot be treated by any means other than transplantation" [5]. Currently the number of patients who would not be considered for completion of the single ventricle pathway is low. The general criteria for elective transplantation are (i) poor systemic ventricular function and (ii) severe atrioventricular valve regurgitation before or after initial palliation. However, the current very restrictive use of transplantation would be less persuasive if a considerably larger donor pool were available through xenotransplantation. This would create a potential space for triple gene-knockout (TKO) porcine heart donors if the demonstrated outcomes in the renal analog were sufficiently encouraging.

Primary Transplantation for Other Selected High-Risk Neonatal Conditions

A few neonatal conditions remain for which current palliative or corrective operations in the setting of progressive circulatory deterioration carry a high hospital mortality. These include neonatal Ebstein's anomaly not amenable to surgical repair with compromised hemodynamics and pulmonary atresia with intact ventricular septum complicated by sinusoids and severe proximal coronary artery stenosis or atresia [1]. Some centers might refer such patients for urgent heart transplantation if a suitable donor were immediately available. However, among non-HLHS patients without prior surgery who are listed for heart transplantation, the wait-list mortality approaches 30% by 3 months [4].

Transplantation After Staged Palliation for Single Ventricle

The third and usually final stage of the three palliative operations for single ventricle is the Fontan operation, which generally occurs between ages 3 and 5 years. In the current era, the late risk of death after the Fontan operation for most forms of single ventricle is quite low. An analysis of the University of Alabama at Birmingham (UAB) experience identified no increase in the hazard, or risk of death, out to 20 years (Fig. 16.3) [6]. However, when the hemodynamic condition deteriorates after the first or second stage of palliation, urgent transplantation or other forms of circulatory support may be the only options to salvage the patient.

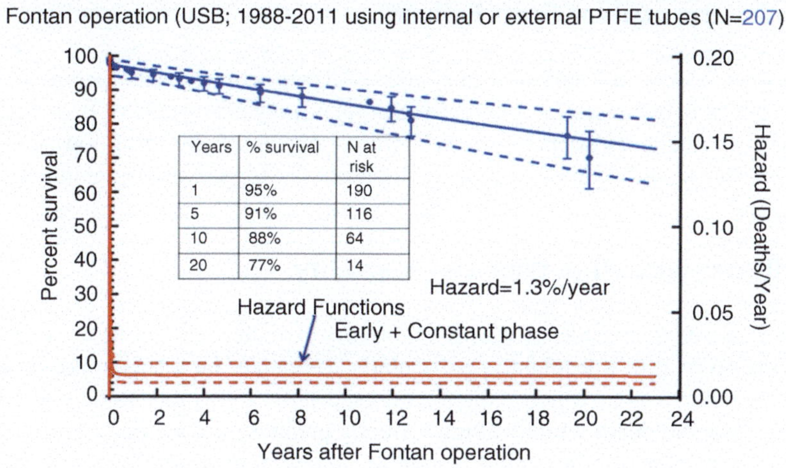

Fig. 16.3 Actuarial and parametric survival after the Fontan operation using polytetrafluoroethylene (PTFE) tubes. The lower red line indicates the instantaneous risk (hazard) of death, in which the rapidly falling early phase merges with a constant phase within the first 3 months. The error bars represent ±1SE. The dashed lines enclose the 70% confidence limits. UAB, University of Alabama at Birmingham. (Reproduced with permission from reference [6])

Another reality that must be recognized is the inferior outcome of transplantation when applied after initial palliative operations (compared to primary therapy). In contrast to the excellent late survival posttransplant for patients undergoing primary heart transplantation for HLHS, a long-term study from the Pediatric Heart Transplant Society revealed a 10-year survival of only 53% after transplantation for HLHS patients with prior palliative surgery [7].

Early graft dysfunction has been a challenge in experimental pig-to-nonhuman primate cardiac transplants, and the challenge would likely be magnified in patients with prior congenital heart surgery, as has been noted in prior studies from the International Society for Heart and Lung Transplantation [8]. Furthermore, neonates are at particularly high risk for early graft dysfunction. If xenotransplantation is going to be applied in the setting of prior complex cardiac operations or primarily in the neonatal period, experimental pig-to-nonhuman primate transplants must demonstrate reproducible good early graft function.

Human Leukocyte Antigen (HLA)-Sensitized Patients

High levels of preformed anti-HLA antibodies against human donors are a predictor of suboptimal outcomes [9]. This may be a pathway for initial xenotransplant application if the likelihood of a suitably matched human donor is low.

Early Posttransplant Graft Failure

When severe early graft failure occurs after pediatric heart transplantation, the outcomes are poor when extracorporeal membrane oxygenator (ECMO) or MCS is needed for circulatory support [8]. When there appears to be little hope for graft recovery, this would be another potential opportunity for xenotransplantation.

Conditions with High Wait-List Mortality

Another avenue to consider is wait-list mortality. Are there situations or conditions in which survival to transplant is particularly dependent on a short time to transplant? In other words, which patients are especially vulnerable to hemodynamic deterioration and death while waiting? Among patients with HLHS who are listed for heart transplant after an initial Norwood operation, the mortality while waiting is high, with nearly 30% of listed patients dying within 3 months without transplant [10]. Another Pediatric Heart Transplant Society study also noted a high wait-list mortality for non-HLHS congenital heart disease in infants [4].

Patients at High Risk with Mechanical Circulatory Support (MCS)

In older children and adults, MCS is the standard therapy for patients who deteriorate while waiting. But there are some patient subsets for whom current methods of MCS are often unsuccessful. The development of suitable devices for infants and small children has been impaired by the relatively limited market for device application compared to older patients and the challenges of miniaturization. Despite these challenges, MCS support plays a major role in supporting pediatric patients to transplantation, with approximately 20% of pediatric patients coming to transplant on MCS [8]. For children older than about 3 years of age, especially with two ventricles, who can receive a small adult continuous flow pump like the HeartWare ventricular assist device (Medtronic), the survival on support is excellent (Fig. 16.4) [11]. Xenotransplantation is not likely to be initially competitive for this group of patients. Furthermore, the rate of transplant is quite high for children on devices [11]. One real opportunity for xenotransplantation, however, is patients who are placed on ECMO emergently and do not come off promptly. If they are also poor candidates for more durable MCS, their outcome is generally poor (Fig. 16.5) [8].

The Berlin Heart Excor is the primary ventricular assist device used in infants and children <3 years of age. Certain patient groups fare poorly with the Berlin heart [12]. The mortality among infants with congenital heart disease who require MCS, and especially with single ventricle, has been generally high and exceeds 90% for neonatal single ventricle patients. Even in a contemporary analysis from a national database, the mortality remains high for congenital heart disease patients

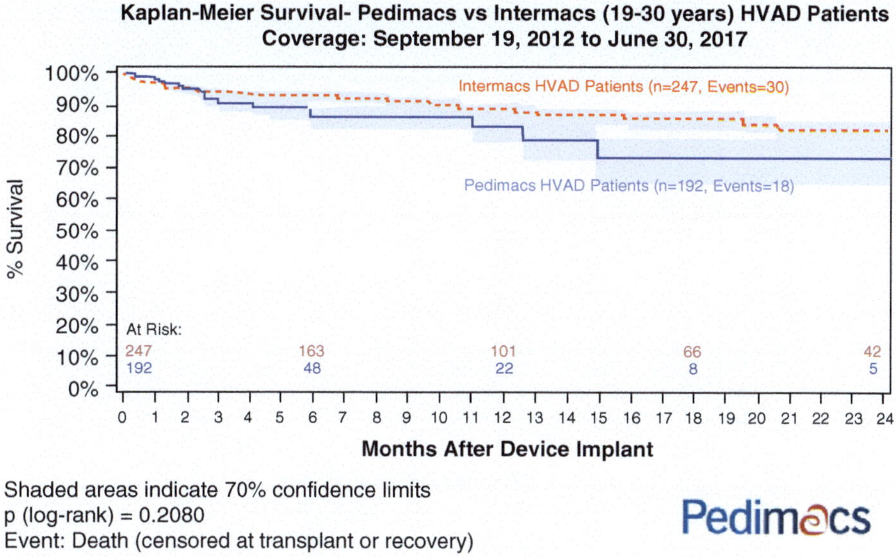

Fig. 16.4 Survival of 193 children (n = 192) (Pedimacs) compared with young adults (aged 19–30 years; n = 247) (Intermacs) implanted with the HeartWare HVAD (Medtronic) from September 19, 2012, to June 30, 2017. (Reproduced with permission from reference [11])

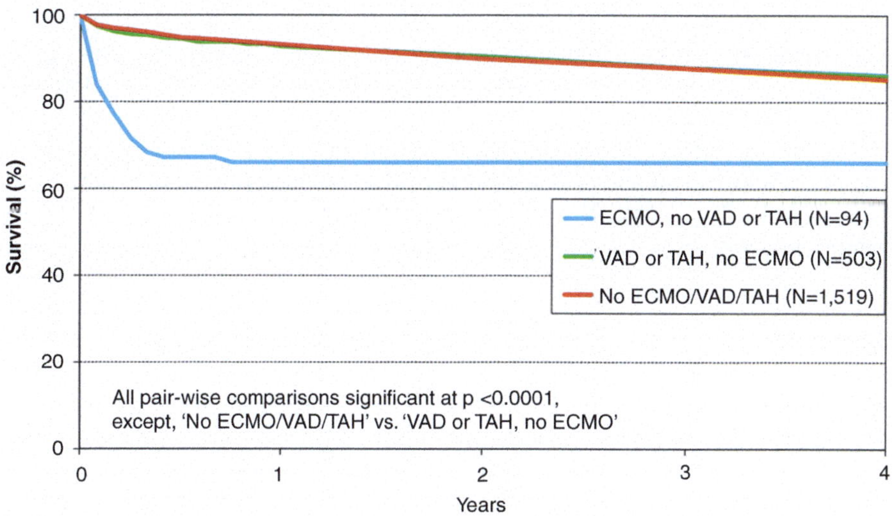

Fig. 16.5 Kaplan-Meier survival posttransplant stratified by mechanical circulatory support pretransplant (transplants: January 2009 to June 2013). (Reproduced with permission from reference [8])

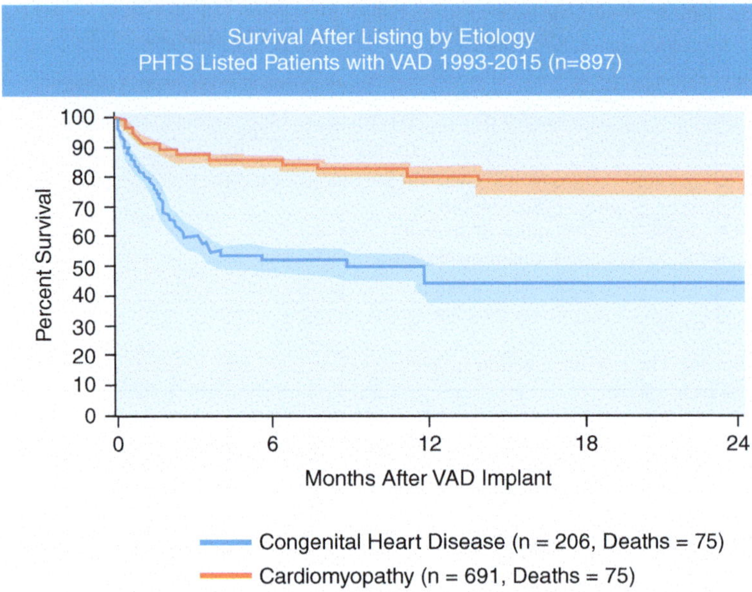

Fig. 16.6 Legend: Kaplan-Meier survival post-implant of ventricular assist device (VAD) while on wait-list, stratified by diagnosis of cardiomyopathy compared to congenital heart disease. (Reproduced with permission from reference [13])

who require MCS if they cannot be transplanted within a month (Fig. 16.6) [13]. Among infants with single ventricle who require MCS, the wait-list mortality exceeds 50% by 3 months.

The National Heart, Lung, and Blood Institute invested >$30 million on a program to develop infant and pediatric durable MCS devices, but to date success has been limited. A multicenter trial to evaluate a pediatric implantable small axial flow pump is ongoing.

Summary

With the considerations outlined above, the following patient cohorts could receive consideration for an initial pediatric xenotransplantation clinical trial, which initially would likely be designed as a "bridge trial" to subsequent human heart transplantation:

(i) Primary heart transplant for HLHS with right ventricular dysfunction, severe atrioventricular valve regurgitation, or coronary anomalies with evidence of circulatory instability
(ii) Primary transplant for high-risk subsets of neonatal Ebstein's anomaly with hemodynamic instability and pulmonary atresia with intact ventricular septum and severe coronary anomalies

(iii) Transplantation after first-stage single ventricle palliation with subsequent poor ventricular function, severe atrioventricular valve regurgitation, and hemodynamic instability (higher risk)
(iv) Congenital heart disease with prior cardiac surgery, severe heart failure, and high sensitization with poor prospects for negative crossmatch
(v) Heart transplant graft failure with need for MCS
(vi) Single ventricle or other congenital heart disease needing MCS in first 2 years of life or with high wait-list mortality

However, there are important barriers to overcome in the present experimental models for such a trial to occur:

(i) Continuous-flow MCS suitable for small infants is under active development with one ongoing clinical trial.
(ii) The results with staged palliation of single ventricle (including HLHS) are currently too good to carry out pig heart xenotransplantation unless major compromise of ventricular function is present.
(iii) Neonates (<1 month of age) and infants with complex congenital heart disease and multiple prior operations are known to be at higher risk for early graft dysfunction, so excellent and reproducible xenograft function must be achieved in the orthotopic pig-to-nonhuman primate model.

References

1. John M, Bailey LL. Neonatal heart transplantation. Ann Cardiothorac Surg. 2018;7:118–25.
2. Chinnocik RE, Bailey LL. Heart transplantation for congenital heart disease in the first year of life. Curr Cardiol Rev. 2011;7:72–84.
3. Cleveland D, Adam Banks C, Hara H, Carlo WF, Mauchley DC, Cooper DKC. The case for cardiac xenotransplantation in neonates: is now the time to reconsider xenotransplantation for hypoplastic left heart syndrome? Pediatr Cardiol. 2019;40:437–44.
4. Guleserian KJ, Schechtman KB, Zheng J, et al. Outcomes after listing for primary transplantation for infants with unoperated-on non-hypoplastic left heart syndrome congenital heart disease: a multi-institutional study. J Heart Lung Transplant. 2011;30:1023–32.
5. Kon AA. Ethics of cardiac transplantation in hypoplastic left heart syndrome. Pediatr Cardiol. 2009;30(6):725–8.
6. Dabal RJ, Kirklin JK, Kukreja M, et al. The modern Fontan operation shows no increase in mortality out to 20 years: a new paradigm. J Thorac Cardiovasc Surg. 2014;148:2517–23.
7. Dipchand AI, Kirk R, Mahle WT, et al. Ten yr of pediatric heart transplantation: a report from the Pediatric Heart Transplant Study. Pediatr Transplant. 2013;17:99–111.
8. Dipchand AI, Rossano JW, Edwards LB, et al. The registry of the International Society for Heart and Lung Transplantation: eighteenth official pediatric heart transplantation report - 2015; focus theme: early graft failure. J Heart Lung Transplant. 2015;34:1233–43.
9. Das BB, Pruitt E, Molina K, et al. The impact of flow PRA on outcome in pediatric heart recipients in modern era: an analysis of the Pediatric Heart Transplant Study database. Pediatr Transplant. 2018;22. https://doi.org/10.1111/petr.13087.
10. Alsoufi B, Mahle WT, Manlhiot C, et al. Outcomes of heart transplantation in children with hypoplastic left heart syndrome previously palliated with the Norwood procedure. J Thorac Cardiovasc Surg. 2016;151:167–74.

11. VanderPluym CJ, Adachi I, Niebler R, et al. Outcomes of children supported with an intracorporeal continuous-flow left ventricular assist system. J Heart Lung Transplant. 2019;38:385–93.
12. Almond CS, Morales DL, Blackstone EH, et al. Berlin heart EXCOR pediatric ventricular assist device for bridge to heart transplantation in US children. Circulation. 2013;27:1702–11.
13. Dipchand AI, Kirk R, Naftel DC, et al. Ventricular assist device support as a bridge to transplantation in pediatric patients. J Am Coll Cardiol. 2018;72:402–15.

Infection in Xenotransplantation: Organ-Source Health and Patient Safety

17

Jay A. Fishman

Abbreviations

PERV Porcine endogenous retrovirus

Introduction: Infection in Transplantation

Infection and cancer are major complications of long-term immunosuppressive therapy used to prevent graft rejection. Infection is derived from environmental exposures in the hospital and the community, from organisms present, often as colonizers or in latent form, in the organ recipient and from organisms carried with the transplanted organ [1–5]. In xenotransplantation, data regarding the microbiology of normal and genetically modified swine are limited. Approaches to the mitigation of the infectious risks of xenotransplantation are based on extrapolation from experience with infection following allotransplantation and on preclinical data developed in studies of immunosuppressed swine and primate xenograft recipients (Table 17.1).

Based on these data, creative strategies have been developed to minimize xenogeneic infectious exposures via screening of source animals and exclusion of potential pathogens (or animals) during animal husbandry. Routine monitoring of xenograft recipients for infection due to both known and unknown pathogens will be complemented, if infectious syndromes emerge, by standard paradigms for

J. A. Fishman (✉)
Transplantation Infectious Disease and Compromised Host Program, Infectious Disease Division and MGH Transplant Center, Massachusetts General Hospital and Harvard Medical School, Boston, MA, USA
e-mail: Fishman.jay@mgh.harvard.edu

D. K. C. Cooper, G. Byrne (eds.), *Clinical Xenotransplantation*,
https://doi.org/10.1007/978-3-030-49127-7_17

Table 17.1 Categories of potential opportunistic infections resulting from clinical xenotransplantation[a]

Common pathogens: community or nosocomially acquired organisms causing infection (e.g., wound infection, pneumonia), specific diagnostic tests generally available, effective therapies available
Opportunistic infections of the immunocompromised host: well-characterized clinical syndromes in human allograft recipients (e.g., cytomegalovirus infection), specific diagnostic tests generally available, effective therapies available
"Traditional zoonosis": well-characterized clinical syndromes in humans (e.g., *T. gondii*), specific diagnostic tests generally available, effective therapies available
"Species specific": incapable of causing infection outside the xenograft (e.g., porcine CMV), some tests available, few standardized tests available for human use
"Potential pathogens": organisms of broad "host range" which *may* spread beyond the xenograft (e.g., adenovirus), few specific tests available, some effective therapies available
"Unknown" pathogens: unknown clinical and microbiological characteristics in vivo in humans (e.g., porcine endogenous retrovirus, PERV), some tests available, some therapies available
New virulence characteristics within the host (e.g., xenotropic viruses)
Not known to be present or pathogenic (e.g., protozoa or retroviruses)
Viral recombinants

[a]Assays must be validated for use in swine and in human samples

management of immunocompromised hosts with infection. As for allotransplantation, there will always remain some irreducible infectious risk associated with transplantation of viable xenograft tissues in graft recipients with organ dysfunction and other medical problems and who are undergoing complex surgical procedures and intensive immunosuppressive therapy.

The risk of infection in transplantation is determined by the semiquantitative relationship between two factors, the "epidemiologic exposures" and a conceptual measure of an individual's susceptibility to infection termed the "net state of immunosuppression." The net state of immunosuppression is largely a function of the intensity of immunosuppressive therapy but also includes metabolic derangements, infection with immunomodulating herpes and other viruses, and technical complications (e.g., devitalized tissues, undrained fluid collections). A decreased risk of infection (e.g., due to donor screening or recipient prophylaxis) increases the tolerability of immunosuppression. The risk factors for allotransplantation have been reviewed elsewhere [1, 2]. The unique features of xenotransplantation result from the microbiology of the nonhuman organ donor and the possibility that greater-than-usual immunosuppressive therapy may be required to prevent graft rejection [5–8].

Xenosis: Which Pathogens?

The term "xenosis" (also "direct zoonosis" or "xenozoonosis") reflects the unique epidemiology of infection due to organisms from a nonhuman source species transmitted with xenogeneic grafts [6, 8–10]. It must be emphasized that *any* organism can potentially cause infection in immunocompromised hosts, so discussion must focus on what are considered the likely pathogens based on experience with allotransplantation (Table 17.1). The microbiological behavior of animal-derived pathogens in the immunosuppressed human host cannot be predicted, and the clinical manifestations of infection are altered by immunosuppression. Various factors may increase the risk of infection in xenotransplantation:

(i) Potential pathogens may be of microbial species previously unappreciated (porcine endogenous retroviruses, PERV) or unexamined (e.g., polyomaviruses) in the source species [11–15].
(ii) Novel clinical syndromes may result from infection with animal-derived pathogens.
(iii) Clinical laboratory assays for organisms from nonhuman species may not be available for use in donor screening or in clinical diagnosis.
(iv) Donor-derived organisms may be nonpathogenic in the native species but cause disease in the new host ("xenotropic organisms") or acquire new characteristics (genetic recombination or mutation) [16–21]. Virulence may increase with passage in a new host (evolutionary adaptation) while diminishing over time in the native host.
(v) As in allotransplantation, incompatibility of transplantation antigens (i.e., MHC antigens) between species may reduce the efficacy of the host's immune response to infection within the xenograft.

As for allotransplantation, keys to the management of infection derived from swine include:

(i) Identification of "likely" pathogens based on experience with related organisms in allotransplant recipients (Table 17.1). In the absence of clinical trials, such predictions are merely educated guesses.
(ii) *Development of sensitive and specific microbiological assays for use in breeding, donor and organ screening, and diagnosis.* Ideally, this would include serological tests and/or measures of T-lymphocyte immunity (e.g., pathogen-specific interferon-gamma release assays) to identify prior exposures and latent infections. In addition, culture systems, microscopic analyses (for parasites), and quantitative molecular assays for use in clinical diagnosis are needed. These must be validated for use with samples from swine and from human xenograft recipients as assays may perform differently in human and porcine sera. Thus far, serological testing for most animal-derived organisms in humans is generally unavailable or unreliable. Serological tests may also be falsely negative in the immunocompromised host. Such assays are available in small numbers of commercial or veterinary programs.
(iii) Identification of therapies appropriate for each pathogen.

With these tools, an "exclusion list" of organisms thought to pose an unacceptable risk to xenograft recipients has been developed as a basis for testing in breeding colonies ("Designated Pathogen-Free Colonies" (Table 17.2) [9, 13, 22–24]). While

Table 17.2 Exclusion list: porcine organisms to consider

Viruses	
Porcine endogenous retrovirus (PERV) A, B, C, AC	Porcine lymphotropic herpesvirus (PLHV)
Porcine adenovirus	Porcine teschovirus
Encephalomyocarditis virus	Rabies virus
Hepatitis E virus	Swine influenza virus
Porcine cytomegalovirus	West Nile virus
Porcine hemagglutinating encephalomyelitis	SARS-Cov-1 and 2
Bacteria	
Mycobacteria spp.	*Shigella*
Pathogenic *E. coli*	*Yersinia*
Campylobacter	*Leptospira* spp.
Salmonella (*choleraesuis*, *typhimurium*)	*Listeria* spp.
Parasites	
Toxoplasma gondii	*Echinococcus* spp.
Cryptosporidium parvum	*Trichinella spiralis*
Strongyloides	*Microsporidium*
Trypanosoma species	

Adapted from [8]

barrier facilities to prevent infection of breeding colonies are essential, it seems likely that it does not matter how such exclusion is achieved if the designated organisms are demonstrably absent from the transplanted organ. Such lists must be dynamic – subject to revision based on experimental and clinical experience and the availability of new therapies.

Source Animal Selection and Exclusion of Likely Pathogens

The need for herd isolation and continuous surveillance of source animals requires meticulous breeding records (including details of nuclear transfer and animal movements) and archiving of specimens (cells and sera) for subsequent use in epidemiological investigations. Microbiological assessments in breeding colonies will be needed for sentinel animals and from the specific animals selected for organ procurement. Swine for xenotransplantation may be bred in "biosecure facilities" to prevent introduction of pig or human pathogens and isolated from other animals, including rodents, insects, and birds, often with care providers gowned and gloved.

We have developed two lists of organisms for consideration in breeding for xenotransplantation. Pig health is assured by standard veterinary practice including routine vaccinations with microbially restricted and mammalian protein-free diets, filtered water, and special housing and avoidance of unnecessary antibiotics (Table 17.3). With the availability of genetic modification of swine (e.g., CRISPR-Cas9) targeting graft rejection, metabolic incompatibilities, or to eliminate endogenous retroviruses, transgenic methods with nuclear transfer are performed in sterile environments, with subsequent embryo transfer to surrogate gilts [25–31].

Table 17.3 Exclusion list: organisms important to swine health status

Viruses
Parvovirus
Porcine circovirus
Porcine delta coronavirus
Porcine diarrhea virus
Porcine reproductive and respiratory virus
Porcine respiratory coronavirus
Porcine sapelovirus 1
Pseudorabies or Aujeszky's disease
Transmissible gastroenteritis virus
Bacteria
Brucella suis
Leptospira spp.
Mycoplasma hyopneumoniae
Salmonella spp.
Parasites
Ascaris suum
Cryptosporidia
Strongyloides ransomi

Adapted from [8]

Based on experience with infections in immunosuppressed human allotransplant recipients and with pig-to-primate xenotransplantation, a second "Designated Pathogen-Free Exclusion List" was developed (Table 17.2). Thus far, infections due to pig-derived pathogens have not been identified in immunosuppressed humans, except for hepatitis E virus (HEV). Regulatory guidance documents exist for clinical trials [10, 32–35]. In practice, these documents require source animal screening to assure animal health and the absence, to the degree possible, of possible pig-derived human pathogens.

Safety in Clinical Trials of Xenotransplantation

Routine Monitoring for Xenogeneic Infection

In immunosuppressed organ recipients, the risks for infection and malignancy are lifelong. Standard pretransplant screening in advance of immunosuppressive therapy is required (Table 17.4). While most donor-derived infections are identified early in the posttransplant course, some infections occur later, often due to immune perturbation by intercurrent viral infection (e.g., cytomegalovirus) or augmented immunosuppression for graft rejection [3, 4, 36–38]. Proof that the source of such infections is swine-derived vs. environmental may be impossible. Based on the technologies applied (e.g., molecular testing, next-generation sequencing), routine samples from recipients might be tested to assure the absence of potential pathogens (e.g., porcine endogenous retrovirus, porcine cytomegalovirus) (Tables 17.5 and 17.6). Similar screens might be applied at times of symptomatic infection or of

Table 17.4 Pretransplant microbiological screening of human xenograft recipients[a]

Name	Testing method(s)
Human immunodeficiency virus, type 1 (HIV-1)	ELISA
Human immunodeficiency virus, type 2 (HIV-2)	ELISA
Hepatitis B virus	Serology
Hepatitis C virus	Serology
Treponema pallidum	Serology
Human cytomegalovirus (CMV)	Serology
Human herpes simplex virus	Serology
Human varicella zoster virus	Serology
Toxoplasma gondii	Serology
Mycobacterium tuberculosis	ELISA (T-spot)

[a]Vaccine status up to date for hepatitis B; hepatitis A; influenza virus; Pneumovax/PCV13; tetanus (Tdap); MMR (measles, mumps, and rubella); varicella zoster virus; if required: meningococcal (including type B), *H. influenzae*; human papillomavirus

Table 17.5 Deployment of microbiological assays in xenotransplantation

Assay type	Screening source animals	Xenograft recipient monitoring	Xenograft recipients – symptomatic infection or increased risk[a]	Healthy contacts of recipient
Cultures (active infection)	X		X	
Serology (past exposures)	X	X	+/–	X
Molecular assay or antigen detection (active infection)		X	X	+/–
Next-generation sequencing (active infection)		X	X	

[a]Increased risk may be associated with treatment of graft rejection or intercurrent viral infection

Table 17.6 Recipient testing (post-xenotransplantation routine)

Virus name – noncommercial testing	Testing method
Porcine endogenous retrovirus (PERV) A, B, C, AC	Qualitative and quantitative (QNAT) nucleic acid testing (NAT); antibody-based tests (serology, ELISA, Western blot)
Porcine lymphotropic herpesvirus type 2 (PLHV-2)	QNAT
Porcine cytomegalovirus (PCMV)	NAT; antibody-based tests
Human cytomegalovirus (HCMV) – per protocol	QNAT
Human Epstein-Barr virus (EBV) – per protocol	QNAT

Adapted from [8]

increased risk (e.g., following treatment of graft rejection). In addition to recipient samples, social or sexual contacts of recipients and source-animal handlers may be considered for inclusion in any monitoring scheme. For this reason, sera and cells from these groups must be archived for future studies.

Recipients of xenografts should have blood samples (sera and cells) obtained and stored at regular intervals. A possible scheme might include serum and leukocyte samples (Table 17.6):

(i) Pretransplant
(ii) Weekly for 1 month postoperatively
(iii) Monthly for 6 months postoperatively
(iv) Quarterly for the first year
(v) Annually for 5 years thereafter

Following periods of fever or of clinical infection (see below), monitoring would be increased to weekly for 1–2 months and then revert to the previous level of surveillance. Samples could be stored on relatives, intimate contacts, and animal handlers every 6 months, with more frequent monitoring (monthly) if the animals or recipients developed signs of infection or were determined to be infected with a xenograft-derived pathogen.

Samples will be used for (i) archiving for future epidemiologic studies (in appropriate storage media for RNA, DNA, cell, and antibody preservation); (ii) NAT testing for PERV (A, B, C, AC), PLHV, and PCMV (if present in donor) and for common human viruses; (iii) cocultivation of peripheral blood leukocytes with permissive human and porcine cell lines for viral detection (including PERV); and (iv) evaluation for any fevers or infectious syndrome per institutional protocols (Tables 17.5 and 17.6).

With periods of fever or of clinical infection, monitoring could increase (e.g., to weekly for 1–2 months and then revert to the previous level of surveillance) depending on the diagnosis obtained. Samples should be stored from social contacts and animal handlers (e.g., every 6 months), with more frequent monitoring (monthly) if the animals or recipients develop signs of infection. Both serologic and molecular assays must be validated for human blood samples.

Management of Xenograft Recipients with Signs of Infection

Organ transplant recipients frequently manifest signs of infection in the form of fever (often without clear source); unexplained leukocytosis; graft dysfunction; respiratory, gastrointestinal, or urinary tract symptoms; sepsis; or abnormal metabolic testing (e.g., hepatitis). Graft rejection and malignancy may present similarly. Most often, these signs and symptoms reflect community-acquired infections or reactivation of latent infections. The risk of xenograft-derived infection requires approaches like those of allograft recipients:

(i) Full microbiological evaluation prior to the initiation of antimicrobial therapy (blood and urine cultures, sputum cultures)
(ii) Radiologic studies and invasive diagnostic testing (needle or surgical biopsies) as appropriate
(iii) Early empiric antimicrobial therapy directed at the most likely pathogens
(iv) Hospital admission with isolation and infectious precautions until further data become available
(v) Universal precautions for all blood samples
(vi) Special testing based on data from the source animals with consideration of both pig-specific pathogen testing and nondirected sequencing of serum samples

The First Recipients

Ideally, initial xenograft recipients would undergo transplantation in the absence of immunosuppression. These might be recipients of porcine skin grafts used for transient wound or burn coverage until sloughed. Such recipients could be assessed for xenogeneic infection locally (at the site of application) and systemically. Significant advances in preclinical studies have demonstrated good xenograft survival using clinically acceptable approaches to immunosuppression. Subsequent recipients requiring immunosuppressive therapy should be free of known infections and not be colonized with antimicrobial-resistant organisms. Infectious risk to the xenograft recipient might be increased by preexisting immunodeficiency states in candidates for xenotransplantation and may mitigate against using xenografts in prior allograft recipients or with underlying immunodeficiencies. Protocols for graft tolerance induction (e.g., stem cell plus organ grafts from the same donor) may avoid the intensive immunosuppression required to maintain graft function in primates but assume systemic spread of pig cells in the recipient with the associated risks of infection and graft-vs.-host disease.

In xenotransplantation, *in the absence of human studies, the absolute risk for infections remains unknown*. Approaches to production and modification of source animals and surveillance in recipients will require adjustment as clinical data emerge. New microbiological assays will be required to screen swine for potential human pathogens and for the diagnosis of pig-specific pathogens in humans. The application of next-generation sequencing technologies to xenograft recipient samples may provide valuable data and another layer of clinical safety. Infections occurring in the xenograft recipient will require early diagnosis and therapy. However, it is unlikely that such opportunistic infections will pose a significant risk to immunologically normal individuals. The recognition that novel organisms may infect xenograft recipients should generate improvements in technologies for the screening of source animals and surveillance of recipients.

Acknowledgment Studies were performed with the generous support of NIH-NIAID 5P01AI45897.

References

1. Fishman JA. Infection in solid-organ transplant recipients. N Engl J Med. 2007;357:2601–14.
2. Fishman JA. Infection in organ transplantation. Am J Transplant. 2017;17:856–79.
3. Fishman JA, Greenwald MA, Grossi PA. Transmission of infection with human allografts: essential considerations in donor screening. Clin Infect Dis. 2012;55:720–7.
4. Fishman JA, Grossi PA. Donor-derived infection–the challenge for transplant safety. Nat Rev Nephrol. 2014;10:663–72.
5. Fishman JA, Patience C. Xenotransplantation: infectious risk revisited. Am J Transplant. 2004;4:1383–90.
6. Fishman JA. Infection in xenotransplantation. BMJ. 2000;321:717–8.
7. Fishman JA, Scobie L, Takeuchi Y. Xenotransplantation-associated infectious risk: a WHO consultation. Xenotransplantation. 2012;19:72–81.
8. Fishman JA. Infectious disease risks in xenotransplantation. Am J Transplant. 2018;18:1857–64.
9. Fishman JA. Infection in xenotransplantation. J Thorac Cardiovasc Surg. 2001;16:363–73.
10. WHO. Second WHO global consultation on regulatory requirements for xenotransplantation clinical trials. 2001.
11. Peretti A, FitzGerald PC, Bliskovsky V, Buck CB, Pastrana DV. Hamburger polyomaviruses. J Gen Virol. 2015;96:833–9. https://www.who.int/transplantation/xeno/report2nd_global_consultation_xtx.pdf?ua=1.
12. Akiyoshi DE, Denaro M, Zhu H, et al. Identification of a full-length cDNA for an endogenous retrovirus of miniature swine. J Virol. 1998;72:4503–7.
13. Fishman JA, Patience C. Xenotransplantation: infectious risk revisited. Am J Transplant. 2004;4:1383–90.
14. Hause BM, Smith C, Bishop B, Stewart C, Simonson R. Complete genome sequence of a porcine polyomavirus from nasal swabs of pigs with respiratory disease. Genome Announc. 2018;6:e00344-18.
15. Denner J. The porcine virome and xenotransplantation. J Virol. 2017;14:171.
16. Coffin JM, Stoye JP, Frankel WN. Genetics of endogenous murine leukemia viruses. Ann N Y Acad Sci. 1989;567:39–49.
17. Frankel WN, Stoye JP, Taylor BA, Coffin JM. Genetic analysis of endogenous xenotropic murine leukemia viruses: association with two common mouse mutations and the viral restriction locus Fv-1. J Virol. 1989;63:1763–74.
18. Rando RF, Srinivasan A, Feingold J, Gonczol E, Plotkin S. Characterization of multiple molecular interactions between human cytomegalovirus (HCMV) and human immunodeficiency virus type 1 (HIV-1). Virology. 1990;176:87–97.
19. Isfort R, Jones D, Kost R, Witter R, Kung HJ. Retrovirus insertion into herpesvirus in vitro and in vivo. Proc Natl Acad Sci U S A. 1992;89:991–5.
20. Javier RT, Sedarati F, Stevens JG. Two avirulent herpes simplex viruses generate lethal recombinants in vivo. Science. 1986;234:746–8.
21. Katz RA, Skalka AM. Generation of diversity in retroviruses. Annu Rev Genet. 1990;24:409–45.
22. Fishman JA. Xenosis and xenotransplantation: addressing the infectious risks posed by an emerging technology. Kidney Int Supple. 1997;58:S41–5.
23. Fishman JA. Infection and xenotransplantation. Developing strategies to minimize risk. Ann N Y Acad Sci. 1998;862:52–66.
24. Fishman JA. The risk of infection in xenotransplantation. Introduction. Ann N Y Acad Sci. 1998;862:45–51.
25. Klymiuk N, Aigner B, Brem G, Wolf E. Genetic modification of pigs as organ donors for xenotransplantation. Mol Reprod Dev. 2010;77:209–21.
26. Hryhorowicz M, Zeyland J, Słomski R, Lipiński D. Genetically modified pigs as organ donors for xenotransplantation. Mol Biotechnol. 2017;59:435–44.
27. Yang L, Guell M, Niu D, et al. Genome-wide inactivation of porcine endogenous retroviruses (PERVs). Science. 2015;350:1101–4.

28. Sachs DH, Galli C. Genetic manipulation in pigs. Curr Opin Organ Transplant. 2009;14:148–53.
29. Phelps CJ, Koike C, Vaught TD, et al. Production of alpha 1,3-galactosyltransferase-deficient pigs. Science. 2003;299:411–4.
30. Niu D, Wei HJ, Lin L, et al. Inactivation of porcine endogenous retrovirus in pigs using CRISPR-Cas9. Science. 2017;357:1303–7.
31. Schaefer KA, Wu WH, Colgan DF, Tsang SH, Bassuk AG, Mahajan VB. Unexpected mutations after CRISPR-Cas9 editing in vivo. Nat Methods. 2017;14:547–8.
32. Hering BJ, Cooper DK, Cozzi E, et al. The International Xenotransplantation Association consensus statement on conditions for undertaking clinical trials of porcine islet products in type 1 diabetes – executive summary. Xenotransplantation. 2009;16:196–202.
33. Hering BJ, Cozzi E, Spizzo T, et al. First update of the International Xenotransplantation Association consensus statement on conditions for undertaking clinical trials of porcine islet products in type 1 diabetes--executive summary. Xenotransplantation. 2016;23:3–13.
34. Food and Drug Administration C, U.S. DHHS. Source animal, product, preclinical, and clinical issues concerning the use of xenotransplantation products in humans. *Guidance for Industry.* 2016.
35. PHS. PHS guideline on infectious disease issues in xenotransplantation. Rockville: Food and Drug Administration, Center for Biologics Evaluation and Research; 2001. http://www.fda.gov/BiologicsBloodVaccines/GuidanceComplianceRegulatoryInformation/Guidances/default.htm.
36. Fishman JA. New technologies for infectious screening of organ donors. Transplant Proc. 2011;43:2443–5.
37. Grossi PA, Fishman JA. Donor-derived infections in solid organ transplant recipients. Am J Transplant. 2009;9(Suppl 4):S19–26.
38. Michaels MG, McMichael JP, Brasky K, et al. Screening donors for xenotransplantation. The potential for xenozoonoses. Transplantation. 1994;57:1462–5.

Histocompatibility Testing for Xenotransplantation

18

Joseph M. Ladowski and Gregory R. Martens

Abbreviations

AMR	Antibody-mediated rejection
Gal	Galactose-α1,3-galactose
GTKO	1,3-Galactosyltransferase gene-knockout
HLA	Human leukocyte antigens
Neu5Gc	N-glycolylneuraminic acid
RBCs	Red blood cells
SLA	Swine leukocyte antigens
β4GalNT2	β1,4 N-acetylgalactosaminyltransferase

Introduction

Antibody-mediated rejection occurs when cells or tissues are transplanted between disparate members of the same species or between members of different species. This immunological insult limits the ability of the donor organ or cells to effectively replace the function of a failing organ. In allotransplantation, e.g., human-to-human transplantation (e.g., blood transfusion and organ allotransplantation), matching the donor tissue to the recipient tissue, with no preformed antibodies in the recipient directed to the donor tissues, avoids rapid antibody-mediated rejection (AMR) of transplanted tissues. Failing to avoid the effect of pre-existing donor-specific antibodies (DSA) often results in acute rejection by activating the complement cascade, resulting in destruction of the organ in minutes to hours.

J. M. Ladowski (✉) · G. R. Martens
Xenotransplantation Program, Department of Surgery, University of Alabama at Birmingham (UAB), Birmingham, AL, USA
e-mail: jladowski@uabmc.edu

D. K. C. Cooper, G. Byrne (eds.), *Clinical Xenotransplantation*,
https://doi.org/10.1007/978-3-030-49127-7_18

Once recognized as a problem in allotransplantation, a number of assays were developed to identify acceptable donor-recipient pairings based on the absence of DSA. These assays mix donor cells with recipient antibodies to probe for complement-fixing IgG and IgM binding to, and killing, the cells through a complement-mediated pathway. These technologies were key in defining the ABO blood group as carbohydrate target antigens in blood and in identifying the human leukocyte antigens (HLA) as major protein antigens in the setting of solid-organ allotransplantation. More recent technological advances have enabled the development of assays that utilize panels of recombinant HLA proteins, with increased sensitivity to detect specific anti-HLA antibodies in each recipient.

The initial problems encountered in the early years of allotransplantation are present in xenotransplantation. The transfer of cells from donor to recipient across the species barrier may result in the rapid AMR of the transplanted tissues. As the pig is the major candidate organ-source animal being pursued in the xenotransplant field, pig antigenicity will be the focus of the following discussion in the context of evaluating the success of genetic engineering in lowering humoral reactivity and in the ability to determine which individuals may be candidates for xenotransplantation as a therapy.

The First Xenoantigens: Carbohydrates

Echoing the early allotransplantation experiments, the initial trials of pig xenotransplantation into nonhuman primates were limited by hyperacute AMR [1–3]. It was eventually determined that carbohydrates containing galactose-α1,3-galactose (Gal) residues attached to glycoproteins or glycolipids via an α1,3 linkage were the first xenoantigens, to which nearly every human has preexisting antibodies [4–6]. The gene producing the galactosyltransferase responsible for producing Gal is nonfunctional in humans, and, likely through exposure to bacterial and viral intestinal flora expressing the sugar, nearly all humans develop antibodies to this carbohydrate [7]. Eliminating Gal expression by disrupting the galactosyltransferase responsible for its production showed markedly reduced antibody binding and prolonged the survival of grafts from minutes to days, but ultimately the grafts would succumb to rejection, and the search continued for additional xenoantigens [8–11].

Further work demonstrated that additional xenoantigenicity could be attributed to a sialic acid variant, present in nearly all mammals except humans [12–15]. Similar to the galactosyltransferase responsible for the Gal linkage, the gene producing the enzyme, cytidine monophosphate N-acetylneuraminic acid hydroxylase (CMAH) that adds Neu5Gc to the underlying carbohydrate structures, has been evolutionarily inactivated in humans, preventing Neu5Gc synthesis and rendering the sialic acid a foreign molecule (e.g., when expressed on pig tissues) to which most humans develop antibodies.

Though not specifically identified, a third major xenoantigen also appears to be carbohydrate-based, as expression of pig β1,4 N-acetylgalactosaminyltransferase (β4GalNT2) on human cells increases their antigenicity [16–19]. Despite humans

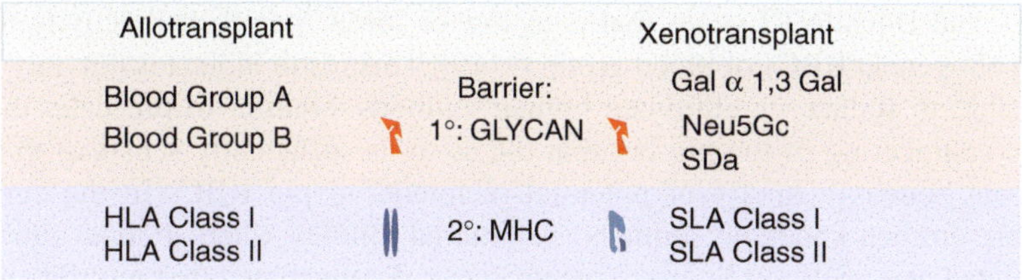

Fig. 18.1 A comparison of known allotransplant antigens and their xenoantigen counterparts

containing a functional β4GalNT2 gene, the vast majority of people appear to have antibodies to pig cells on which it is expressed. Deletion of all three genes (GT, CMAH, and β4GalNT2) eliminates expression of their corresponding glycans and has allowed the production of novel strains of pig to which approximately 30% of people lack natural antibodies [19].

Discussed at length below, the swine leukocyte antigens (SLA), homologs of the HLA that serve as key antigens in allotransplantation, are becoming increasingly appreciated as proteinac xenoantigens are recognized by potential pig xenograft recipients (Fig. 18.1).

The history of allotransplantation and recent pig-to-nonhuman primate xenotransplantation models suggest that the use of pigs with reduced humoral antigenicity, and the identification of recipients with virtually no anti-pig antibodies, will be essential to the success of xenotransplantation. Pig organs that do not express these three carbohydrate xenoantigens are likely to form the basis of pig organ grafts that will be capable of functioning long term in humans. What follows is a brief review of several assays that have been used to screen human blood for the presence of anti-pig antibodies against a variety of xenoantigens and cell types.

Human Antibodies Against Genetically Engineered Pig Red Blood Cells (RBCs)

RBCs are a convenient target with which to evaluate how genetic engineering may alter the expression of xenoantigens in a pig. Venipuncture and low-speed centrifugation allow large numbers of RBCs to be easily isolated without euthanasia of the animal. Additionally, wild-type pig RBCs express the three major known carbohydrate-based xenoantigens (Gal, Neu5Gc, and Sda), and these antigens are eliminated upon inactivation of the appropriate genes. Flow cytometry has been used to examine antibody binding to pig RBCs where, following co-incubation of pig RBCs with heat-inactivated human sera, fluorescent anti-human immunoglobulin secondary reagents enable bound antibodies to be detected [20, 21]. This assay has been used to demonstrate that gene-editing does reduce human IgM and IgG binding to pig cells.

These flow cytometric analyses have shown that approximately 30% of people exhibit humoral reactivity toward pig RBCs isolated from pigs lacking the GGTA1,

CMAH, and B4GalNT2 genes that is as low or lower than their reactivity toward human allogeneic RBCs of blood group O [19]. This result indicates not only that it is possible to reduce or eliminate human antibody binding to pig cells through genetic engineering of the pig but that the assay is sufficiently sensitive to reveal additional xenoantigenicity of potential recipients to pig RBCs in the currently available antigen-knockout animals. Additional studies where bound antibodies were eluted and analyzed by mass spectroscopy demonstrated that multiple human immunoglobulin isotypes interacted with these cells, including IgM, IgG1, IgG2, IgG3, and IgG4 [20]. RBCs also enable simple evaluation of antibody-initiated cytotoxicity by measuring the release of hemoglobin into the medium following incubation of cells with human serum and exogenous complement.

Despite their utility, using RBCs to examine antibody reactivity with xenoantigens presents two challenges: (i) ongoing access to pigs of the desired genotypes is required, given that RBCs cannot be expanded in culture and do not tolerate frozen storage well, and (ii) being a relatively simple cell type, RBCs do not present all xenoantigens that may be expressed on other cells or tissues in the pig. For example, as discussed below, they do not express the potentially antigenic SLA molecules.

Human Antibodies Against Pig Peripheral Blood Mononuclear Cells (PBMCs)

PBMCs have long been used as target cells in crossmatching assays in allotransplantation to identify compatible donors and recipients. They have also been a key tool in the analysis of human and primate antibody reactivity with various genetically modified pigs. They suffer from many of the same benefits and drawbacks as RBCs in that they are easy to obtain from blood but require continued access to genetically modified pigs. PBMCs differ from RBCs in that they can be stored frozen for the long term and express SLA in addition to Gal, Neu5Gc, and Sda-derived carbohydrates, though the freezing and thawing process of PBMCs increases the background levels of antibody binding compared to fresh cells for, at this time, unknown reasons. PBMCs have been used repeatedly in defining the contribution of all four xenoantigens to human-anti-pig humoral reactivity, typically with a flow cytometry-based assay where fluorescent anti-immunoglobulin reagents are used to detect binding to the target cells.

Similar to RBCs, PBMC-based crossmatch experiments suggest that approximately 30% of people do not have antibodies that react with cells taken from pigs engineered to lack the three major glycan xenoantigens [19]. With careful selection of phenotyping reagents and fluorescent immunoglobulin-detecting reagents, analyzing PBMCs in flow cytometry-based experiments affords the opportunity to evaluate multiple cell populations (B cells, T cells, and monocytes) for xenoantigenicity. PBMCs are also useful as targets in complement-mediated cytotoxicity assays that can be used to probe for the presence of cytotoxic antibodies (Fig. 18.2).

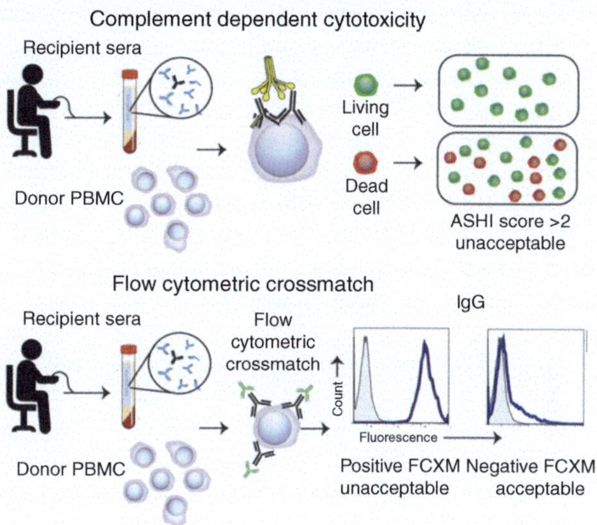

Fig. 18.2 Illustrations of the complement-dependent cytotoxicity assay and the flow cytometry crossmatch - two histocompatibility assays crucial to demonstrating the presence of a xenoantigen

Endothelial Cells as Targets in Crossmatching Assays

Endothelial cells are more challenging to isolate than either RBCs or PBMCs in that surgical access to the vasculature is required to isolate them. However, once obtained, they can be convenient cell targets because they express the major known xenoantigens and can be expanded in cell culture, frozen for future use, analyzed by flow cytometry and cytotoxicity assays, immortalized, and cloned to create homogenous cell populations. In contrast to RBCs and PBMCs, endothelial cells must be grown on solid supports, requiring assays such as ELISAs to monitor antibody binding while the cells remain attached, or they must be removed from the culture dish. SLA class II expression provides an additional caveat in that endothelial cells lose their class II expression during cell culture, though this can be restored with interferon-gamma treatment [22].

Swine Leukocyte Antigens (SLA): A Protein Xenoantigen

In humans, the highly polymorphic nature of the major histocompatibility complex (MHC), including the HLA proteins, causes people to frequently develop anti-HLA antibodies following exposure and sensitization to transplanted organs, blood product transfusions, and/or pregnancy [23]. Developing tests that evaluate the antibody repertoire of potential recipients for reactivity against HLA on potential donor organs/cells has been critical to the successful pairing of donors with recipients possessing low risks of acute AMR. As mentioned previously, SLA proteins are the homologs of HLA molecules, and consequently human and pig class I and class II molecules share approximately 75–80% amino acid sequence identity, and structural analyses of class I HLA and SLA crystal structures are very similar (SLA class II molecular structures are not currently available for comparison) [24–26].

Given their similarities, it has been hypothesized that SLA and HLA will share common epitopes that can drive cross-species reactivity between human anti-HLA antibodies and their SLA counterparts. A number of studies provided evidence in support of this concept. Oostingh and colleagues examined reactivity of pig serum IgG with PBMCs from 23 different pigs [27]. Prior to examining the interaction, the sera were incubated with wild-type pig RBCs, which do not express SLA, in an effort to deplete anti-glycan antibodies and minimize the presence of non-SLA antibodies (e.g., to the three known glycans) that could interfere with SLA antibody detection.

These experiments yielded several key observations. First, the presence of anti-HLA antibodies in a person's serum increased the frequency with which that individual demonstrated anti-pig reactivity. Second, anti-pig reactivity varied depending on which pig "donor" cells were tested, and a serum could demonstrate strong reactivity against one pig while having little to no reactivity toward a second pig having different SLA alleles. This appears analogous to the situation in allotransplantation where reactivity of a recipient's antibodies with donor cells can vary as the HLA haplotypes change.

Another study supporting the presence of SLA class I-reactive antibodies in human sera again used wild-type pig RBC absorption to minimize the assay background and again found evidence of SLA class I-specific IgG in patients who also expressed anti-HLA antibodies [28]. Serum samples were further absorbed on human platelets, which express SLA class I, but not class II, to reduce cross-reactivity with pig PBMCs, and further indicated that SLA class I was indeed the antigenic target [29]. Despite the findings of the described studies, the evidence supporting SLA as a xenoantigen was conflicting as, although some groups arrived at the same conclusions using similar assays, others demonstrated no evidence of human antibody binding to SLA proteins [30–34]. Given the diversity of pigs used among all of the studies and the polymorphic nature of SLA, it is theoretically possible that the lack of SLA reactivity in some studies arises from the varying SLA haplotypes analyzed. Given the high background of pig cells prior to the development of double- and triple-gene knockout pigs, another potential explanation includes the signal of anti-glycan antibodies dwarfing or overwhelming the signal provided by anti-SLA antibodies, giving a false-negative correlation between anti-HLA antibodies and SLA. Several of the studies that showed a positive correlation attempted to diminish carbohydrate-reactive immunoglobulins through wild-type RBC absorption or other means as a method to unmask anti-SLA antibodies.

Further studies have been performed to examine SLA class I reactivity with increasing molecular detail. The first used human monoclonal antibodies, most of which were IgM with known reactivity against specific HLA class I proteins [35]. As was seen in the studies using RBC-depleted sera, these HLA-reactive antibodies showed variable reactivity with PBMCs expressing different SLA haplotypes. Because amino acid residues which comprise the HLA epitopes targeted by these monoclonal reagents were known, SLA sequence analyses were performed to determine if they contained the same or similar epitopes.

Although many of the reactive SLA alleles seemed to share epitopes with the reactive HLA alleles, a shared epitope on SLA did not guarantee reactivity, suggesting that additional amino acids in the pig proteins also contributed to various epitopes. The reactivity of human immunoglobulin with SLA class I has also been evaluated by comparing the reactivity of serum antibodies with pig PBMC that either expressed or lacked SLA class I proteins [19, 36]. This approach revealed the presence of IgG, and occasionally IgM, in patient sera that had been sensitized to HLA. Human sera from HLA-sensitized transplant wait-list patients were incubated with pig PBMCs and washed, and bound antibodies were eluted and re-probed on beads coated with single antigen HLA class I proteins (Fig. 18.3). This study

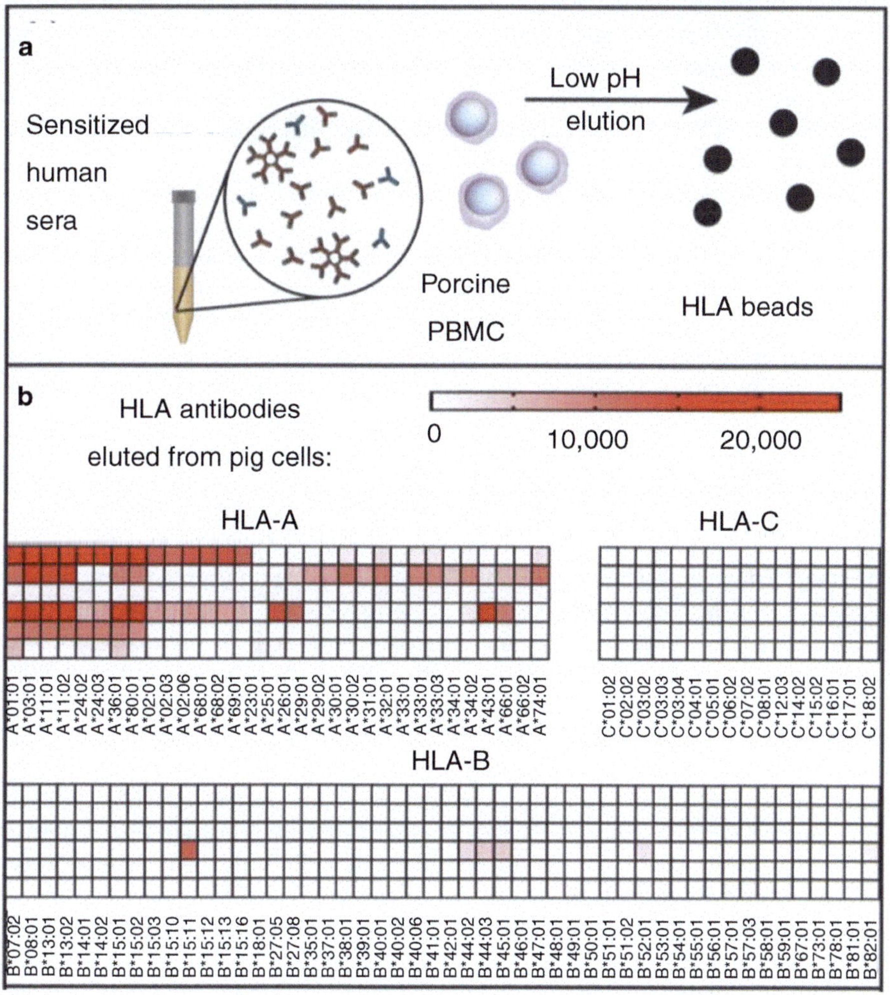

Fig. 18.3 A schematic demonstrating the mechanism behind (**a**) binding and eluting human sera to porcine PBMC and (**b**) rebinding to single antigen HLA beads to determine the possible amino acids responsible for MHC antigen cross-reactivity

demonstrated that only a subset of HLA-specific IgG in a given serum reacted with pig cells. This supports the idea that common epitopes in HLA and SLA are responsible for species cross-reactivity.

To extend these observations, cells expressing individual SLA class I alleles were mixed with sera from various HLA-sensitized patients. Amino acids which made up the epitope were predicted on the basis of HLA allele reactivity. Several sera were found with potential reactivity to the amino acid, lysine, at position 144 (144 K). Mutating this amino acid in HLA-A3 to glutamine was used to demonstrate that 144 K was responsible for some or all of the antibody reactivity with HLA-A3. Binding of this serum to SLA class I allele 1*12, which has a 144 K residue, was also performed. In two of three sera, IgG interaction with the SLA molecule was confirmed, and mutating 144 K to Q diminished this reactivity. In the third serum, despite HLA-A3 reactivity being strongly dependent on 144 K, no binding was observed against SLA-1*12 [37]. These data agree with the monoclonal antibody experiments described above – where shared epitopes appear to drive SLA class I reactivity with human antibodies. However, interspecies differences may complicate epitope identification for some patients.

SLA class II faces some unique complications compared to class I. The protein expression is more ubiquitous than HLA class II, being found on a greater variety of cells, but the level of expression is variable and diminishes with cellular culture and passage. As a result of these challenges, previous studies examining the potential cross-species reactivity of SLA class II-specific antibodies on pig PBMCs relied on indirect measurements [28, 29]. Given that SLA class II expression is more variable than class I in pigs, it is possible that the sensitivity of these indirect assays was insufficient to consistently identify reactivity on pig PBMCs.

This challenge was addressed initially by introducing the transcription regulator MHC Class II transactivator (CIITA) into fibroblasts, which normally do not express SLA class II [38]. CIITA drives transcription of class II genes and therefore made it possible to isolate SLA class II-positive fibroblasts, providing an assay to directly measure and compare human IgG binding on cells that were either uniformly class II-positive or negative. Using this assay, it was possible to detect IgG bound to SLA class II.

To extend these results and examine reactivity with single SLA class II proteins, class II-negative cells were transfected with cDNA expressing ten different SLA class II proteins [39]. Probing these ten cell lines with different human sera showed strong evidence of IgG and IgM directed to SLA proteins in human sera. In addition, similar to findings from allotransplantation and the SLA class I studies, most sera demonstrated allelic specificity even in the small number of SLA class II tested.

A putative epitope, based on the HLA class II reactivity pattern of the human serum, was determined to be an arginine at position 55 in the beta chain of class II. These sera cross-reacted with an SLA-DQ molecule containing this same amino acid, and mutations of that residue successfully reduced human antibody binding to the pig antigen.

While SLA class I and class II do appear to react with some human antibodies, the allelic specificity of the interaction suggests that it may be possible in many cases to find pig strains with SLA haplotypes that do not elicit anti-SLA humoral responses in people.

Conclusion

The field has come a great distance from the initial xenotransplant experiments that largely resulted in acute graft failure. The organ survival time from in vivo pig-to-nonhuman primate studies utilizing organs from genetically engineered pigs with the three glycan gene disruptions is encouraging, and it is generally agreed that, on this background, pig-to-human trials can proceed. Careful patient selection, utilizing precise histocompatibility tests to screen potential recipients for preformed anti-pig antibodies, provides hope that the first trial will be a success and open the door to a new age in transplantation.

References

1. Lexer G, Cooper DK, Rose AG, et al. Hyperacute rejection in a discordant (pig to baboon) cardiac xenograft model. J Heart Transplant. 1986;5:411–8.
2. Cooper DK, Human PA, Lexer G, et al. Effects of cyclosporine and antibody adsorption on pig cardiac xenograft survival in the baboon. J Heart Transplant. 1988;7:238–46.
3. Dalmasso AP, Vercellotti GM, Fischel RJ, Bolman RM, Bach FH, Platt JL. Mechanism of complement activation in the hyperacute rejection of porcine organs transplanted into primate recipients. Am J Pathol. 1992;140:1157.
4. Good AH, Cooper DKC, Malcolm AJ, et al. Identification of carbohydrate structures that bind human antiporcine antibodies: implications for discordant xenografting in man. Transplant Proc. 1992;24:559–62.
5. Oriol R, Ye Y, Koren E, Cooper DK. Carbohydrate antigens of pig tissues reacting with human natural antibodies as potential targets for hyperacute vascular rejection in pig-to-man organ xenotransplantation. Transplantation. 1993;56:1433.
6. Galili U. The α-gal epitope (Galα1-3Galβ1-4GlcNAc-R) in xenotransplantation. Biochimie. 2001;83:557.
7. Damian RT. Molecular mimicry: antigen sharing by parasite and host and its consequences. Am Nat. 1964;98:129.
8. Chen G, Qian H, Starzl T, et al. Acute rejection is associated with antibodies to nonGal antigens in baboons using Gal-knockout pig kidneys. Nat Med. 2005;11:1295.
9. Yamada K, Yazawa K, Shimizu A, et al. Marked prolongation of porcine renal xenograft survival in baboons through the use of alpha1,3-galactosyltransferase gene-knockout donors and the cotransplantation of vascularized thymic tissue. Nat Med. 2005;11:32.
10. Kuwaki K, Tseng Y-L, Dor FJMF, et al. Heart transplantation in baboons using alpha1,3-galactosyltransferase gene-knockout pigs as donors: initial experience. Nat Med. 2005;11:29.
11. Tseng Y-L, Kuwaki K, Dor FJMF, et al. Alpha1,3-Galactosyltransferase gene-knockout pig heart transplantation in baboons with survival approaching 6 months. Transplantation. 2005;80:1493.
12. Bouhours D, Pourcel C, Bouhours JE. Simultaneous expression by porcine aorta endothelial cells of glycosphingolipids bearing the major epitope for human xenoreactive antibodies (gal alpha 1-3gal), blood group h determinant and n-glycolylneuraminic acid. Glycoconj J. 1996;13:947–53.
13. Zhu A, Hurst R. Anti-N-glycolylneuraminic acid antibodies identified in healthy human serum. Xenotransplantation. 2002;9:376–81.
14. Burlak C, Wang ZY, Chihara RK, et al. Identification of human preformed antibody targets in GTKO pigs. Xenotransplantation. 2012;19:92.
15. Burlak C, Paris LL, Lutz AJ, et al. Reduced binding of human antibodies to cells from GGTA1/CMAH KO pigs. Am J Transplant. 2014;14:1895.

16. Byrne GW, Stalboerger PG, Du Z, Davis TR, McGregor CGA. Identification of new carbohydrate and membrane protein antigens in cardiac xenotransplantation. Transplantation. 2011;91:287.
17. Byrne GW, Du Z, Stalboerger P, Kogelberg H, McGregor CGA. Cloning and expression of porcine β1,4 N-acetylgalactosaminyl transferase encoding a new xenoreactive antigen. Xenotransplantation. 2014;21:543.
18. Estrada JL, Martens G, Li P, et al. Evaluation of human and non-human primate antibody binding to pig cells lacking GGTA1/CMAH/β4GalNT2 genes. Xenotransplantation. 2015;22:194.
19. Martens GR, Reyes LM, Butler JR, et al. Humoral reactivity of renal transplant-waitlisted patients to cells from GGTA1/CMAH/B4GalNT2, and SLA class I knockout pigs. Transplantation. 2017;101:e86.
20. Wang ZY, Martens GR, Blankenship RL, et al. Eliminating xenoantigen expression on swine RBC. Transplantation. 2017;101:517.
21. Wang ZY, Burlak C, Estrada JL, Li P, Tector MF, Tector AJ. Erythrocytes from GGTA1/CMAH knockout pigs: implications for xenotransfusion and testing in non-human primates. Xenotransplantation. 2014;21:376.
22. Steimle V, Siegrist C, Mottet A, Lisowska-Grospierre B, Mach B. Regulation of MHC class II expression by interferon-gamma mediated by the transactivator gene CIITA. Science. 1994;265:106.
23. Tinckam KJ, Chandraker A. Mechanisms and role of HLA and non-HLA alloantibodies. Clin J Am Soc Nephro. 2006;1:404.
24. Pratt K, Sachs DH, Germana S, et al. Class II genes of miniature swine. II. Molecular identification and characterization of B (beta) genes from the SLAc haplotype. Immunogenetics. 1990;31:1.
25. Hirsch F, Germana S, Gustafsson K, Pratt K, Sachs DH, LeGuern C. Structure and expression of class II alpha genes in miniature swine. J Immunol. 1992;149:841.
26. Lunney JK, Ho C-S, Wysocki M, Smith DM. Molecular genetics of the swine major histocompatibility complex, the SLA complex. Dev Comp Immunol. 2009;33:362.
27. Oostingh GJ, Davies HFS, Tang KCG, Bradley JA, Taylor CJ. Sensitisation to swine leukocyte antigens in patients with broadly reactive HLA specific antibodies. Am J Transplant. 2002;2:267.
28. Varela ID, Mozo PS, Cortes AC, Blanco CA, Canedo FV. Cross-reactivity between swine leukocyte antigen and human anti-HLA-specific antibodies in sensitized patients awaiting renal transplantation. J Am Soc Nephrol. 2003;14:2677.
29. Taylor CJ, Tang KGC, Smith SI, White DJG, Davies HFS. HLA-specific antibodies in highly sensitized patients can cause a positive crossmatch against pig lymphocytes. Transplantation. 1998;65:1634.
30. Bartholomew A, Latinne D, Sachs DH, et al. Utility of xenografts: lack of correlation between PRA and natural antibodies to swine. Xenotransplantation. 1997;4:34.
31. Wong BS, Yamada K, Okumi M, et al. Allosensitization does not increase the risk of xenoreactivity to α1,3-galactosyltransferase gene-knockout miniature swine in patients on transplantation waiting lists. Transplantation. 2006;82:314.
32. Hara H, Ezzelarab M, Rood PPM, et al. Allosensitized humans are at no greater risk of humoral rejection of GT-KO pig organs than other humans. Xenotransplantation. 2006;13:357.
33. Naziruddin B, Durriya S, Phelan D, et al. HLA antibodies present in the sera of sensitized patients awaiting renal transplant are also reactive to swine leukocyte antigens. Transplantation. 1998;66:1074.
34. Zhang Z, Hara H, Long C, et al. Immune responses of HLA-highly-sensitized and non-sensitized patients to genetically engineered pig cells. Transplantation. 2018;102:e195–204.
35. Mulder A, Kardol MJ, Arn JS, et al. Human monoclonal HLA antibodies reveal interspecies crossreactive swine MHC class I epitopes relevant for xenotransplantation. Mol Immunol. 2010;47:809.
36. Reyes LM, Estrada JL, Wang ZY, et al. Creating class I MHC–null pigs using guide RNA and the Cas9 endonuclease. J Immunol. 2014;193:5751.

37. Martens GR, Ladowski JM, Estrada J, et al. HLA Class I-sensitized renal transplant patients have antibody binding to SLA Class I epitopes. Transplantation. 2019;103:1620.
38. Ladowski, J.M., Reyes, L.M., Martens, G.R., et al. Swine leukocyte antigen (SLA) class II is a xenoantigen. Transplantation. 2018;102(2):249–54.
39. Ladowski JM, Martens GR, Reyes LM, et al. Examining the biosynthesis and xenoantigenicity of Class II swine leukocyte antigen proteins. J Immunol. 2018;200:2957.

Part VI

Regulatory, Economic, and Social Aspects of Clinical Trials of Xenotransplantation

Xenotransplantation: The FDA Perspective

19

Winson W. Tang and Judith Arcidiacono

Abbreviations

FDA US Food and Drug Administration

Introduction

The US Food and Drug Administration (FDA) is responsible for the regulatory oversight of a wide range of products, including cell and gene therapy products regulated by the Office of Tissues and Advanced Therapies (OTAT) in the Center for Biologics Evaluation and Research (CBER). FDA oversight of animal products including genetically altered animals resides within the Center for Veterinary Medicine (CVM). When genetically altered animals are used to produce human medical products, including xenotransplantation products, such as genetically modified or nongenetically modified xenografts, CBER and CVM collaborate on the assessment of source animals. The CVM review is focused on the safety of the regulated article in the animals, whereas the CBER focuses its review on the safety and efficacy of the xenograft.

Guidance on the regulatory approaches for xenotransplantation products regulated by CBER and genetically altered animals regulated by CVM can be found at https://www.fda.gov/downloads/BiologicsBloodVaccines/GuidanceComplianceRegulatoryInformation/Guidances/Xenotransplantation/UCM533036.pdf and https://www.fda.gov/downloads/AnimalVeterinary/GuidanceComplianceEnforcement/GuidanceforIndustry/UCM113903.pdf.

W. W. Tang · J. Arcidiacono (✉)
U.S. Food and Drug Administration, Silver Spring, MD, USA
e-mail: Judith.Arcidiacono@fda.hhs.gov

D. K. C. Cooper, G. Byrne (eds.), *Clinical Xenotransplantation*,
https://doi.org/10.1007/978-3-030-49127-7_19

The choice of pigs as a source animal for human xenotransplantation is because their organs, tissues, and cells are similar in size and function to those of humans. The meeting held at the University of Alabama at Birmingham (UAB) focused on pig-to-human kidney transplantation. To initiate a first-in-human clinical trial, an extensive clinical development program for the use of xenotransplantation products is needed to ensure patient and public safety and to ensure that the xenograft functions as expected.

The development program for xenotransplantation products must address several key issues:

(i) A consistent, reproducible, well-controlled, adequately scaled animal husbandry and manufacturing process incorporating characterization of the product's critical quality attributes
(ii) Preclinical testing incorporating proof-of-concept (POC) and toxicology studies that provide scientific rationale and support for the proposed use of the product
(iii) Clinical evidence of effectiveness and clinical evidence of an acceptable safety profile weighed against the benefit of the therapy

Consideration of each of these key areas at an early stage may improve the overall efficiency of product development.

The public health risks to xenotransplantation can be mitigated by appropriate animal husbandry that includes animals bred from closed herds of known origin, maintenance of animal health, and facility maintenance. Procedures for quarantine and transport of source animals and a program for screening for infectious disease should be in place. Production of xenografts require Current Good Manufacturing Practices that include appropriate procedures, reagents, and test methods; controls for tracking, labeling, and cross contamination; and appropriate conditions for processing, storage, and shipping. Process validation, adventitious agent testing, and product characterization are also required.

In cases where results of some testing for whole-organ xenografts may not be available prior to transplantation, some testing may be conducted prior to organ harvest. For example, identity testing and potency assay testing may be conducted on samples obtained from whole-organ biopsy or surrogate samples, such as adjacent tissue.

Prior to initiation of a clinical trial, a sponsor will need to provide adequate data generated from pharmacology and toxicology studies to establish that it is reasonably safe to conduct the proposed clinical investigation. The results of these studies provide data critical to (i) establish the scientific rationale and biologic plausibility of the proposed approach (i.e., demonstration of proof-of-concept); (ii) identify and characterize potential local and systemic toxicities, including the time frame for onset (i.e., acute vs. long term), incidence, severity, and transient or chronic nature of the findings; (iii) support subject eligibility criteria; and (iv) identify physiologic/toxic parameters to help guide appropriate clinical monitoring. Some additional early considerations for preclinical development may include selecting a biologically relevant animal model, the use of a clinically applicable immunosuppressive regimen, the use of appropriate control groups, and assessment of in vivo safety and activity. The preclinical studies should mimic the clinical situation as closely as possible to help guide the design of the clinical trials.

The primary objective of a Phase 1 first-in-human study is to assess the safety of the xenograft. However, co-primary or secondary objectives often include assessments of biologic activity to help guide the subsequent development program. The sponsor should consider the target indication, interpretability issues, and the risk of the study procedure when selecting the study population for a first-in-human study. The risk of xenotransplantation is an unacceptable risk when administered to normal healthy volunteers, and thus the sponsor should enroll a study population of subjects with life-threatening disease for which there are no other therapeutic options.

When possible, first-in-human studies should be conducted in individuals who can understand and consent to the study procedures and risks and thus should exclude children. However, there may be potential exceptions, such as infants with congenital heart diseases associated with early mortality. The number of subjects in the first study should be limited to avoid exposing many subjects to the potential risk while allowing for collection of preliminary evidence of safety. At the same time, consideration should be given to the power of the sample size to rule out adverse events that occur at a clinically meaningful rate.

A first-in-human study should also include an intersubject staggering treatment interval to avoid concurrent exposure of multiple subjects to the xenograft. The staggering interval should be of sufficient duration to allow for the monitoring of acute and subacute. The study should also prospectively define stopping rules, usually based on a number or frequency of specific adverse events, that if triggered would temporarily halt the study, pending a safety review. It should include a safety monitoring plan that can capture early, intermediate, and delayed adverse events that may be expected, based on preclinical and clinical data, as well as on theoretical concerns.

Finally, the duration of follow-up for an individual subject prior to the assessment of the primary efficacy and safety endpoints will also need to be determined.

There are additional physiologic incompatibilities that may render pig-to-human xenotransplantation impractical [1, 2]. These safety concerns are specific to the organ that will be transplanted, and the kidney will be used as the reference organ in this discussion. The kidney serves many important homeostatic functions, including the clearance of uremic toxins. For xenotransplantation to fulfill its promise, the porcine kidney must replace many if not all the functions of the normal kidney.

For example, the kidney is an important endocrine organ that secretes erythropoietin (EPO). However, porcine EPO is only 80% homologous to its nonhuman primate counterpart and does not support erythropoiesis in the nonhuman primate [1]. One potential solution is to administer recombinant human EPO. However, the presence of porcine EPO may engender an immune response that generates anti-EPO antibodies that have the potential to neutralize recombinant EPO and precipitate aplastic anemia. Therefore, the development program should include a monitoring and mitigation plan in the event of cross-reactive EPO antibody development.

The kidney is also an important organ for the acute control of calcium and phosphate balance via the parathyroid hormone (PTH)-vitamin D axis. The active form of vitamin D in humans, 1,25-dihydroxyvitamin D, depends upon 1-alpha hydroxylation within the kidney. Therefore, it would be important to ascertain the effect of porcine 1-alpha-hydroxylase enzyme on the synthesis of human vitamin D. In addition, PTH promotes tubular reabsorption of calcium while inhibiting phosphate reabsorption. However, in a transgenic pig-to-cynomolgus monkey bilateral nephrectomy model, serum calcium levels were within the normal range although hypophosphatemia was noted suggesting that the activity of a nonhuman primate PTH may not be entirely normal within a porcine xenograft. Therefore, clinical studies should include rigorous protocol(s) for evaluating calcium and phosphate balance, including the need for long-term studies on their effect on bone metabolism.

Fluid and sodium balance are maintained by an interplay between dietary intake and bodily excretion. Although there are losses through the skin and gastrointestinal tract, the kidney is the major regulator of sodium and water balance. The kidney generates over 180 L of filtrate per day, and much of the filtered water and electrolyte must be resorbed. The maintenance of homeostasis is dependent upon the interactions of three hormonal axes – aldosterone, natriuretic peptides, and vasopressin. The latter is secreted by the posterior hypothalamus in response to hyperosmolality and leads to the reabsorption of solute-free water through increased transcription of the water channel Aquaporin-2. By comparison, sodium balance depends upon the activity of the renin-angiotensin-aldosterone axis and a family of natriuretic peptides.

Simplistically, the two systems serve to counteract the effect of the other such that aldosterone increases sodium reabsorption while the natriuretic peptides promote natriuresis. The compatibilities between the porcine kidney and the human recipient are unknown, although there are reports to suggest that they may not be compatible. For example, porcine renin, which is normally produced by the juxtaglomerular cells of the glomerulus in humans, does not cleave human angiotensinogen and thus would not be expected to promote aldosterone synthesis. Therefore, the preclinical and clinical development program should design studies to assess the intricacies of fluid and electrolyte homeostasis.

A nonselective urinary loss of protein has been reported in a transgenic pig-to-cynomolgus monkey bilateral nephrectomy model [1]. Both globulin and albumin were detected in the urine by electrophoresis suggesting protein loss via the glomerulus.

Although this may represent subacute/chronic rejection, it may also reflect the appropriate generation of an immune response to the foreign constituents of the porcine kidney. Thus, the potential exists for glomerular diseases that would be considered autoimmune in a "normal" human to develop via either direct binding of host antibodies to antigens present on the porcine kidney (in situ immune complex formation) or deposition of circulating immune complexes. However, under this scenario, it would be an appropriate immune response to a foreign antigen. Therefore, care must be taken when assessing a renal biopsy, and the Banff classification may not be appropriate for interpreting xenograft rejection.

Patients with end-stage renal disease often have comorbidities that require medical therapy. Some of these include medications that may affect proteins produced by the porcine kidney, for example, renin blockers. Others may act directly on the porcine kidney to exert their effect such as inhibitors of the sodium-glucose cotransporter (SGLT2 inhibitors). Still other drugs are metabolized by the kidney, such as morphine and paracetamol. Finally, the clearance of many small molecule drugs is dependent on renal clearance. Thus, it is likely that pharmacokinetic/pharmacodynamic studies may need to be conducted to demonstrate that the porcine kidney replicates these properties of the human kidney.

Finally, the immune response to infectious agents is a carefully orchestrated combination of innate and adaptive response to the invading pathogen. However, the latter depends upon antigen presentation in the context of HLA. The porcine kidney obviously lacks HLA. Therefore, in the event of an infection localized to the porcine kidney such as pyelonephritis, it is unclear if the human host will be able to mount an adaptive immune response to the pathogen.

The advancement of clinical xenotransplantation relies heavily on an appropriate clinical development plan that includes a robust preclinical program, the development of suitable animal herds, and a clinical trial design with built-in protections for the xenograft recipient and the public. Developers of xenotransplantation products are encouraged to interact with the FDA early in the development process to ensure that regulatory requirements can be met. Developers may utilize the INTERACT Program to obtain preliminary informal advice (https://www.fda.gov/vaccines-blood-biologics/industry-biologics/interact-meetings-initial-targeted-engagement-regulatory-advice-cber-products). It is recommended that early consultations with the FDA include both CBER and CVM.

Acknowledgment The authors would like to acknowledge Dr. Allen Wensky, FDA, CBER, Office of Tissues and Advanced Therapies (OTAT), and Dr. Marc Cavaille-Coll, FDA, CDER, Office of New Drugs, for their critical review of this manuscript.

Disclosures Winson Tang coauthored this manuscript during his tenure as a Medical Officer at the FDA, Center for Biologics Evaluation and Research, OTAT. Judith Arcidiacono is an International Regulatory Expert and the lead for policy on xenotransplantation at FDA/CBER/OTAT.

References

1. Soin B, Smith KG, Zaidi A, et al. Physiological aspects of pig-to-primate renal xenotransplantation. Kidney Int. 2001;60:1592–7.
2. Iwase H, Klein E, Cooper DKC. Physiologic aspects of pig kidney transplantation in nonhuman primates. Comp Med. 2018;68:332–40.

Xenotransplantation: A Payer's Perspective

20

Jon Friedman

Introduction

There are two major cost drivers for patients with end-stage organ disease who need a lifesaving organ transplant. The first is the anticipated cost of the medical expense of maintaining the patient on the wait-list (e.g., expenses associated with hemodialysis or ventricular assist devices) and of the transplant itself. The second is the unexpected cost of treating the complications and hospitalizations associated with pretransplant care.

High Costs of Transplants and Wait-List Maintenance

There is a high cost to maintaining transplant patients on a wait-list. In a commercial population, the cost of pre-kidney transplant maintenance hemodialysis can easily approach $260,000 per year per covered life. An even more impactful example is the bridge-to-heart transplant implantation of a ventricular assist device (VAD) with an average cost of roughly $700,000 (not including an additional $30,000–580,000 per year for maintenance) [1].

In kidney transplantation, the cost of the evaluation and wait-list phases exceeds the cost of the actual transplant, especially in deceased donor transplants (driven mostly by dialysis costs, as noted above and in Fig. 20.1). Compared to the high cost of maintenance dialysis, renal transplantation is a more cost-effective treatment for end-stage renal disease.

J. Friedman (✉)
Medical Benefit Management, Optum, Eden Prairie, MN, USA
e-mail: jon.friedman@optum.com

D. K. C. Cooper, G. Byrne (eds.), *Clinical Xenotransplantation*,
https://doi.org/10.1007/978-3-030-49127-7_20

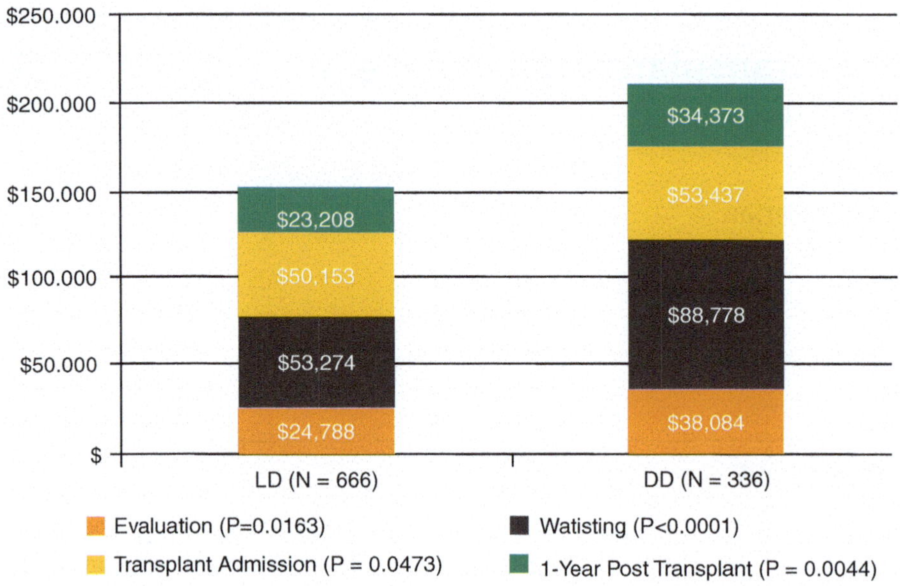

Fig. 20.1 Total medical cost per patient by phase. Living (LD) versus deceased donor (DD) transplant costs

Strategies for Reducing Transplant Costs

Any procedure, medication, or intervention that helps avoid these high maintenance costs and promotes earlier transplantation should be attractive to the payer and patient communities. The number of recipients needing a transplant far exceeds the available number of human donor organs, which drives up overall cost in this population. However, because of the increasing donor pool and higher transplantation rates, preemptive transplants are becoming a cost-reduction strategy for payers.

Another cost-reduction strategy that could help payers avoid or eliminate waitlist maintenance costs is the use of xenografts. Xenotransplantation could contribute significantly to achieving improved outcomes (quality) at a better unit cost (value).

If xenotransplantation becomes a mainstream therapy for end-stage organ failure, the xenograft could be used in a role similar to a "living donor" organ as a preemptive renal transplant in patients with no identified living human donors. Compared to the potentially long wait times for a deceased donor transplant, this strategy could avoid all or most of the costs of hemodialysis.

A preemptive renal transplant reduces overall healthcare costs in an end-stage renal disease patient by approximately $504,000, inclusive of the transplant

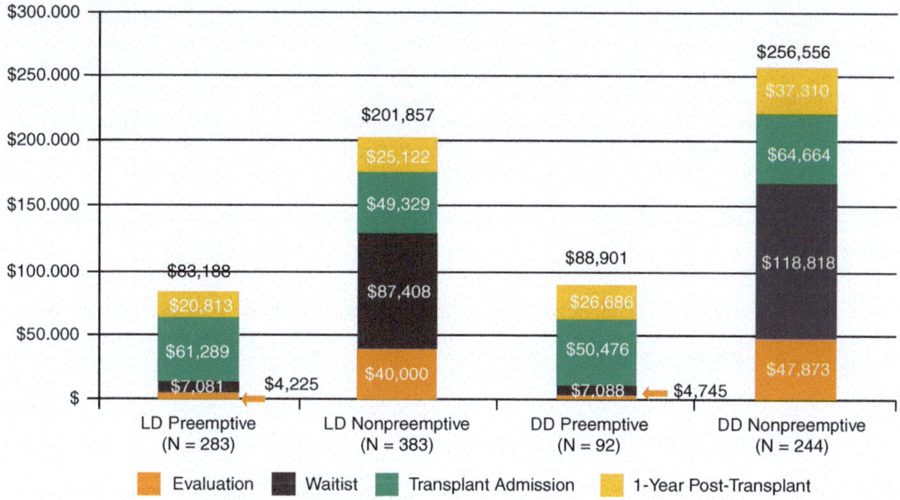

Fig. 20.2 Total medical cost per patient by phase. Preemptive transplant advantage in medical expense. (*DD* deceased donor; *LD* living donor)

(Anderson, D., Optum Insight Natural History of Disease internal data). Most of the savings are generated by eliminating costs in the evaluation and wait-list phases, plus avoidance of "crash and burn" uremic admissions (Fig. 20.2).

Potential Impact of Xenotransplantation

In summary, xenotransplantation has the potential to positively impact transplant outcomes and costs. Clinically, xenotransplantation can shorten wait times and thus ensure that the patient undergoes kidney transplantation while still in good health (except for renal failure), which can lead to a reduction in adverse posttransplant clinical outcomes and thus improve survival rates. Financially, xenotransplantation can favorably affect multiple factors involved in total cost of care.

Xenotransplantation is one of many possible solutions that could improve transplant clinical and fiscal outcomes. Other possible advances include expanding the organ supply by accepting HCV+ donors and using bariatric surgery as a bridge to transplantation in the end-stage renal disease population with obesity.

Given the sheer number of possible pig organs, xenotransplantation has the potential to make a large impact in the field of transplantation. However, there is still a need for well-designed clinical trials with conclusive findings of long-term safety and efficacy to gain general acceptance of xenotransplantation and to obtain possible coverage by payers.

Jon Friedman, MD, FAST, serves as the Chief Medical Officer of Optum's Medical Benefit Management group. Optum is a leading information and technology-enabled health services business dedicated to helping make the health system work better for everyone. With more than 135,000 people worldwide, Optum delivers intelligent, integrated solutions that help to modernize the health system and improve overall population health. Optum is part of UnitedHealth Group (NYSE:UNH). For more information, visit www.optum.com.

Reference

1. Levitsky J, Formica RN, Bloom RD, et al. The American Society of Transplantation consensus conference on the use of hepatitis C viremic donors in solid organ transplantation. Am J Transplant. 2017;17:2790–802.

Public Perceptions Toward the Clinical Trials of Organ Xenotransplantation

21

Wayne Paris, Chace Mitchell, Zoie Werkheiser, Alan Lipps, David K. C. Cooper, and Luz A. Padilla

Abbreviations

UAB	University of Alabama at Birmingham
WHO	World Health Organization
XTx	Xenotransplantation

Introduction

Given the increasing possibility of clinical trials, it is important to fully explore and understand how psychosocial concerns and theological beliefs might influence the public's acceptance of xenotransplantation (XTx) as a clinical therapy. The

W. Paris (✉)
School of Social Work, Abilene Christian University, Abilene, TX, USA

Xenotransplantation Program, Division of Transplantation, Department of Surgery, University of Alabama at Birmingham (UAB), Birmingham, AL, USA
e-mail: wdp10a@acu.edu

C. Mitchell · Z. Werkheiser
Xenotransplantation Program, Division of Transplantation, Department of Surgery, University of Alabama at Birmingham (UAB), Birmingham, AL, USA

A. Lipps
School of Social Work, Abilene Christian University, Abilene, TX, USA

D. K. C. Cooper
Xenotransplantation Program, Division of Transplantation, Department of Surgery, University of Alabama at Birmingham (UAB), Birmingham, AL, USA

L. A. Padilla
Department of Epidemiology and Surgery, Pediatric and Congenital Heart Center of Alabama, Children's Hospital of Alabama, Birmingham, AL, USA

D. K. C. Cooper, G. Byrne (eds.), *Clinical Xenotransplantation*,
https://doi.org/10.1007/978-3-030-49127-7_21

importance of psychosocial factors in allotransplantation has been well-established [1]. The theological and ethical questions that XTx raises for society and future recipients have also been partially addressed [2]. The questions yet to be answered are whether our current level of psychosocial and theological knowledge is adequate and whether public support is strong enough for XTx to be considered as a realistic clinical therapy when a transplant is needed.

The potential psychosocial problems that may be associated with XTx should be identified to the extent possible. A failure to do so now increases the chances that they cannot or will not be addressed adequately when clinical trials begin. Just as an understanding of theological perspectives is important in preparation for moving XTx from the laboratory into the clinic, so too is the broader question of whether or not the "public" is supportive of the prospect of pig-to-human organ transplantation. There is little or no point in continuing scientific advancement if the public will never accept XTx as a possible bridge to allotransplantation or as a definitive alternative.

In an attempt to address these matters, in this brief chapter, we will review the World Health Organization (WHO) guidelines for programs preparing for clinical trials and provide a brief analysis of the published psychosocial and theological literature in relation to XTx.

Public Attitudes Toward Xenotransplantation

The importance of public comment and understanding toward XTx cannot be overestimated. Tallacchini reported that, through the efforts of the Health Canada 2000 project, Canadian citizens introduced a new strategy to deal with the scientific uncertainties of XTx and its unknown risks [3]. In her view:

> ...the Canadian public consultation took a critical stance towards the so-called "knowledge deficit model" (Wynne, 1992), according to which citizens are afraid of science and technology because they do not know enough, and they tend to agree with scientists when they learn more. Instead, the Canadian consultation showed that citizens who were unaware of xenotransplantation tended to accept it, and they became more scared and reluctant the more knowledge they were acquiring about the procedures and risks.

The WHO has contributed to the conversation (about what defines "relevant voices" that need to be heard in regulating XTx) through the 2008 Changsha Communiqué [4], the Geneva Report on the Second WHO Global Consultation of Regulatory Requirements for XTx [5], and the recent 2019 Changsha Communiqué [6]. This does not mean that exploring public attitudes and beliefs is a "one-size-fits-all" legitimization process. Rather, to be fully aware of the level of the public's risk acceptance prior to initiation of clinical XTx, local "relevant voices" need to be consulted through a multilevel process that includes patients; health professionals; religious, business, and academic leaders; lay citizens; etc. This allows those potentially affected by the proposed technology to have an opportunity to present their views, question or challenge the views of others, and have their questions or challenges answered prior to initiation of any clinical trial [7].

The importance of a failure to fully engage the public has historical precedent. In 2002 and 2004, public consultations in Australia were flawed in both their design and process. By preemptively suggesting a desired outcome for XTx to be "allowed to proceed," they failed in their ability to meaningfully engage and involve the citizens [8]. This approach resulted in a complete moratorium on clinical trials of animal-to-human organ transplantation in Australia until 2009.

This does not suggest that relevant involvement of the public is not without difficulties. Clearly, the issues and questions raised by XTx require a certain level of sophistication and experience by the public at large to be fully appreciated.

What Cook suggested in 2011 [8] remains relevant today:

> While difficulties remain with public consultation and participation processes, and tense relations between the public and science continue, these should not be reasons to abandon meaningful two-way dialogue. In public consultation, we need to move away from an imposed, standardized and top-down model of privileging scientific knowledge over all other forms of knowledge and towards consultation that genuinely includes and engages the public, and values their input on an equal status with the scientific point of view. These different positions and viewpoints need to be respected.

Recent Surveys at the University of Alabama at Birmingham (UAB)

The challenges associated with serious public engagement have proven to be germane to the current experience at UAB. As part of a multilevel assessment (e.g., patient, physician, nurse, and public attitudes) consistent with WHO guidelines, it was determined that non-Caucasian UAB patients were almost six times less likely to support XTx than Caucasian patients [9]. Given that (i) most of the non-Caucasian patients in the survey were African-American and (ii) approximately two thirds of patients awaiting kidney transplantation at UAB are African-American and (iii) in the historical context of the US Public Health Service Tuskegee (Alabama) Syphilis Study [10], this finding necessitates an additional level of exploration and research prior to initiation of clinical trials.

Simply stated, this observation highlights the need for an additional level of expertise, in this case to be provided by inclusion of the National Center for Bioethics in Research and Health Care at Tuskegee University, to further explore this subject and help the medical team to better understand the African-American patients' concerns about participating in "experimental medicine." This observation regarding non-Caucasian attitudes to XTx was not anticipated but clearly reflects the importance of a program's commitment to WHO expectations in regard to inclusion of the public's opinion, which is a necessary part of planning before initiating a clinical trial.

The significance of being aware of the public's opinion can also be illustrated through a brief review of the historical context of the "psychosocial" literature regarding XTx. In 2004, Hagelin described public opinion survey results [11] as follows:

> Overall, there was no overwhelming support for xenotransplantation, but over time it seemed as if lower proportions oppose it. Proportional differences in support and opposition between geographical regions remain. Opinions to xenotransplantation depend on many socio-economic factors. The influence of gender, education, and religion on opinions about xenotransplantation were similar as to what is usually the case with other related issues, like the use of animals in biomedical research and other biotechnology/genetic engineering application.

One observation from our surveys and focus groups is that, although the public at large is supportive of XTx, patients awaiting organ transplantation and those who have a close family member awaiting transplantation, e.g., mothers of infants with complex congenital heart disease, were especially positive and pragmatic in their attitudes to XTx. If there was no realistic therapeutic alternative, a pig organ transplant would be welcomed.

Is Our Understanding of the Factors that Influence the Public Any More Definitive Today?

In an attempt to answer this question, the UAB XTx program is in the process of conducting a meta-analysis of the "psychosocial" literature relating to XTx (1985–2019). Meta-analysis is a well-known tool to insure that medical treatments are based on the best available empirical data, but it can also prove to be helpful to establish the relationship between two variables – in this case, the differences between those who are or are not in favor of XTx.

PubMed and Cumulative Index of Nursing and Allied Health databases were searched from 1985 through 2019 for studies specifically related to patient, nursing, and physician attitudes to XTx. A total of 51 studies were identified. Of these, 41 surveyed patients, 9 surveyed nurses, and 1 surveyed physicians. After excluding abstracts, articles published in foreign languages, and those that could not be located through the university online library services, a total of 19 papers were available for meta-analysis.

In a preliminary meta-analysis of the studies being reviewed, cross-sectional designs with considerable independent variable heterogeneity were employed [12–31]. The majority of studies reported that >50% of those surveyed supported XTx [13–15, 17–21, 24–28, 30, 31], with a range from a low of 37% [16] to a maximum of 83% [24].

Several variables were found to be associated with a more favorable attitude toward XTx, as measured by odds ratio (OR; the statistic that quantifies the strength of the association between two events) (Table 21.1). These included (i) a subject's personal experience with transplantation (e.g., through a family member or friend), (ii) the perceived benefit from the procedure, (iii) a partner's positive attitude toward medical treatment, (iv) the subject's engagement with, and acceptance of, biotechnology, (v) a higher level of education, (vi) a positive attitude toward deceased human organ donation, and (vii) a younger age [32].

Table 21.1 Preliminary meta-analysis of factors found to increase support for XTx

Factors	Odds ratio	Significance
Personal experience with transplantation	16.8	$p < 0.00$
Perceived benefit of the procedure	9.8	$p < 0.00$
Partners' positive attitude toward medical treatment	5.6	$p < 0.00$
Engagement with biotechnology	2.6	$p < 0.00$
Higher education level	2.4	$p < 0.00$
Positive attitude toward deceased human organ donation	2.2	$p < 0.00$
Younger age	1.2	$P < 0.02$

Factors Influencing Attitudes to Xenotransplantation

Given a lack of available statistical data for comparison by meta-analysis, the roles that theological beliefs, a knowledge of genetic engineering of pigs, and certain other factors may play in determining a willingness to consider XTx are difficult to evaluate, and the above results should be considered preliminary. In general, the literature reports a majority of those surveyed are supportive of the procedure, but there is pronounced variability in attitudes, influenced by such factors as country of origin, religious beliefs, and potential concern about the risks to the public health [30].

At best, our understanding of the potential influence of numerous factors is rudimentary in nature. For example, after soliciting the opinions of Jewish, Christian, and Muslim theologians, one recent report [33] concluded:

> The consideration of theological beliefs presents XTx programs with serious and complex views to consider. As evidenced by the existing literature, theologian opinions are not always consistent with those of potential patients…. theologians themselves do not always agree as to the viability of XTx, or as to the rationale appropriate to making the decision about the procedure. Nevertheless, the important takeaway is that a theologically informed XTx program is one that has the greatest potential to maximize the benefit to their future patients, and more likely to have broad public support.

From this report [33], however, it would seem that the Christian and Jewish religions accept the concept of XTx, though the Moslem community may have some reservations, though these are modest and would not automatically exclude XTx as a form of lifesaving therapy. The views of adherents of other religions vary considerably. To some Buddhists, the belief that animals should be protected would prevent them from availing themselves of the procedure [34]. Hinduism is a decentralized religion absent one standard set of beliefs; and research has shown that some are totally opposed to the use of animals or animal products for the treatment of human disease which would preclude XTx, while others do allow the donation and receipt of human tissue and would potentially be amenable to the procedure [35]. For non-monotheistic religions practiced in some parts of the world, e.g., Japan, allotransplantation is limited due to the lack of widespread public acceptance of organ donation after brain death. Although Japanese organ donation rates have gradually been increasing since the 1997 law (and amended in 2009), allowing for organ retrieval from brain-dead heart-beating donors was implemented [36]. Presently the majority of allotransplants performed in Japan are from living related donors. The progress of basic XTx research within the experimental laboratory has recently been recognized by increasing the level of research funding for islet XTx, although there has not been a corresponding willingness of the Japanese government to regulate or implement clinical trials [37].

The complexity of attitudes toward XTx was further addressed by Amin and colleagues [30] when they summarized their findings as follows:

> …stakeholders' attitudes to xenotransplantation as a means of treating human disease and restoring critical functions in untreatable patients is complex and thus should be viewed as a multi-dimensional process. Attitudes to xenotransplantation were determined predominantly by two direct predictors: the specific application-linked perceptions of their benefits and perceived moral concerns. Stakeholders' attitudes to xenotransplantation also involved intricate relationships between other factors, such as perceived risk, engagement, attitude to nature, and religiosity.

Unfortunately, to measure the strength of association between a specific factor and willingness to consider XTx (or not) is complicated by the innumerable ways in which the questions have been asked. For example, the role of religious beliefs was asked three different ways (i.e., religious attitudes, religious affiliation, and whether the subject's religion supported XTx). These questions result in answers that are not directly comparable. Researchers would facilitate our ability to develop a clearer profile of who does, or does not, support XTx by reporting statistics that allow computation of the results (e.g., means, standard deviations, sample sizes, correlation matrices, etc.). At best, based on the above analysis, all one can report is that there is great variability about the role that religious beliefs, attitudes to genetic engineering of pigs, and other factors play within individual patient populations.

Comment

The reality is that the views and beliefs of potential XTx recipients and families, and the staff that will be caring for them, have not been widely explored to the extent necessary for individual programs to initiate clinical trials. For potential patients, the procedure raises individual religious, cultural, or psychological issues that need to be better understood. The broader question about XTx is what it means to be "human" from the perspective of the patient, family, and community, rather than from that of the medical profession [2]. The current literature is limited in its ability to provide a clear impression of the psychosocial factors that need to be considered for an individual program to have an appropriate level of understanding.

The preponderance of the existing literature is primarily from an individual professional, cultural, geographic, or philosophical perspective and, although helpful, does not provide the breadth of understanding necessary to fully appreciate the attitudes toward XTx of patients and of the broader community. The closer we come to the clinical application of XTx, the more important it is to incorporate and fully analyze local opinions and attitudes toward the procedure, as outlined by the WHO guidelines.

Acknowledgment The authors would like to thank the numerous UAB medical personnel, patients, and the Birmingham, Alabama, residents who freely gave of their time and completed surveys or attended community focus groups.

Conflict of Interest The authors declare no conflicts of interest.

References

1. Paris W, Thompson S, Brawner N, et al. Medico-social aspects. In: Cooper DKC, Miller LW, Patterson GA, editors. The transplantation and replacement of thoracic organs. Hingham: Kluwer Academic; 1996. p. 137–40.
2. Paris W, Jang K, Colsch L, et al. Psychosocial challenges of xenotransplantation: the need for a multidisciplinary, religious, and cultural dialogue. Xenotransplantation. 2016; https://doi.org/10.1111/xen12263.
3. Tallachini M. Medical technologies and EU law: the evolution of regulatory approaches and governance. In: Cremona M, editor. New technologies and EU law. Oxford, New York: Oxford University Press; 2017. p. 9–37.
4. World Health Organization. https://www.who.int/transplantation/xeno/en/. Accessed 15 June 2019.
5. World Health Organization. https://www.who.int/transplantation/publications/en/. Accessed 15 June 2019.
6. Third WHO Global Consultation on regulatory requirements for xenotransplantation clinical trials. *IXA Update*. https://www.tts.org/ixa/news/ixa-quarterly-updates/2809-ixa-update. Accessed 19 May 2019.
7. Young IM. Inclusion and democracy. Oxford: Oxford University Press; 2000.
8. Cook PS. What constitutes adequate public consultation? Xenotransplantation proceeds in Australia. J Bioeth Inq. 2011;8:67–70.
9. Padilla L, Jang K, Cleveland D, et al. Patient attitudes towards xenotransplantation in preparation for clinical trials. Paper presented at the International Xenotransplantation Association's 2019 Congress; October 10–13, 2019; Munich, GDR.
10. Centers for Disease Control. U.S. Public Health Service syphilis study. https://www.cdc.gov/tuskegee/timeline.htm. Accessed 15 June 2019.
11. Hagelin J. Public opinion surveys about xenotransplantation. Xenotransplantation. 2004;11:551–8.
12. Coffman KL, Sher L, Hoffman A, et al. Survey results of transplant patients' attitudes on xenografting. Psychosomatics. 1998;39:379–83.
13. Schlitt J, Brunkhorst R, Haverich A, Raab R. Attitude of patients toward transplantation of xenogeneic organs. Langenbeck's Arch Surg. 1999;384:384–91.
14. Deschamps JY, Roux FA, Gouin E, Sai P. Reluctance of French patients with type 1 diabetes to undergo pig pancreatic islet xenotransplantation. Xenotransplantation. 2005;12:175–80.
15. Long T, Wray J, Myers L, Banner N. Attitudes of potential candidates for heart and heart-lung transplantation to xenotransplantation. Prog Transplant. 2002;12:280–28.
16. Kranenburg LW, Kerssens C, Ijzermans JN, Zuidema W, Weimar W, Busschbach JJ. Reluctant acceptance of xenotransplantation in kidney patients on the waiting list for transplantation. Soc Sci Med. 2005;61:1828–34.
17. Rios A, Conesa C, Ramirez P, et al. Attitude toward xenotransplantation among residents. Transplant Proc. 2005;37:4111–6.

18. Martinez-Alarcon L, Rios A, Conesa C, et al. Attitude toward xenotransplantation in kidney and liver patients on the transplant waiting list. Transplant Proc. 2005;37:4107–10.
19. Idvall M. The xenotransplantation narratives of nine type 1 diabetic patients with renal failure. Xenotransplantation. 2006;13:509–11.
20. Rubaltelli E, Burra P, Sartorato V, et al. Strengthening acceptance for xenotransplantation: the case of attraction effect. Xenotransplantation. 2008;15:159–63. https://doi.org/10.1111/j.1399-3089.2008.00474.x.
21. Rios A, Martinez-Alcaron L, Sanchez J, et al. Population study about the attitude toward organ xenotransplantation in the population from Eastern Europe resident in Southeast Spain. Int J Artif Organs. 2010;33:30–9.
22. Rios A, Martinez-Alcaron L, Ayala-Garcia MA, et al. Level of acceptance of a clinical solid organ xenotransplantation program among personnel in organ transplant-related services in Spanish, Mexican, and Cuban hospital centers. Transplant Proc. 2010;42:222–7. https://doi.org/10.1016/j.transproceed.2009.11.007.
23. Abalovich A, Wechsler C, Lara S, Bervottini M. Pig islet xenotransplantation acceptance in a Latin-American diabetic population. Xenotransplantation. 2010;17:263–6. https://doi.org/10.1111/j.1399-3089.2010.00598.x.
24. Martinez-Alarcon L, Rios A, Pons JA, et al. Attitudes toward islet cell and tissue xenotransplantation among kidney and liver patients on the transplant waiting list. Transplant Proc. 2010;42:3098–101. https://doi.org/10.1016/j.transproceed.2010.05.071.
25. Rios A, Febrero B, Lopez-Navas A, et al. Xenotransplantation of cells for diabetes: is there a good acceptance rate among Spanish adolescents? Transplant Proc. 2011;43:55–7. https://doi.org/10.1016/j.transproceed.2010.11.007.
26. Martinez-Alarcon L, Rios A, Pons JA, et al. Attitudinal study of organ xenotransplantation in patients on the kidney and liver transplant waiting list in a country with a high rate of deceased donation. Xenotransplantation. 2011;18:168–75. https://doi.org/10.1111/j.1399-3089.2011.00673.x.
27. Fernandez RF, Bucchi C, Navarro P, Beltran V, Borie E. Bone grafts utilized in dentistry: an analysis of patients' preferences. BMC Med Ethics. 2015;16:71. https://doi.org/10.1186/s12910-015-0044-6.
28. Valente P, Gomes C, Tomada N. Small intestinal submucosa grafting for Peyronie disease: outcomes and patient satisfaction. Urology. 2017;100:117–24. https://doi.org/10.1016/j.urology.2016.09.055.
29. Febrero B, Rios A, Lopez-Navas A, et al. Teenagers and their future role in transplantation: an analysis of their attitudes toward solid organ xenotransplantation. Transplant Proc. 2018;50:526–9. https://doi.org/10.1016/j.transproceed.2017.09.062.
30. Amin L, Hashim H, Mahadi Z, Che NA, Ismail K. Determinants of stakeholders' attitudes to xenotransplantation. Xenotransplantation. 2018;25:e12430. https://doi.org/10.1111/xen.12430.
31. Martinez-Alarcon L, Rios A, Santaines-Borreda E, et al. Student nurses at Spanish universities and their attitude toward xenotransplantation. Xenotransplantation. 2019;26:e12507. https://doi.org/10.1111/xen.12507.
32. Mitchell C, Werkheiser Z, Lipps A, Cooper DKC, Paris W. A meta-analysis of patients, nurses, and physicians' opinions about xenotransplantation. Paper presented at the International Xenotransplantation Association's 2019 Congress; October 10–13, 2019; Munich, GDR.
33. Paris W, Seidler J, FitzGerald K, Padela AI, Cozzi E, Cooper DKC. Jewish Christian and Muslim theological perspectives about xenotransplantation. Xenotransplantation. 2018; https://doi.org/10.1111/xen.12400.
34. Sykes M, d'Apice A, Sandrin M, et al. Position paper of the ethics committee of the international xenotransplantation association. Xenotransplantation. 2003;10:194–203.
35. Jenkins ED, Yip M, Melman L, Frisella MM, Matthews BD. Informed consent: cultural and religious issues associated with the use of allogenic and xenogenic mesh products. J Am Col Surg. 2009;210:402–10.

36. Saito K, Takahara S, Nakagawa T, et al. Obstacles of non-heart beating donor kidney transplantation in Japan to date and future perspectives. Transplant Proc. 2013;45:2866–70.
37. Kobayashi T, Miyagawa S. Current activity of xenotransplantation in Japan. Xenotransplantation. 2019; https://doi.org/10.1111/xen.12487.

Part VII

Summation

22 What Did the Workshop Achieve?

Devin E. Eckhoff and Guerard Byrne

Abbreviations

AMR	Antibody-mediated rejection
FDA	US Food and Drug Administration
HLA	Human leukocyte antigen
SLA	Swine leukocyte antigen
TKO	Triple-knockout
VAD	Ventricular assist device

Introduction

The Department of Surgery at the University of Alabama at Birmingham (UAB) has attracted several leaders in the field of xenotransplantation research to its faculty. Laboratory studies have progressed rapidly to the point where serious consideration is being given to initial clinical trials, particularly of genetically engineered pig kidney transplantation in patients with end-stage renal disease. Members of the Xenotransplantation Program at UAB thought that progress had advanced to the point where bringing together scientists and researchers in the field was required to (i) determine the next steps necessary to initiate a clinical trial, (ii) determine how

D. E. Eckhoff (✉)
Xenotransplantation Program, Department of Surgery, University of Alabama at Birmingham, Birmingham, AL, USA

G. Byrne
Department of Surgery, University of Alabama at Birmingham, Birmingham, AL, USA

Institute of Cardiovascular Science, University College London, London, UK

D. K. C. Cooper, G. Byrne (eds.), *Clinical Xenotransplantation*,
https://doi.org/10.1007/978-3-030-49127-7_22

patients would be selected, and (iii) suggest what appropriate monitoring for rejection and infection would be necessary. Thus, the *Pathway to Clinical Xenotransplantation* Workshop was held at UAB on March 21 and 22, 2019.

The purpose was to discuss the current status of xenotransplantation research and to consider what steps are required to safely initiate a clinical trial. More than 100 scientists and physicians attended, with speakers from UAB and other distinguished academic institutions, as well as participants from the US Food and Drug Administration (FDA, which regulates such trials) and the National Institutes of Health (NIH, which has funded some of the research). The invited speakers gave stimulating and thought-provoking talks ranging from kidney xenotransplantation to the economics of xenotransplantation.

The potential impact of xenotransplantation is immense, not only in organ transplantation but in tissue and cell transplantation. The conference focused primarily on the transplantation of pig kidneys and hearts, but xenotransplantation may ultimately play a role in the treatment of diverse conditions that include diabetes (where pig islet transplantation may be lifesaving) and Parkinson's disease (where transplantation of pig neuronal cells could alleviate many of the symptoms). At the close of the symposium, there was enthusiasm among the attendees for beginning a clinical trial.

Obstacles still existing before conducting a clinical trial include (i) determining the potential for disease transmission and reactivation, (ii) defining the optimal genetically engineered pig source, (iii) the appropriate histocompatibility testing of patients, and (iv) the selection of the appropriate candidates for a pilot trial. The chapters in this book were written by experts who presented at the Workshop (augmented by other experts who did not) and include discussions of xenotransplantation immunology, genetic engineering of pigs, biosecure pig housing, potential infectious risks, and social, religious, and economic aspects of xenotransplantation.

Summary and Conclusions for the Workshop on *Pathways to Clinical Xenotransplantation*

The Workshop was organized into five sessions covering major topics in xenotransplantation. After each major session, the presenters, and sometimes other invited experts, held an open panel discussion with the audience. These discussions ranged across many of the challenges faced with bringing xenotransplantation into the clinic. Most notable were the panel discussions following the presentations on (i) patient evaluation and selection and (ii) the regulatory and economic aspects of clinical xenotransplantation.

The opening session was devoted to a review of the current status of xenotransplantation research and focused on primate studies, presented by three UAB faculty members – A. Joseph Tector, David K.C. Cooper, and Christopher G.A. McGregor.

Session 1: Xenotransplantation in Nonhuman Primates – The Present Position

Joe Tector, MD, PhD, presented his personal view of the steps needed to take xenotransplantation into the clinic. He highlighted that xenotransplantation has the potential to alleviate the organ donation shortage. The primary immunologic barrier has been determined to be xeno-reactive antibodies. Advances in genetic engineering now provide a means of eliminating the xenoantigens on the endothelial surface of the pig organ, thus removing the antibody barrier. There are three primary antigens – galactose-α1,3-galactose [Gal], N-glycolylneuraminic acid [Neu5Gc], and Sda [encoded by β1,4Nacetylgalactosylaminyl transferase] – that have been identified and deleted from pigs (triple-knockout [TKO] pigs). Many humans have no or minimal antibody binding or serum cytotoxicity to TKO pigs. Kidneys transplanted from these pigs, in combination with T cell costimulation blockade targeting the CD40/CD154 pathway, have resulted in prolonged survival in nonhuman primate recipients.

David Cooper, MD, PhD, emphasized that the barriers to clinical xenotransplantation not only are antibody-mediated but also involve complement and coagulation dysfunction, inflammation, and innate and adaptive immune cell activity. He stated that over the last 5 years, there has been a significant improvement in the overall success of pig-to-nonhuman primate kidney transplantation with survival frequently reaching over 200 days. The keys to success have been the incorporation of costimulation blockade targeting the CD40/CD154 pathway and utilizing TKO pigs with additional insertion of "protective" human transgenes targeting differences in complement, coagulation, and inflammation between pigs and primates.

He also presented data showing that there could be problems with overgrowth of transplanted pig kidneys, which could be inhibited by deleting the gene for growth hormone receptors. Sometimes the primates develop a hypovolemia-dehydration syndrome, where the serum creatinine increases and which is associated with low arterial and venous pressures. The syndrome can be reversed simply by the intravenous infusion of normal saline. The cause of the syndrome remains uncertain but may be associated with an inability of pig renin to cleave primate angiotensinogen. This may impair vasoconstriction and fluid retention. Theoretically, if this hypothesis is correct and the condition is problematic, the syndrome could be corrected by insertion into the pig of the gene for human renin.

Christopher McGregor, MB, BS, gave a synopsis of the remaining barriers to successful pig cardiac xenotransplantation. Some of the similarities of the ideal "donor" are shared both by kidney and cardiac xenotransplantation. Genetic engineering of the organ-source pigs has the potential to remove most of these barriers. He drew attention to the difficulties in obtaining adequate function of pig hearts in the hours immediately following orthotopic transplantation in nonhuman primates, a condition termed "perioperative cardiac xenograft dysfunction" (PCXD), the cause of which remains uncertain, though it does not appear to be associated with the immune response.

Additionally, it was discussed that cardiac transplantation is a lifesaving organ transplant and, unlike renal transplantation, there is no bridge, like dialysis, to support cardiac patients. Ventricular assist devices can serve as a bridge in selected patients but make the transplants technically more challenging and have the potential for sensitization to human leukocyte antigens (HLA). Thus, finding clinically appropriate patients may be more difficult in the case of cardiac xenotransplantation.

Session 2: Donor Genetics and Potential Infectious Risks

This session focused on three main areas in developing and monitoring genetically engineered pigs and in compatability testing. David L. Ayares, PhD, CEO and CSO of Revivicor, a biotechnology company, discussed genetic engineering of the organ-source pigs. Jay Fishman, MD, discussed consideration of the organ-source pig's health status and potential infectious risks associated with xenotransplantation. Matthew Tector, PhD, covered the potential relationship between histocompatibility testing and xenotransplantation.

Dr. Ayares discussed the several techniques he has developed or used to produce genetically modified pigs. Currently, pigs are now available in which expression of the three known xenoantigens against which humans have natural (or preformed) antibodies has been deleted. This has greatly reduced the risk of graft rejection after pig organ transplantation into nonhuman primates. In addition, the pigs may express up to six "protective" human transgenes, such as human complement- and/or coagulation-regulatory genes. When bred and housed in an isolated, biosecure facility, the potential organ-source pigs should be free of all potentially pathogenic microorganisms. Although the potential risks associated with the transfer of porcine endogenous retroviruses will remain uncertain until clinical trials are initiated, the data available to date from in vitro and in vivo studies in humans and animals suggests that the risks are small.

Dr. Fishman emphasized that the diagnosis of infection is more difficult in an immunocompromised host because of diminished signs of inflammation. Dual infections or even multiple infections are common. There is a potential broader range of pathogens, and the toxicity of drugs used in xenotransplantation may be undetermined. Because of its unique potential risk, xenotransplantation will require the archiving of blood and tissues from both the organ-source pig and the human recipient. Dr. Fishman concluded that the likely infectious risk is not greater than following allotransplantation, but is not negligible.

Dr. Matthew Tector discussed the immunologic barriers to xenotransplantation and how appropriate histocompatibility testing can be used to minimize the risk of immunologic graft loss. There was some discussion over whether patients who are highly sensitized to human leukocyte antigens (HLA) should be excluded from the initial clinical trial (because there is some evidence of cross-reactivity between anti-HLA antibodies with swine leukocyte antigens [SLA]). It was suggested that HLA-sensitized patients, in whom the T cell response to a pig organ might be increased, should be avoided, though others believed that, if a negative crossmatch was obtained, there would be no greater risk.

It was agreed, however, that these patients, for whom a suitable allograft is often difficult to identify, may ultimately benefit most from the opportunity to receive a pig xenograft. Studies are progressing in which the expression of SLA in the organ-source pig can be modified by genetic engineering that may allow every HLA-sensitized patient to receive a pig graft against which he/she expresses no antibodies.

Session 3: Antibody-Mediated Allotransplant Rejection – Lessons for Xenotransplantation

Drs. Stuart Knechtle, Robert Montgomery, Stanley Jordan, and Jeffrey Platt discussed their perspectives on how lessons learned from antibody-mediated rejection (AMR) of an allograft, associated with either HLA or ABO incompatibility, might impact xenotransplantation.

Stuart Knechtle, MD, has developed and characterized a primate preclinical model of allosensitization. His model of allosensitized monkey kidney allotransplantation shared many similarities to pig-to-nonhuman primate kidney xenotransplantation. The genetically engineered pig offers an opportunity to select pigs that do not express antigens against which humans have natural (preformed) antibodies. This is in contrast to AMR of an allograft that is associated with the presence of donor-specific anti-HLA antibodies present in the recipient before the transplant. To prevent T cell-dependent de novo antibody formation, it will be essential to administer sufficient immunosuppressive therapy to prevent new antibodies from developing. In this respect, the novel agents that inhibit the secondary signal of T cell activation (known as costimulation blockade, specifically of the CD40/CD154 pathway) appear to be more efficacious than conventional immunosuppressive agents.

Robert Montgomery, MD, discussed the prevention and treatment of acute AMR in allotransplantation and the importance of crossmatching in HLA highly sensitized patients. He discussed the three types of AMR – (i) the anamnestic response that occurs in pre-sensitized patients in the first 3 months tends to be severe; (ii) the second type of AMR is associated with persistent preformed antibody after desensitization where donor-specific antibodies are present before transplantation, and that occurs in the first 3 months and tends to be mild or moderate; and (iii) lastly, there is AMR from the development of de novo donor-specific antibodies in unsensitized patients. This can occur at any time posttransplantation and can be mild-to-moderate. The first two types of AMR respond well to current therapies, but the third form does not and tends to be chronic.

Stanley Jordan, MD, discussed the various treatments used for AMR, including intravenous immunoglobulin (IVIg) therapy, plasma exchange, rituximab, complement inhibitors, e.g., eculizumab, and the novel use of IgG-degrading enzymes derived from *Streptococcus pyogenes* (IdeS) for HLA highly sensitized patients. It is thought that the lessons learned from desensitization could be directly applicable to AMR occurring in xenotransplantation.

Jeffrey Platt, MD, pointed out that several components of the immune barrier to xenotransplantation are similar to those relevant to ABO-incompatible allografts. He proceeded to provide insights drawn from the successful transplantation of ABO-incompatible hearts in newborn infants, which he suggested were particularly pertinent to the application of xenotransplantation in infants or, indeed, in all human recipients.

Although many of the approaches summarized by the speakers in this session have proved invaluable in organ allotransplantation, in xenotransplantation, we have the ability to genetically modify the organ "donor" rather than just "treat" the recipient. The problem of natural antibody has largely been resolved by depletion of the relevant glycan xenoantigens from the pig. The presence in the recipient of anti-HLA antibodies that cross-react with SLA is likely to be resolved by deleting and replacing (i.e., mutating) the target antigens in the pig. For example, modifying SLA expression has been demonstrated by Tector's group to have potential to overcome the problem of cross-reactivity. An organ-source pig with minimal preformed reactivity, suitable for the broadest range of potential patients, including those with HLA sensitization, would be one that includes glycan elimination and site-specific SLA mutation. This would be a preferable approach to eliminating SLA expression in the pig, which would render it immunocompromised, possibly resulting in a risk of infectious complications and/or decreased survival.

Session 4: Patient Evaluation and Selection for First Clinical Trials of Kidney or Heart Xenotransplantation

The selection of candidates for the first clinical trial of xenotransplantation was the focus on the second day of the workshop. Jayme Locke, MD, PhD, discussed the pros and cons of kidney transplantation, and Christopher McGregor and James K. Kirklin, MD, discussed heart transplantation, with input from transplant nephrologist, Robert Gaston, MD. Dr. Jay Fishman discussed potential infectious risks to the xenograft recipient.

Dr. Locke emphasized the magnitude of the organ donor shortage and that efforts to increase the supply of deceased human organs are not keeping up with the demand. There are several classes of candidates that would potentially benefit from participating in a clinical trial of xenotransplantation. Patients with end-stage renal disease who are likely to die before a kidney from a deceased human donor becomes available were considered to be the optimal group who might be offered xenotransplantation. However, patients with diseases known to recur quite rapidly in a kidney allograft, e.g., focal segmental glomerulosclerosis, were also a group that might benefit from xenotransplantation, though there is currently no evidence whether the disease will or will not recur in a pig kidney graft. A third group discussed was patients with renal failure in countries where the resources are insufficient to provide chronic dialysis for all patients and yet the expertise to carry out kidney transplantation successfully is available. South Africa was given as an example of such a country, but there are many others. Without a kidney from a deceased or living human donor, many patients face an early death. The logistics of carrying out an initial clinical trial in such a country, however, could be challenging.

Dr. Locke indicated her preference that the first candidate be an older patient in generally good health, except for renal failure, but with a predictably long wait-time and little real prospect of obtaining a deceased human donor organ. This type of patient (i) would be viewed as an acceptable candidate with a potentially good outcome, (ii) would benefit from receiving a xenograft when an allograft was unlikely, and (iii) would still have the option of returning to dialysis if the xenograft should fail. There was some support for this view, but others indicated that, at their own institutions, aggressive use of deceased human donor kidneys had reduced the waiting period for this sort of older patient and that an allograft would be preferred.

Dr. McGregor discussed the similar problem of an inadequate availability of human hearts. This was currently partially being addressed by the implantation of ventricular assist devices (VADs), but pig heart transplantation might be preferable. He suggested that patients with amyloid disease might be those who might first be considered for pig cardiac transplantation because they were unsuitable for VAD support.

Dr. Kirklin discussed the progress that has been made in recent years in the function of VADs and other forms of mechanical support of a failing heart. Because of the availability of these devices, the selection of patients for a pig heart transplant is possibly more difficult than of those for pig kidney transplantation. Nevertheless, VADs are still associated with considerable morbidity, and there are no forms of mechanical support that are truly effective in babies and small children. Thus, xenotransplantation may well be justified in the pediatric age group.

A straw poll of the audience clearly favored kidney xenotransplantation as the first clinical application.

Dr. Fishman emphasized that the potential infectious risks to a patient with a xenograft would be those seen commonly in patients will allografts (primarily associated with the detrimental effects of immunosuppressive therapy), rather than those that might be related to transfer of a microorganism with the pig organ. Physicians in the field of infectious disease have considerable experience in the management of immunosuppressed patients with allografts and should be well-qualified to treat similar infectious complications in patients with pig xenografts.

Session 5: Regulatory, Economic, and Social Aspects of Clinical Trials of Xenotransplantation

Karl Kraebber, MS, discussed the regulatory aspects of xenotransplantation and the requirements for a facility for the breeding and housing of the genetically modified pigs that will be used for the first clinical trial. Much progress has been made in building such a biosecure facility in Birmingham, and this should meet the US Food and Drug Administration (FDA) requirements for a clinical trial. He briefly summarized how this facility would function.

Winson Tang, MD, of the US FDA drew attention to the need in any clinical trial to ensure patient and public safety. In regard to xenotransplantation, this clearly related to the potential for infection. He also stressed that the sponsor of a trial should demonstrate that the xenograft would function as expected.

The FDA concurred that kidney xenotransplantation was likely to be the first clinical application and that several companies were already in pre-pre-IND discussion with the FDA. He emphasized that the first-in-human studies were primarily safety trials and that the FDA's current risk-benefit analysis would limit selection of the first patients to those who were very ill, at risk of imminent death, and unlikely to receive an allograft. Using the kidney as an example, he stressed the several potential physiologic incompatibilities between pigs and humans.

For cardiac xenotransplantation, adult patients with amyloid disease, and pediatric patients with severe congenital heart disease not suitable for three-stage palliation, were discussed as potential first candidates for xenotransplantation. For each type of patient, especially pediatric patients with complex malformations who had possibly undergone previous operations, there was concern that primary cardiac xenograft function would need to be optimized before a clinical study could be considered. Dr. Tang indicated that, in general, the FDA preferred that any new therapy is first shown to be safe and effective in adults, who can understand and consent to the required procedures and potential risks, before being applied to younger patients. Therefore, the trial should not include children unless a special case can be made for their participation. Nevertheless, there may be an exception to this for diseases, such as hypoplastic left heart syndrome, not suitable for three-stage palliation, which occur only in children and are likely to be fatal. Early consultation with the FDA was recommended.

There was an extended discussion about the required interval between treating the first and second patients. Currently, regulators set this interval at 1 year for any individual study. This interval, which was perceived as excessive by most of the researchers present, might be shortened if data, such as patient perception of risk, could be provided.

It was clear throughout these discussions that the regulatory perspective for xenotransplantation was enthusiastic but that the risk-benefit analysis of the first clinical trial would be focused on the patient.

Jon Friedman, MD, the chief medical officer of Medical Benefit Management, Optum Specialty Management, a company involved in facilitating the financing of healthcare, gave the final presentation. In view of the relative "unknowns" of xenotransplantation, e.g., cost of the pig organs and future immunosuppressive drug therapy, etc., predictions of the cost associated with organ xenotransplantation are difficult. Nevertheless, he provided data indicating that, when organ xenotransplantation became commonplace, it would probably reduce the costs of healthcare of patients with end-stage organ failure, particularly, for example, those requiring chronic dialysis.

Summary: How Close Are We to Clinical Xenotransplantation

The *Pathways to Clinical Xenotransplantation* Workshop explored the logistic, regulatory, immunologic, and ethical barriers to the first clinical trial in xenotransplantation. Experts from throughout the USA attended and participated in spirited and productive debate. The advances that have been made in understanding AMR and the tools that have been developed to produce genetically modified pigs have proceeded at a rapid pace. Significant work has been carried out on minimizing the infectious risk, and, with proper monitoring, xenotransplantation should present a similar risk to allotransplantation. The requirements for the facilities to house the pigs, and monitoring both the animals and the human recipients, have been rigorously defined. The preclinical work in nonhuman primates shows survival of life-supporting pig kidneys frequently functioning for more than 200 days.

Although more preclinical work could always be done, workshop participants felt that the primary barrier will be to meet the regulatory requirements by the FDA. The general and enthusiastic consensus was that this could be achieved within the next several years.

Index

D. K. C. Cooper, G. Byrne (eds.), *Clinical Xenotransplantation*,
https://doi.org/10.1007/978-3-030-49127-7

The manufacturer's authorised representative in the EU is Springer Nature Customer Service Centre GmbH, Europaplatz 3, 69115 Heidelberg, Germany. If you have any concerns regarding our products, please contact ProductSafety@springernature.com

Printed and bound by CPI Group (UK) Ltd, Croydon, CR0 4YY
30/09/2026
02225328-0005